Praise

A FIELD GUIDE TO
FROM ALASKA TO CALIFORNIA

"Having fished the Pacific coast my entire life, I've never seen anything that even comes close to this incredible book. Clean, straightforward and easy to read. The consummate guide to our waters. The dead-nuts-on accurate and absolutely beautiful illustrations alone are worth the price of admission."

—*Michael Fowlkes, Executive Producer, Fox Sports/WFN Inside Sportfishing*

"Val Kells is one of the most extraordinary fish illustrators I've ever seen. Her keen attention to detail and exquisite aesthetic sense are truly astounding. This outstanding guide to the fishes of the West Coast that she has assembled with Luiz Rocha and Larry Allen is the ultimate guide to the piscine wonders of the deep. It will be the standard for decades to come."

—*Ray Troll, Trollart*

"Destined to become *the* classic guide to fishes of the Pacific coast for generations of fish aficionados."

—*Milton Love, Research Biologist, Marine Science Institute, University of California, Santa Barbara*

"Fish geeks rejoice! Thanks to Val Kells' beautiful illustrations and unrelenting attention to detail, the West Coast SCUBA diving community has a terrific new book to use in their quest to identify those pesky mystery fish encountered on their dives. Sculpin 'a-fishy-onados' will be especially thrilled with the extensive list of species represented."

—*Janna Nichols, Reef Environmental Education Foundation (REEF)*

"An up-to-date, well-organized, and beautifully illustrated compact field guide to fishes of the Pacific Northwest covering shallow (<200 m) coastal brackish and marine fishes. Organized by orders and families, with descriptions of features, habitats, and biology and wonderful, detailed color illustrations of each species. This will become the new reference for the region."

—*Gregor M. Cailliet, Professor Emeritus and Director, Pacific Shark Research Center, Moss Landing Marine Laboratories, California State University*

"With all the protected species, anglers need to make positive IDs of the fish they catch. This will be a great resource for West Coast anglers. Illustrations are very clear, and the script is concise but detailed. Every sportfishing and private boat should have a copy aboard."

—Peter Gray, President, Let's Talk Hook-up

"Ed Ricketts would have been proud of the dedication that went into this field guide. Quality illustrations and easy-to-read text make this book a must-have for any fisherman, diver, or aquarist."

—Randy Hamilton, Vice President of Husbandry, Monterey Bay Aquarium

"*A Field Guide to Coastal Fishes: From Alaska to California* is a most welcome addition to the books on fishes of the West Coast. Identifying features of nearly 700 species are primarily based on details of coloration, beautifully illustrated by Val Kells' vibrant color paintings that seem to jump off the page. This book and *A Field Guide to Coastal Fishes: From Maine to Texas*, also illustrated by Kells, provide complete coverage of all shallow-water marine fishes from the continental United States. Professional ichthyologists and amateur fish enthusiasts alike will want to own copies of these beautifully illustrated guides to the country's marine fishes."

—Phillip A. Hastings, Professor and Curator of Marine Vertebrates, Scripps Institution of Oceanography

"Val Kells, Luiz Rocha, and my good friend Larry Allen have demonstrated their passion for and knowledge of Pacific marine fishes in this book. *A Field Guide to Coastal Fishes: From Alaska to California* is the result of an outstanding collaboration that will prove to be an asset to anglers and marine biologists alike."

—Bart Hall, Producer, Fred Hall Shows, Inc.

"Kells, Rocha, and Allen have put together a fantastic field guide illustrated with accurate detailed paintings as well as current and relevant information on features, habitat, and biology. This will be a great resource for fish biologists, fishers, and enthusiasts. Clearly, decades of experience and knowledge are wrapped up in this guide—can't wait to add it to my library."

—Chris Lowe, Professor of Marine Biology, California State University, Long Beach

A FIELD GUIDE TO COASTAL FISHES

FROM ALASKA TO CALIFORNIA

A Field Guide to Coastal Fishes

FROM ALASKA TO CALIFORNIA

Val Kells
Luiz A. Rocha · Larry G. Allen

Johns Hopkins University Press
Baltimore

Printed in Canada on acid-free paper
4 6 8 9 7 5 3

Johns Hopkins University Press
2715 North Charles Street
Baltimore, Maryland 21218-4363
www.press.jhu.edu

Library of Congress Cataloging-in-Publication Data

Names: Kells, Valerie A., author, illustrator.
Title: A field guide to coastal fishes : from Alaska to California /
Val Kells, Luiz A. Rocha, Larry G. Allen.
Description: Baltimore : Johns Hopkins University Press, 2016. | Includes
bibliographical references and index.
Identifiers: LCCN 2015936213 | ISBN 9781421418322 (pbk. : alk. paper) |
ISBN 1421418320 (pbk. : alk. paper)
Subjects: LCSH: Marine fishes--Pacific Coast (U.S.)--Identification. |
Marine fishes--British Columbia--Pacific Coast--Identification. |
Fishes--Pacific Coast (U.S.)--Identification. | Fishes--British Columbia--
Pacific Coast--Identification. | Fishes. | Marine fishes. |
British Columbia--Pacific Coast. | United States--Pacific Coast.
Classification: LCC QL628.A4 K45 2016 | DDC 597.177
LC record available at https://lccn.loc.gov/2015936213

A catalog record for this book is available from the British Library.

To my sons, Dave and Drew
—VK

To my wife, Claudia,
and my children, Gabriel and Sophia
—LR

To my ichthyological mentor,
Mike Horn
—LA

Contents

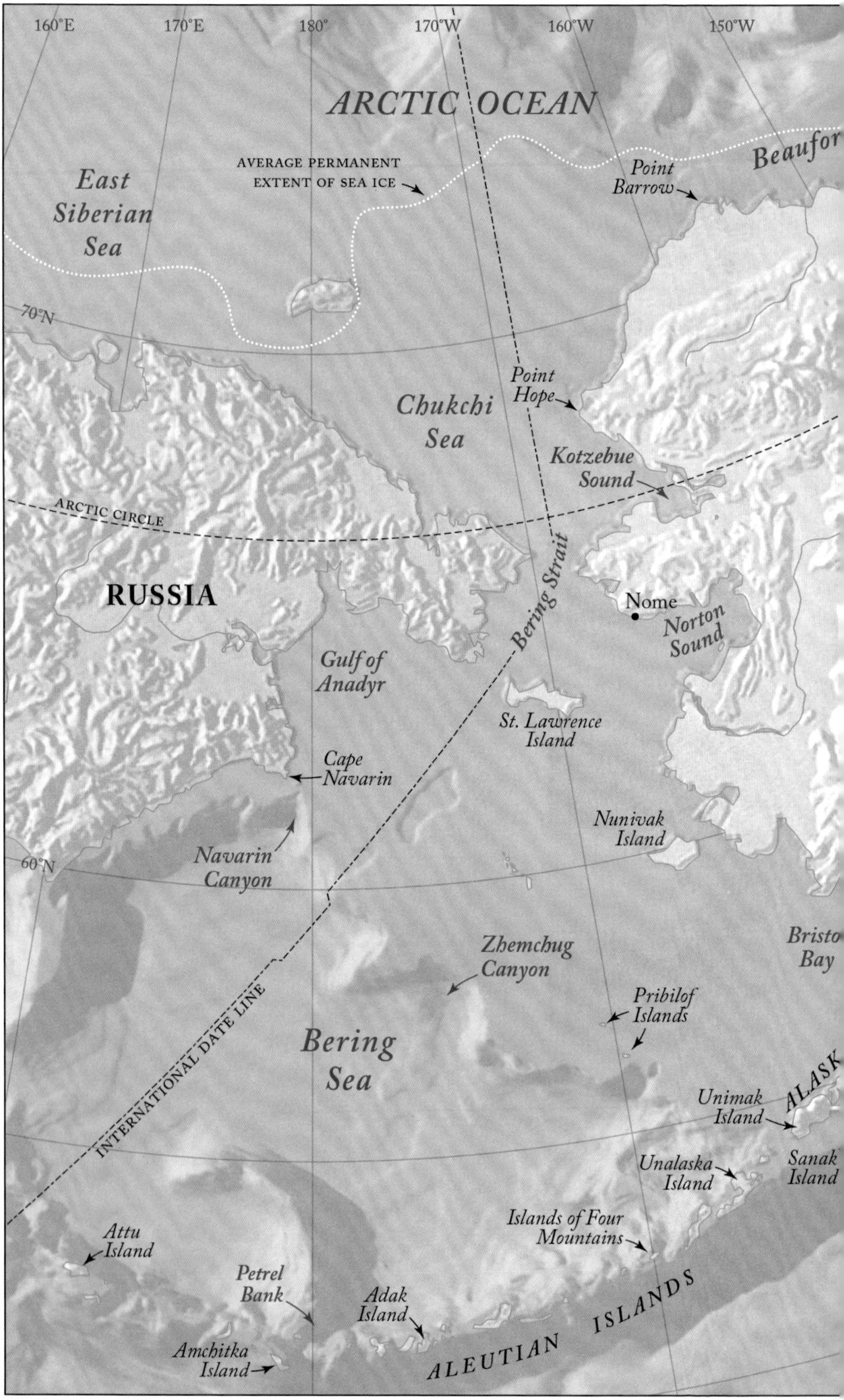
160°E
170°E
180°
170°W
160°W
150°W
ARCTIC OCEAN
AVERAGE PERMANENT EXTENT OF SEA ICE
East Siberian Sea
Point Barrow
Beaufor
70°N
Point Hope
Chukchi Sea
Kotzebue Sound
ARCTIC CIRCLE
Bering Strait
RUSSIA
Nome
Norton Sound
Gulf of Anadyr
St. Lawrence Island
Cape Navarin
Nunivak Island
Navarin Canyon
60°N
Zhemchug Canyon
Bristo Bay
Pribilof Islands
INTERNATIONAL DATE LINE
Bering Sea
Unimak Island
ALASK
Sanak Island
Unalaska Island
Islands of Four Mountains
Attu Island
Petrel Bank
Adak Island
Amchitka Island
ALEUTIAN ISLANDS

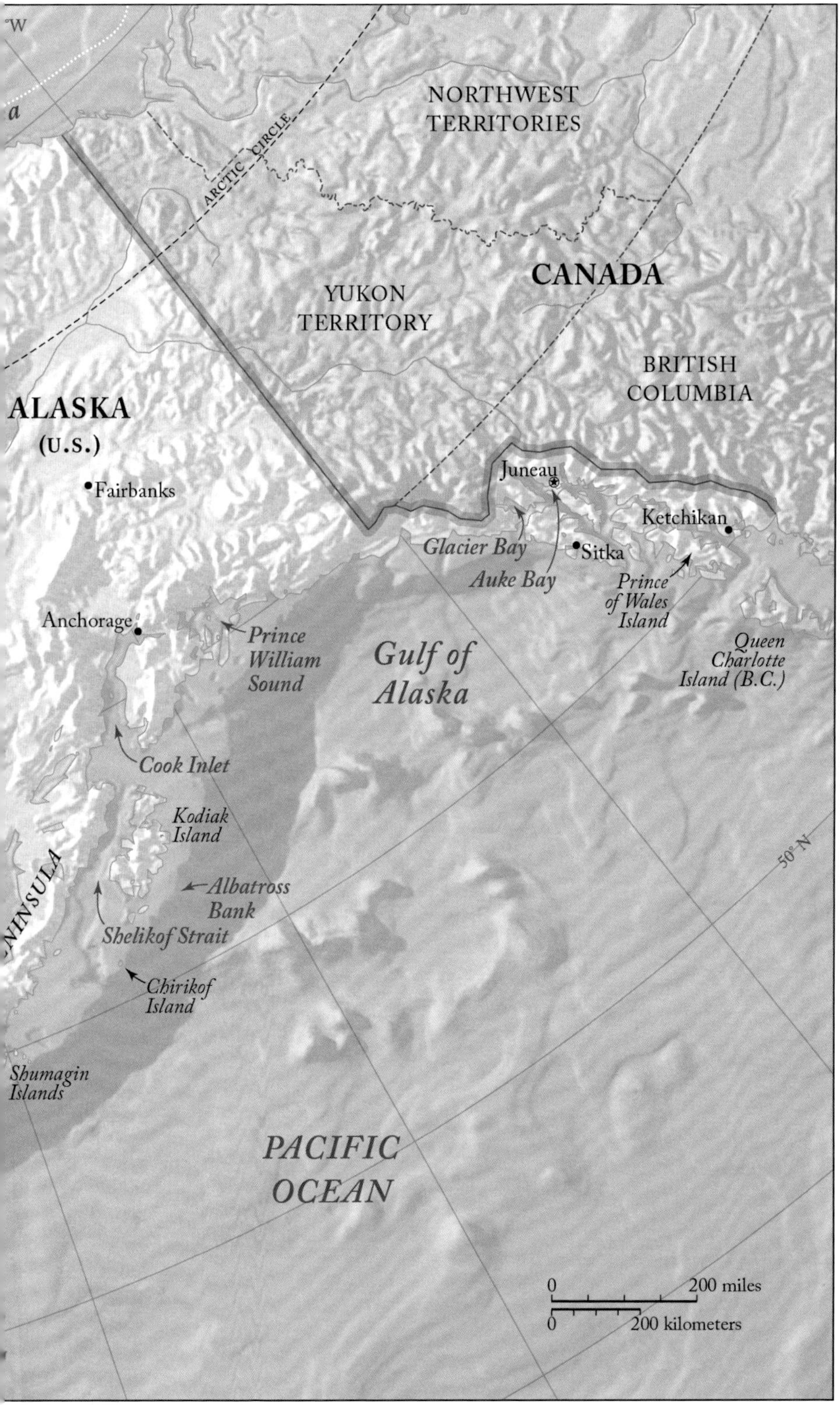

°W
NORTHWEST
TERRITORIES
ARCTIC CIRCLE
CANADA
YUKON
TERRITORY
BRITISH
COLUMBIA
ALASKA
(U.S.)
Fairbanks
Juneau
Ketchikan
Glacier Bay
Sitka
Auke Bay
Prince
of Wales
Island
Anchorage
Prince
William
Sound
Gulf of
Alaska
Queen
Charlotte
Island (B.C.)
Cook Inlet
Kodiak
Island
Albatross
Bank
Shelikof Strait
Chirikof
Island
Shumagin
Islands
50° N
PACIFIC
OCEAN
0
200 miles
0
200 kilometers

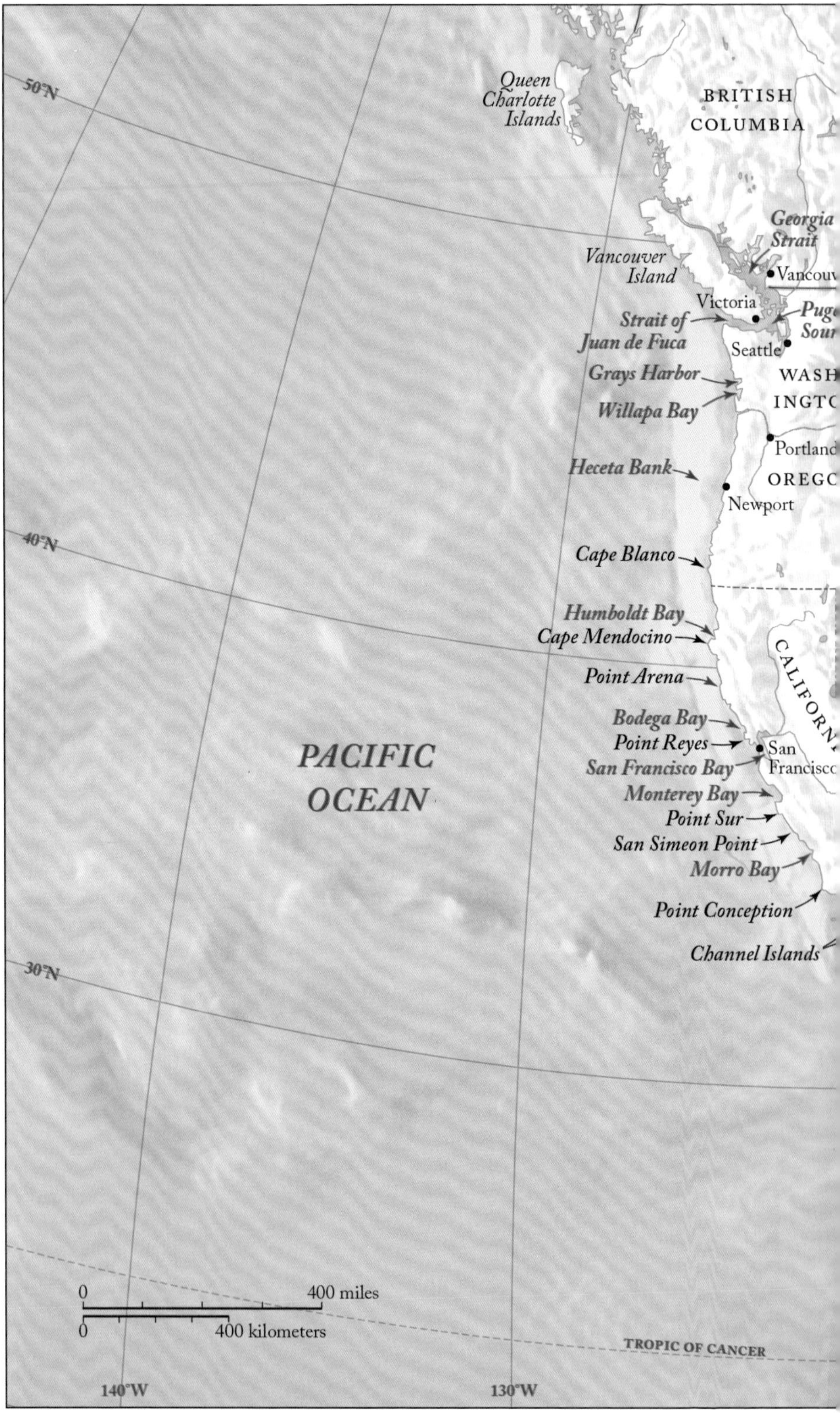
Queen Charlotte Islands
BRITISH COLUMBIA
50°N
Georgia Strait
Vancouver Island
Vancouv
Victoria
Strait of Juan de Fuca
Pug
Sour
Seattle
Grays Harbor
WASH
INGTO
Willapa Bay
Portland
Heceta Bank
OREGO
Newport
40°N
Cape Blanco
Humboldt Bay
Cape Mendocino
CALIFORN
Point Arena
Bodega Bay
Point Reyes
San Francisco
San Francisco Bay
PACIFIC OCEAN
Monterey Bay
Point Sur
San Simeon Point
Morro Bay
Point Conception
Channel Islands
30°N
0
400 miles
0
400 kilometers
TROPIC OF CANCER
140°W
130°W

CANADA
LBERTA
SASKATCHEWAN
MANITOBA
ONTARIO
CKY MOUNTAINS
MONTANA
NORTH DAKOTA
MN
WI
SOUTH DAKOTA
IDAHO
IOWA
Columbia River
WYOMING
NEBRASKA
UNITED STATES
MO
KANSAS
UTAH
COLORADO
VADA
AR
OKLAHOMA
NEW MEXICO
ARIZONA
Angeles
TEXAS
an Diego
Tijuana
BAJA CALIFORNIA
Gulf of California
Punta
Cabo San Lázaro
Cabo San Lucas
MEXICO
110°W

Acknowledgments

In the spring of 2011, *A Field Guide to Coastal Fishes: From Maine to Texas* was published and immediately embraced by divers, fishermen, students, researchers, teachers, naturalists, and scientists. One such scientist was Luiz Rocha, who at the time worked at the University of Texas Marine Science Institute. He sent Val the following note:

> "Hey Val, Congratulations again on the book, fantastic work—I just got my copy. Anyhow, I am moving to take the job of Curator of Fishes at the California Academy of Sciences in San Francisco. So, let's get together if you get bored and decide to draw some west coast fishes!
> Cheers, Luiz"

Val quickly seized the opportunity to collaborate on a new and necessary book for West Coast communities. The partnership of Kells and Rocha was established and Larry Allen joined the team as the third coauthor in August 2011. Production on *Alaska to California* began soon thereafter.

∞

No book is solely the result of the author's work. A successful and comprehensive book results from the collaboration and effort of many people. We are blessed with a wide network of friends, associates, and colleagues—many of whom graciously and freely gave their time and support.

We would like to thank the following persons for contributions of information, knowledge, referrals, and assistance: Steve Adams, Nature Conservancy; John Butler, National Marine Fisheries Service; John Caruso, Tulane University; Judith Connor, Monterey Bay Research Institute; John Fong, California Academy of Sciences; Robert N. Lea, California Department of Fish and Game; Milton Love, University of California, Santa Barbara; Ann Matarese, NOAA Federal; John McCosker, California Academy of Sciences; Katherine Mecklenberg, California Academy of Sciences; Royden Nakamura, California Polytechnic State University; D. Ross Robertson, Smithsonian Tropical Research Institute; Mark A. Steele, California State University, Northridge; and Duane Stevenson, NOAA Federal and Burke Museum.

We also gratefully acknowledge the California Academy of Sciences and California State University, Northridge, for generously underwriting Val's watercolor paper and the production of the range maps.

For providing valuable photographic reference necessary to illustrate obscure or poorly documented species, or for directing us toward sources of photo reference, we extend deep gratitude to: Martini Arostegui; Karin Fletcher; Bernie Hanby; Nick Harling; Craig Hoover; Dan Ituarte; Gregory Jensen; Andy Lamb; Ami Latker; Kevin Lee; Ted Pietsch; Joseph Tomelleri; Ray Troll; John T. Snow; Scott Stevenson; and most appreciably, Steve Wozniak. Steve's enthusiasm for this project was outstanding and his extensive photographic library saved the day on many occasions.

Acknowledgments

We thank our many friends and family who gave their unwavering encouragement and friendship throughout the long production of this book. Many people touched this project in ways small to very large. They include: Bob Baydush and Susan Smith; Lane Becken; the Browning-Lynch family; Julie Neff Carew; Bruce Cauthen; Don and Dorothy Chan; Candice Cobb; Cole Columbus; Joasha Dundas; Tommy and Sarah Everett; Pieter Folkens; Jessie Grove; Jeff and Sarah Haas; Terri Kirby Hathaway; Lisa Huizenga Hogan; Betsey Evans Kates; John Kells; Ron and Peg Kernehan; Kate and Stathis Koniditsiotis; Leslie Lanier; Katherine and Elizabeth McDonald; Betty, Casey, Georgia, Jo, Leigh Middleditch III, and Leigh Middleditch, Jr.; the entire Migliaccio family; Norman Miller; Mary Michaela Murray; Elizabeth Neff; Robin Noonan; Tim O'Brien; Carson Oldham; Donald J. Orth; Jean and David Patteson; Rosemary and Charles Robinson; Mike Roth; Heidi Sonen; Allen and Melinda Sutton; Jason Woodle; and Bob Yates. Cheers!

A large number of online friends were supportive and enthusiastic about this project. It was a great source of motivation to read encouraging notes from people we've known for years, and others we have yet to meet in person. These notes particularly helped propel Val through the mountains and valleys of Rockfishes and Sculpins. For taking the time, we send our thanks . . . you know who you are!

The range maps were created with great care by Bob Cronan of Lucidity Information Design. IUCN Red List designations were compiled by Garrett Leidy. Drew Middleditch helped carefully proofread the text. Thanks for your excellent work, Bob, Garrett, and Drew.

A special shout-out goes to Bradley Bryant for telling Val, "Go for it, Miss Val! It's meant to be!" And to Janna Nichols for her tenacious help locating images of the elusive Puget Sound Sculpin. This little fish was so rarely encountered many believed it was a phantom.

We owe an enormous amount of gratitude to the directors, editors, and staff of Johns Hopkins University Press. In particular, we thank our editor, Vincent J. Burke, PhD, and our copyeditor, Debby Bors. Debby spent many hours meticulously refining the manuscript. Her keen eyes and hard work added a final, priceless polish. Vince had the faith, vision, and trust to champion our project. With unwavering support, he helped to nurture it from concept to reality. Thank you, Vince—this book is another dream come true.

Finally, we would like to express our deepest thanks to our families. This book was an enormous undertaking made much easier by their steadfast patience, encouragement, and understanding.

Thank you, Andrew, Dave, and Drew. Thank you, Claudia, Gabriel, and Sophia. Thank you, Teresa.

—Val, Luiz, and Larry

About This Book

We developed this book to fulfill the need for a comprehensive, current, and compact field guide to fishes of the Pacific northwest and as a companion to *A Field Guide to Coastal Fishes: From Maine to Texas*. We hope that this book is as well received as *Maine to Texas*, and does even more good work.

Area and Species Covered

The species included are brackish and marine fishes that are encountered from the northern border of Alaska to the southern border of California. This area generally extends from the intertidal zone to depths of about 660 feet (200 meters). We provide identification and natural history information for most fishes we know to have stable populations within this range of coverage. This includes all native and several non-native fishes that spend all or part of their adult lives in marine waters.

We also describe species that are predominantly freshwater inhabitants but are also found in low-salinity waters. For comprehensiveness, some deep-water species and their families are included. Rare species and those generally occurring below 660 feet have been mostly omitted. Other poorly documented species were excluded for lack of information such as specimen photographs or video clips for live color reference.

Many of the fishes that occur from Alaska to California also occur at other, often distant, locations. Strays and waifs may be found in areas outside of the species' typical range. Wherever possible, those locations are noted in the text. Most of the fishes found in the southern waters of the western United States also occur off of Baja California and adjacent waters; thus the range map includes this broader area.

Names and Sequence of Species

The Latin and common names of the families in this book follow those presented in *Fishes of the World*, fourth edition, by Joseph S. Nelson. The common names of species are taken from *Common and Scientific Names of Fishes from the United States, Canada, and Mexico*, seventh edition, published by the American Fisheries Society (AFS), Special Publication 34. With a few exceptions, Latin names, authority, and date follow Eschmeyer's most recent online *Catalog of Fishes*. The orders and families of fishes are organized in phylogenetic sequence following *Fishes of the World*. The inclusion of Scaridae in Labridae and splitting of Serranidae and Epinephelidae follow the most recent research. Sequences of species within each family follow the alphabetical order of genus and species names, rather than the alphabetical order of common names, with Pipefishes being the only exception. For example, the following Rockfishes are listed as:

Rougheye Rockfish - *Sebastes aleutianus*
Pacific Ocean Perch - *Sebastes alutus*
Kelp Rockfish - *Sebastes atrovirens*

The first letter of each word in single- and multiple-word names is capitalized, except after a hyphen, unless that word requires capitalization as a proper noun. This is in accordance with recent changes adopted by the American Fisheries Society and as published in *Common and Scientific Names of Fishes from the United States, Canada, and Mexico*, seventh edition, Special Publication 34, AFS. Although we elected to use the most recently accepted AFS common names for the individual species, other commonly used local names or those accepted by the Food and Agriculture Organization are also mentioned wherever possible. Whenever we encountered errors or conflicting information in regard to Latin or common names, we made appropriate corrections and inclusions based on the most recently published documentation.

Organization and Presentation

We have arranged this book into three primary sections: Introduction, Families, and Species. These sections are supported by supplemental materials, which include a glossary of terms, a list of additional resources, descriptions of rare species also in the area, and an index.

The Introduction provides an overview of the evolution, diversity, and features of fishes. It also includes information that will help the user identify fishes. Each family section describes, in concise terms, each of the 157 families of fishes that are found along the coast from Alaska to California. A printable Key to Families presented in this book is available online at www.press.jhu.edu.

The species section is the largest in this book and includes descriptions of 705 individual species, based on the most recently published scientific information available. An additional 34 rare species are described in the appendix. A condensed summary of range and habitat is provided for each species. To save room, names of states are abbreviated. For example, California reads: CA. North, South, East, and West are also sometimes abbreviated and read, respectively: N, S, E, W. The biologic description provides a brief summary of the species' behavior, diet, and/or ecology, where room allowed. The depths provided are approximate maximum recorded depths, although many fishes may be more common at shallower depths. The lengths given for each species are the approximate maximum recorded adult total length.

Each account is accompanied by a large, full-color illustration of the species. Each fish is shown in living color, as it would appear in hand, or at the surface in clear water. While no two fish of the same species are exactly alike, the illustration intends to closely represent the species as one might observe it. Great care was taken to accurately portray the correct placement and proportion of anatomical features. All illustrations show the adult fish unless otherwise noted. The illustrations are presented in a "size relative" fashion, meaning those in the same genus and on the same page are shown at a size relative to the longest fish in the genus. Juveniles can be much smaller than adults; they are shown larger than size relative for clarity. Each illustration provides identifying anatomical and color features. All are shown facing left or from above with all fins displayed. The only exception is the presentation of right-eyed flatfishes, which are shown facing right.

Diversity and Classification

Fishes comprise the most diverse group of vertebrates on Earth. There are currently over 32,000 known living species of fishes and many thousands of others that have become extinct in the 500 million years since the early fish ancestors first swam in the seas. Over 800 marine fish species live along the northwest Pacific coast of the United States, inhabiting bays, inlets, estuaries, rocky reefs, and rocky, sandy, and kelp-lined shores. These species range in size from the Whale Shark, which may grow to 50 feet, to the Arrow Goby, which as an adult grows to about 2.3 inches in length. Pacific Northwest coastal fishes also include species that represent the most primitive of ichthyofauna—like the Pacific Hagfish—to the most highly evolved forms, such as the Balloonfish.

The scientific classification of fishes is, and will likely always be, an ongoing process subject to debate and change as new information unfolds. However, many scientists divide fishes into five recognized classes: Myxini, the Hagfishes; Petromyzontida, the Lampreys; Chondrichthyes, the Cartilaginous Fishes; Actinopterygii, the Ray-finned Fishes; and Sarcopterygii, the Lobe-finned Fishes. Of these five classes only the Lobe-finned Fishes do not occur along the western United States.

The table below shows how three representative fishes are classified in the three classes that occur along the western United States:

	Pacific Hagfish	**Whale Shark**	**Southern Clingfish**
Kingdom:	Animalia	Animalia	Animalia
Phylum:	Chordata	Chordata	Chordata
Class:	Myxini	Chondrichthyes	Actinopterygii
Order:	Myxiniformes	Orectolobiformes	Perciformes
Family:	Myxinidae	Rhincodontidae	Gobiesocidae
Genus:	*Eptatretus*	*Rhincodon*	*Rimicola*
Species:	*stoutii*	*typus*	*dimorpha*

The "jawless" fishes—Hagfishes and Lampreys—are truly primitive. They lack true jaws, do not have paired fins, and exhibit a simple, cartilaginous skeletal structure. They have a single nostril located on the top of the head. Their form of locomotion is simple and eel-like. The simplicity of the jaws in this class of fishes limits them to rasping prey. However, this characteristic has not prohibited them from succeeding. Through scavenging and parasitizing, jawless fishes have survived, evolved, and prospered for hundreds of millions of years.

The Cartilaginous fishes of the Class Chondrichthyes—the familiar sharks, skates, and rays as well as the less familiar chimaeras—are more structurally advanced than jawless fishes, even while lacking true bones. They have true jaws, and their nostrils are located on both sides of the head, generally under the snout. The skull and jaws are constructed of large single units, rather than multiple pieces as seen in the Ray-finned Fishes. Cartilaginous Fishes lack a swim bladder and most rely on large oily livers for buoyancy. Unlike jawless fishes, they possess paired fins: the pectoral and pelvic fins.

Another differentiating feature is that all cartilaginous species practice internal fertilization and the females produce either egg cases or live young.

The largest and most diverse class of fishes is, by far, the familiar Ray-finned fishes of the Class Actinopterygii. This group possesses a bony, rather than cartilaginous, skeleton. Like the Cartilaginous fishes, bony fishes also possess true jaws. However, their jaws are composed of many small bones rather than large cartilaginous units. The skull is also a complex structure of small bones. The wide array of jaw and tooth types in this class has spawned a large variety of feeding systems, including biting, crushing, filter-feeding, sucking, picking, and scraping. Nostrils are found on the upper part of both sides of the head. Bony fishes usually have swim bladders, many of which are complex in structure. They also possess paired fins, but in some the pectoral and pelvic fins may be absent. Their methods of reproduction are wide and varied.

Adaptations to Life in Water

Living completely submerged in water presents a host of challenges. Fishes need to regulate the amount of salt and water in their bodies and extract dissolved oxygen from the water. Their senses are adapted to aquatic life. Beyond the basics of sight and smell, most fishes have a lateral line, a sensory organ that is highly developed to detect the minutest motions in the water. Fishes have also developed many ways to communicate with each other under water. Some grind their pharyngeal teeth, while others grunt by manipulating the air bladder. Some fishes have light-producing organs that they may use to locate each other or attract prey in the darkness.

Water can be over 900 times more dense than air. Many fishes cruise through this dense solution by undulating the body, caudal peduncle, and tail to create forward thrust. Others move by flapping, fanning, or sculling the fins. Eels move much like snakes, winding their way through the water and over the bottom. Skates undulate their pectoral-fin lobes, whereas rays flap them. Many fishes have developed ingenious forms of locomotion. Searobins "crawl" across the bottom by "walking" with their free pectoral-fin rays. Remoras, while able to swim freely, have an adapted dorsal fin that forms a suction disk. This disk allows the Remora to "hitch a ride" on its host. Some fishes hardly swim at all. The Grunt Sculpin spends most of its life hopping and crawling along high-energy, wave-swept rocky bottoms.

Identifying Fishes

All fishes change shape and most change color as they develop. Juveniles can be drastically different from their adult counterparts. Adults of the same sex and species often have subtle differences. Numerous fishes are sexually dimorphic, meaning males and females are different in color and form. Many fishes change color and pattern depending on the time of day, time of migration, diet, depth, mood, or breeding phase. Some change color when they are hunting; others change color to appear intimidating. In addition, almost every fish changes color when it is caught, when it is in distress, or after it has died.

Introduction

Depending on the subject at hand, identification can be an easy or daunting task. Some species of fishes are so unique that they do not resemble any other and are thus easy to identify. Others are so similar in appearance that only subtle nuances distinguish one from another. Even though the variety and changes in the color of fishes is great, observed colors and patterns are the most common tools used for identification.

Below are several examples of commonly observed color patterns in Pacific coast fishes.

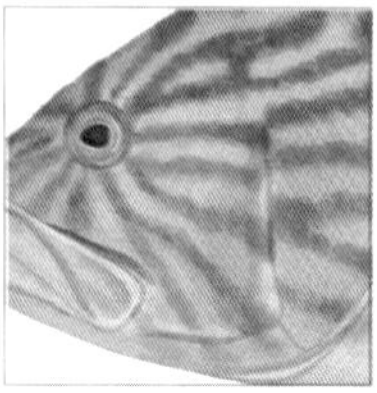

dark lines radiating from eyes

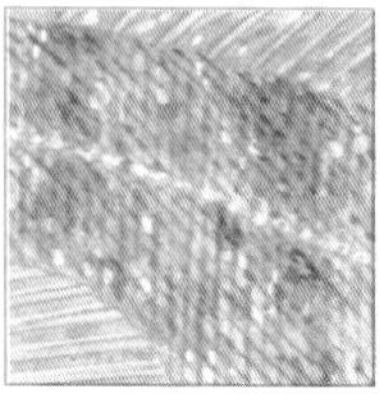

mottled pattern

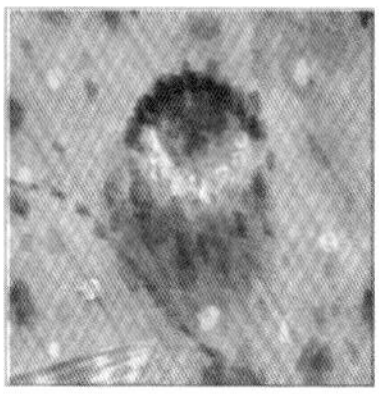

eye spot

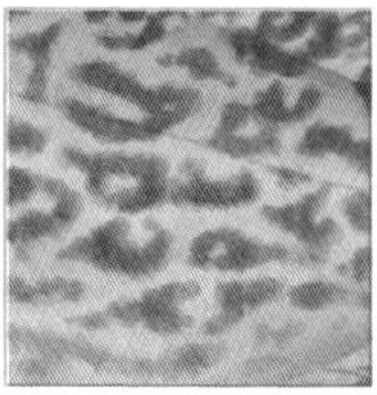

vermiculating pattern

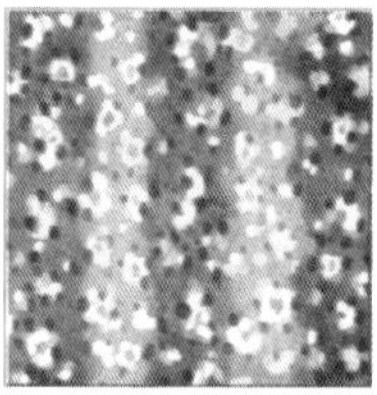

alternating pale and dark bars

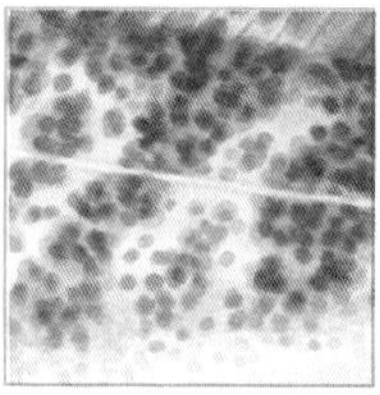

spots forming oblique bars

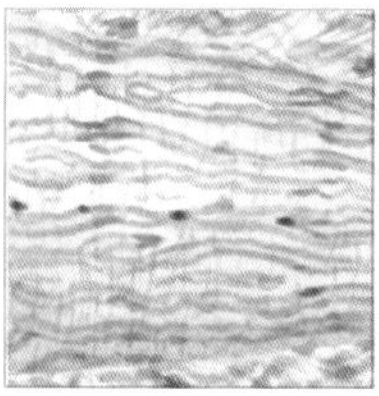

dark, wavy stripes

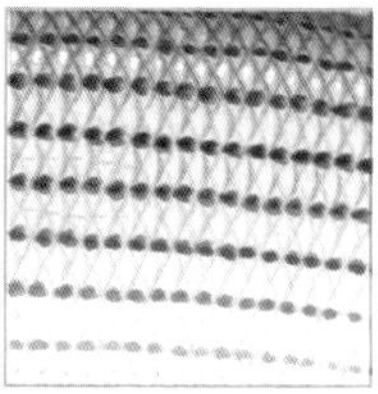

dark spots forming stripes along scales

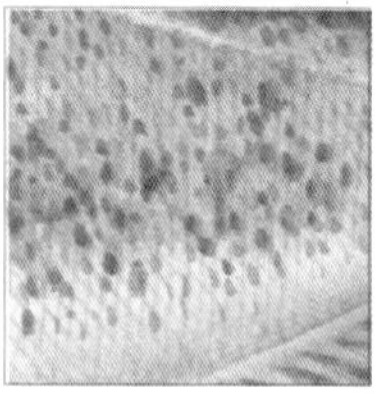

dark speckles

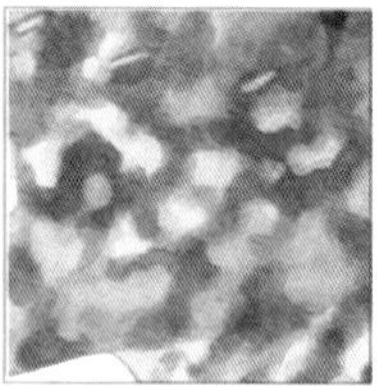

irregular spots and blotches

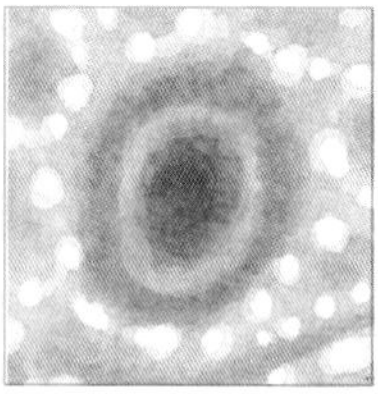

large, dark ocellated spot

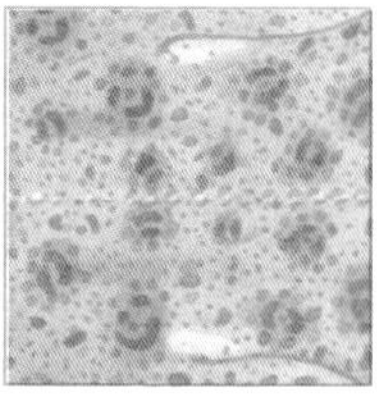

rosettes

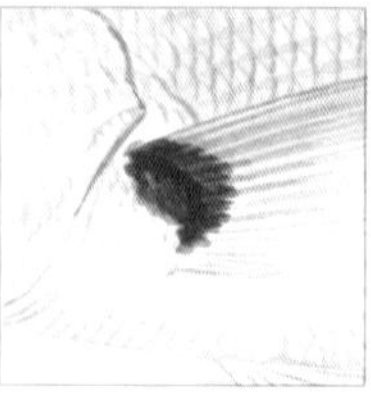

dark blotch on pectoral-fin base

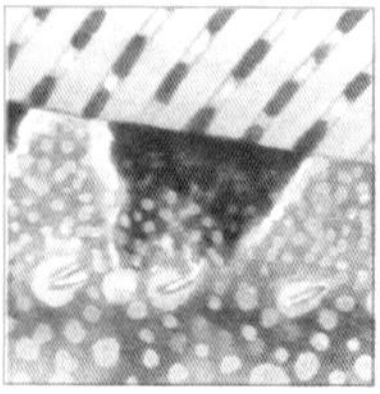

dark saddle

banded pectoral fin

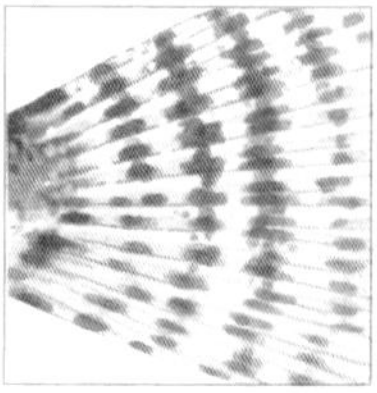

banded caudal fin

When a fish is not identifiable by either color or pattern, the fish's anatomy can help to secure an identification. Shape, size, and placement of anatomical features vary from one species to the next and thus distinguish one from another. The following illustrations show the primary external features of several cartilaginous and bony fishes that are commonly used as tools for identification.

Cartilaginous Fishes

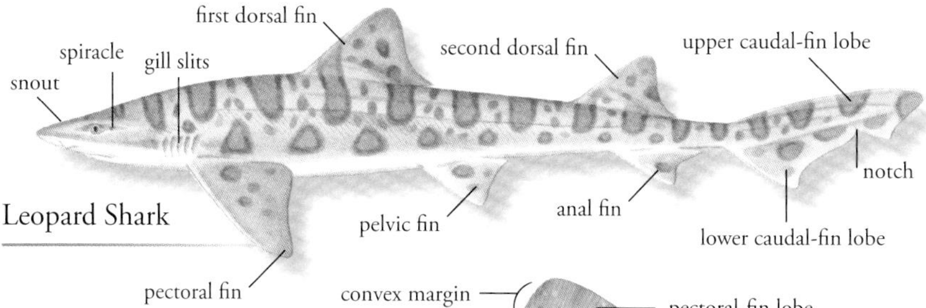

Leopard Shark

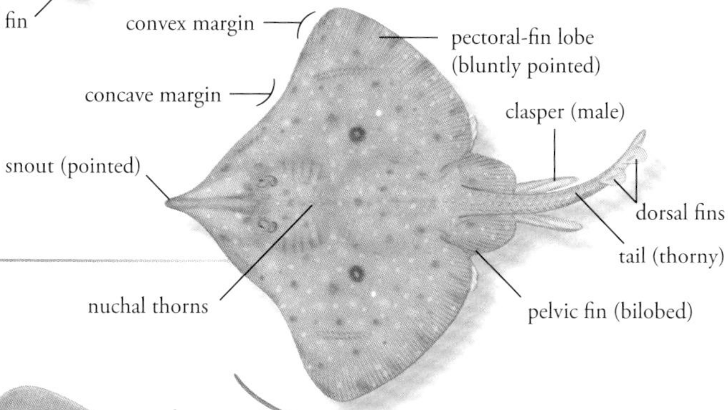

Longnose Skate

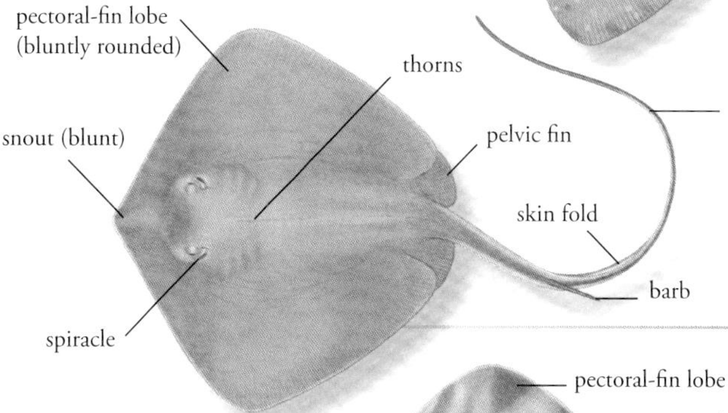

Diamond Stingray

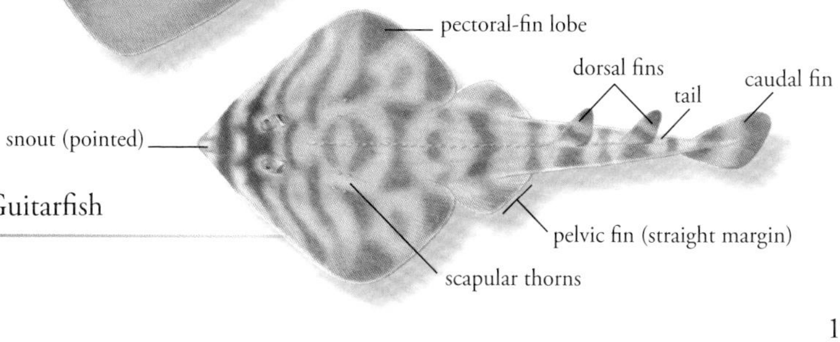

Banded Guitarfish

Bony Fishes

Stone Scorpionfish

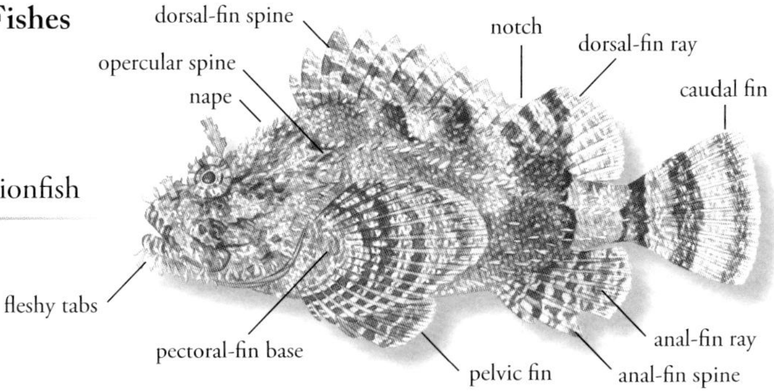

Bonyhead Sculpin

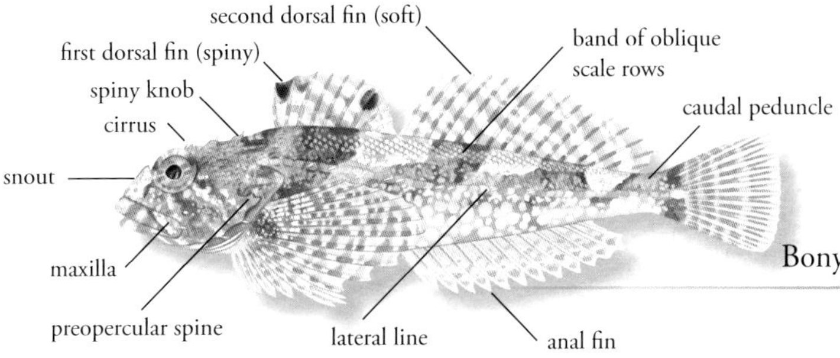

Bigeye Trevally

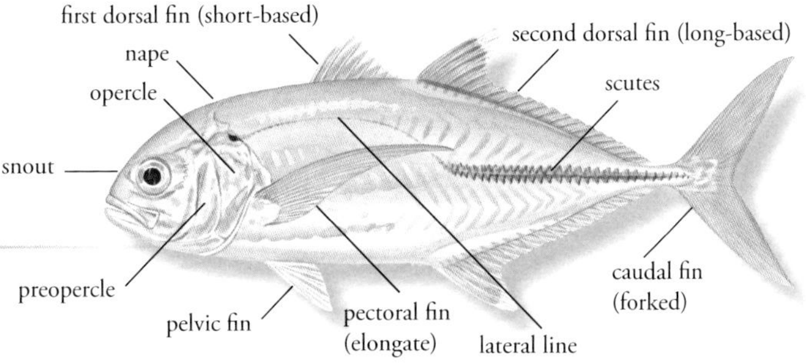

Redtail Surfperch

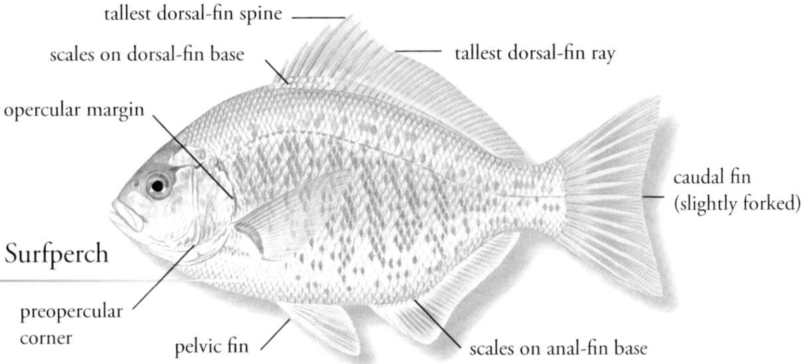

Lengths and proportions will also help in differentiating one fish from another. Total lengths are used in this book. To determine total length, measure from the tip of the snout or the tip of the lower jaw to the tip of the caudal fin. If the caudal fin is forked, press the upper and lower tips toward each other. The lengths provided in the accounts give the reader a general idea of how big a species may become. The illustration below shows positions of specific identifying features and indicates lengths and depth.

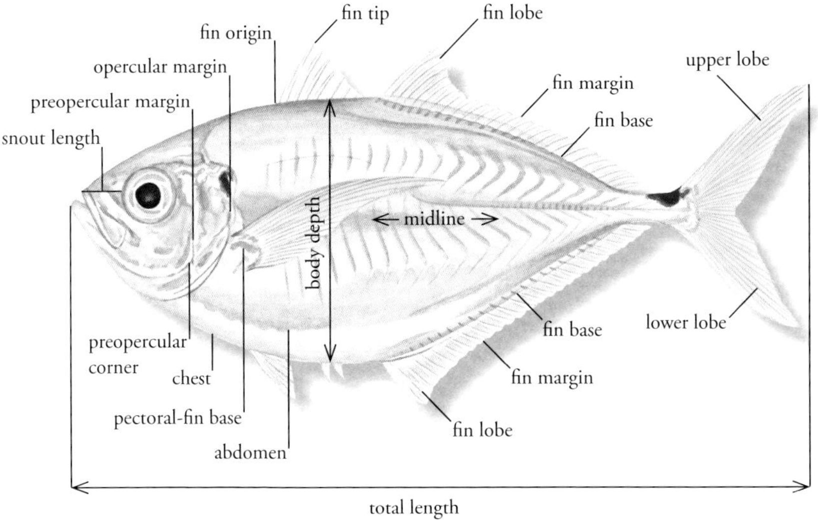

The overall shape of a fish's body can also help in identification. Shapes of the fishes presented in this book are in profile, as a fish would appear from the side. Proportions of a fish's depth relative to its length are important identifiers. A fish is said to be "deep-bodied" when the measurement of depth is great relative to length, such as in the Popeye Catalufa. A fish is said to be elongate when its depth is small relative to length, such as in the Longnose Lancetfish. Some examples of profile are shown below.

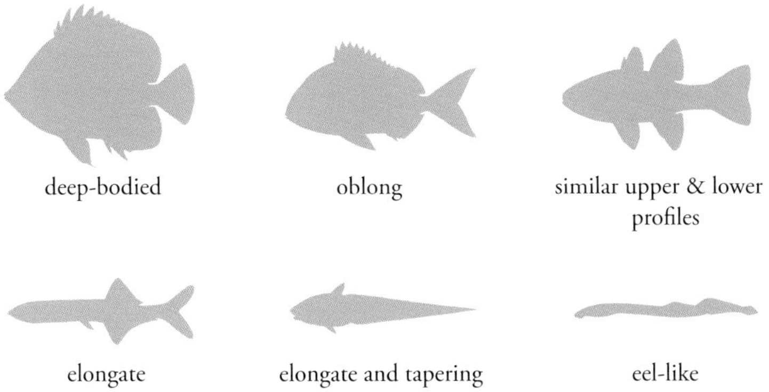

The cross-sectional shape of a fish is the shape of the body as it appears head on. A fish that is flattened from side to side is said to be laterally compressed. A fish that is flattened from top to bottom is described as flattened or depressed. Below are some simplified cross-sectional views.

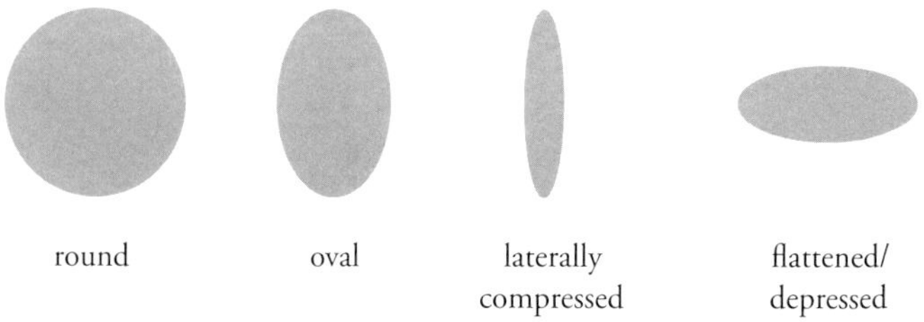

Sometimes it is necessary to go further and count spines, rays, or scales to determine the identity of a fish. When counting spines and rays, the norm is to count anteriorly (front) to posteriorly (toward tail). Even if a spine is very small or directed forward, it is still counted. Additionally, if the last ray on the dorsal or anal fin is split to a unified base, it is still counted as a single ray.

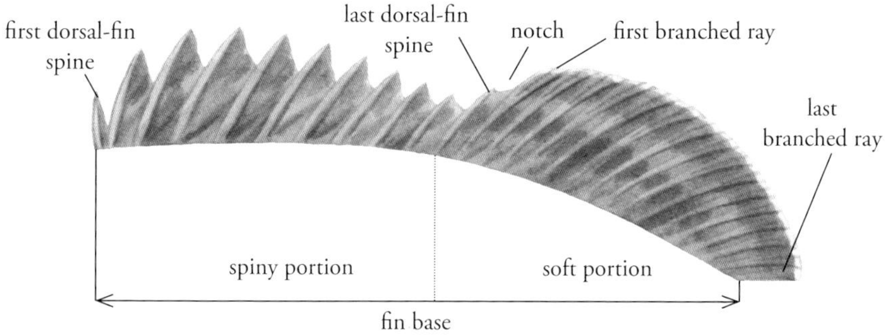

There are times when lateral-line scales need to be counted. These are counted from the first pored, lateral-line scale behind the opercle to the last pored scale that corresponds to the crease in the caudal peduncle when the caudal fin is moved from side to side. At other times, scales above the lateral line are counted. These are counted from the highest arch of the lateral line diagonally backward to the base of the dorsal fin.

The habitat, geographic area, and depth range may also help in identification. For example, the blue-sided and blue-blotched forms of Blue Rockfish, *Sebastes mystinus,* have very similar appearances and features. However, their ranges barely overlap. When using depth as an identifier, it should be noted that many recorded depths are those taken by trawl or line. These records reflect the deepest point of the trawl or line and not necessarily the deepest level at which a fish may swim. Depth records may also reflect seasonal migrations or the preferred depth of juveniles or adults.

Conservation

Many people have devoted countless hours to the conservation of our precious marine environments. It cannot be overstated how important it is to reduce, and possibly reverse, the harm due to run-off, pollution, destruction of habitat, and overfishing. If each person takes one small step toward conservation and preservation, the overall affect could be tremendous.

Many fishes included in this book are currently Threatened or Endangered. The IUCN (International Union for Conservation of Nature) Red List of Threatened Species contains an evaluation of the extinction risk of global plants and animals. This information plays a significant role in guiding conservation activities and serves to monitor changes in the conservation status of species. Of the many fish species that are monitored, sharks are one of the most vulnerable. They have been heavily exploited and are largely misunderstood and misrepresented. Sharks grow slowly, have a long gestational period, and do not produce a large quantity of offspring. Therefore, it is very difficult for them to recover from the depletion they have suffered.

On the positive side, there are several examples of population recovery due to conservation. One is the Giant Sea Bass. This enormous fish was severely overfished in the early 1900s. California passed legislation to prohibit taking this fish (with few exceptions) in 1981, and Mexico protected it in 1992. While it has made significant recovery, it is still listed as Critically Endangered on the IUCN Red List. The White Seabass was also in severe decline until laws restricted recreational catches and prohibited gill-netting this fish. Aided by captive breeding, this fish is now listed as Least Concern.

We encourage our readers to educate themselves about the fishes they encounter. Most fishes have defense mechanisms meant to protect them from other fishes. Many fishes will defend themselves if threatened. A venomous fish would rather be left alone than use its precious venom. We also encourage our readers to

- Handle fishes gently.
- Please respect seasonal and catch limits.
- Obtain proper permits and licenses, as many states use sales revenues to fund conservation and enforcement.
- Do not release non-native fishes into any open body of water.
- Whenever possible, practice catch-and-release and respirate the fish before releasing it back into the water.
- Release native live bait back into the water.
- Dispose of waste properly.
- Anchor in designated areas.
- Participate in fisheries management by reporting tagged fishes.
- Use circle hooks to help prevent the fish from swallowing the hook.

In short, if we care for fishes and their environment, they will be here in the future for us and future generations to admire.

FAMILIES

Order MYXINIFORMES - Hagfishes

Family Myxinidae - Hagfishes

Hagfishes are finless, scaleless, and eel-shaped. They possess a single nostril above the mouth. True jaws are absent. The mouth is circled by a series of barbels and bears rows of keratinized, rasp-like teeth. Eyes are rudimentary and lack lenses and irises. Hagfishes are bottom-dwelling scavengers that occur worldwide in marine and estuarine waters. Three species in the area. Page 60.

Order PETROMYZONTIFORMES - Lampreys

Family Petromyzontidae - Northern Lampreys

Lampreys are long, scaleless, and eel-shaped. There is a single nostril above the disk-shaped, jawless mouth. The mouth consists of a series of horny teeth arranged in a circular pattern. A series of gill pouches connects separately to a row of external openings. Dorsal fins are paired on the posterior half of the body. Lampreys are anadromous or land-locked in northern temperate waters. Adults are parasitic or non-parasitic. Three species in the area. Page 60.

Order CHIMAERIFORMES - Chimaeras

Family Chimaeridae - Shortnose Chimaeras

Shortnose Chimaeras have a large head with a blunt snout. Lateral-line canals form grooved lines over the snout and head and along the body. First dorsal fin is erect, with a spine that is toxic in some. The tail tapers to a pointed tip. Shortnose Chimaeras occur on or near deep muddy bottoms of the Atlantic and Pacific oceans. They swim by flapping broad, wing-like pectoral fins. Females lay paired egg cases. One species in the area. Page 60.

Order HETERODONTIFORMES - Bullhead Sharks

Family Heterodontidae - Bullhead Sharks

Bullhead Sharks are small and stout-bodied, with two spined dorsal fins. The snout is short and "pig-like." Mouth is located ventrally and has rows of sharp front teeth, plus flattened rear teeth used for crushing prey. Head is squarish with raised ridges over eyes. Bullhead Sharks are found scattered circumglobally in shallow temperate to tropical seas. One species in the area. Page 62.

Order ORECTOLOBIFORMES - Carpet Sharks

Family Rhincodontidae - Whale Shark

The Whale Shark is the largest fish on Earth. The mouth is wide, with five rows of very long gill plates. The eyes are small and just behind the mouth. Three dorsal ridges are present, with the lowest ridge becoming a strong keel at the caudal peduncle. The first dorsal fin is larger than the second, and the caudal fin is very tall. Whale Sharks occur circumglobally in warm seas. They filterfeed mostly on plankton. One species in this family. Page 62.

Order LAMNIFORMES - Mackerel Sharks

Family Odontaspididae - Sand Tiger Sharks

Sand Tiger Sharks are stout, with a conical, depressed snout. The large mouth extends beyond the eyes and contains protruding teeth. Gill slits are low on the body and anterior to the pectoral fins. Dorsal, pelvic, and anal fins are similar in size. Pectoral fins are relatively small. They occur worldwide in warm marine waters. Sand Tiger Shark embryos feed on lesser developed embryos, fertilized eggs, and unfertilized eggs before birth. One species in the area. Page 62.

Family Mitsukurinidae - Goblin Shark

The Goblin Shark has an elongate and flattened snout and highly protrusible jaws. The teeth are long and slender. Dorsal fins are low and similar in shape. Caudal fin is long and low. Body is elongate and soft. Goblin Sharks are scattered circumglobally in temperate seas. They occur in deep water over continental shelves, upper slopes, and around seamounts to about 4,200 ft. Rarely encountered at the surface. May use modified snout to detect prey, which they quickly grab with their specialized jaws. One species in this family. Page 62.

Family Megachasmidae - Megamouth Shark

The Megamouth Shark is large, with a broad, rounded head and a wide, gaping mouth. Jaws contain numerous rows of small, hooked teeth. Gill slits are long, with the last two over the pectoral-fin base. Pectoral fins are long and pointed. First dorsal fin is taller than the second. Caudal fin is long and low. They occur circumglobally in tropical to warm temperate seas. Found coastally to well offshore over very deep water. Near surface at night, in deep water during the day. Megamouth Sharks filter-feed on plankton using modified gill rakers. First discovered in 1976, with only 24 known specimens known up to 2004, it is poorly understood. One species in this family. Page 64.

Family Alopiidae - Thresher Sharks

Thresher Sharks have a round, streamlined body and a very long, asymmetrical caudal fin with upper lobe almost as long as body. The head is short and the snout is pointed. The mouth is small. The fourth and fifth gill slits are above the pectoral-fin base. Pectoral fins are as long as or longer than the head. Second dorsal fin and anal fin are very small. Thresher Sharks occur worldwide in tropical to cold seas. The elongated tail is used to disrupt schooling fishes and to stun prey. Three species in the area. Page 64.

Family Cetorhinidae - Basking Shark

The Basking Shark is very large and stout. The snout is pointed and may be long. Eyes are relatively small. The large mouth opens to reveal five long gill plates. Gill slits nearly circle the throat and are anterior to pectoral fins. The caudal peduncle possesses a distinct keel. Basking Sharks live worldwide in temperate seas and are usually found near the surface. They swim slowly through the water with their mouths agape. Plankton is captured in the numerous gill rakers. One species in this family. Page 64.

Order LAMNIFORMES - Mackerel Sharks, *cont.*

Family Lamnidae - Mackerel Sharks

Mackerel Sharks have a round body and a conical snout. The mouth is large, with triangular-shaped teeth. Gill slits are anterior to pectoral fins. Gill rakers are absent. Second dorsal and anal fins are small. Caudal peduncle is heavily keeled. Caudal-fin lobes are similarly sized. Mackerel Sharks are found worldwide in tropical to temperate seas. They have specialized circulatory systems that keep them warmer than the ambient water temperature. Ancestors of this family reached a length of 65 ft. Three species in the area. Page 66.

Order CARCHARHINIFORMES - Ground Sharks

Family Scyliorhinidae - Catsharks

Catsharks are typically small and cigar-shaped with a low, asymmetrical caudal fin. The snout is short and depressed. The mouth extends behind the posterior margin of the small, oval to slit-like eyes. The first dorsal fin is small and originates above or posterior to the origin of the pelvic fins. Catsharks are bottom-dwelling in circumglobal temperate to tropical marine waters from intertidal areas to about 6,500 ft. This is a large and diverse family with 65 known species. Three species in the area. Page 66.

Family Triakidae - Hound Sharks

Hound Sharks are slender with a low, asymmetrical caudal fin, a depressed snout, and slit-like eyes. Gill slits are small and spiracles are close to the eyes. Dorsal and ventral fins are similarly shaped. Hound Sharks are primarily demersal and occur worldwide in warm temperate and tropical seas, rarely in fresh water. They occur from near shore to the outer continental shelf and feed on invertebrates and bony fishes. Five species in the area. Page 68.

Family Carcharhinidae - Requiem Sharks

The Requiem Shark family is diverse and has about 50 species with several common traits. The head is neither flattened nor laterally expanded. Eyes are circular or oval with well-developed nictitating membranes. Spiracles are absent or very small. The mouth is usually large and extends well beyond the eyes. The caudal fin is long and asymmetrical, with rippled or undulating dorsal margin. Requiem Sharks are circumglobal in a wide variety of marine to freshwater habitats. They are strong and active. Seven species in the area. Page 70.

Family Sphyrnidae - Hammerhead Sharks

Hammerhead Sharks are moderately slender with a greatly flattened and laterally expanded head. The eyes and nostrils are on the outer margins of the modified head. Spiracles are absent. The first dorsal fin is tall. Hammerhead Sharks occur worldwide in warm marine to brackish waters. The flattened head creates more surface area for increased electrochemical perception. The position of the eyes enhances vision. Most hammerhead sharks are shy and difficult to approach. Three species in the area. Page 72.

Order HEXANCHIFORMES - Sixgill sharks

Family Chlamydoselachidae - Frill Sharks

Frill Sharks are moderately large and serpent-like. The mouth is broad and gaping, with rows of three-cusped teeth. There are six gill slits, with the first continuous around the throat. There is a single, low dorsal fin that is much smaller than the anal fin. Caudal fin is long and low. Frill Sharks occur over continental and insular slopes in scattered worldwide seas. They are occasionally found near shore. This family has one described species and possibly a second undescribed species. One species in the area. Page 74.

Family Hexanchidae - Cow Sharks

Cow Sharks are slender to stout in shape and are distinguished by the presence of six or seven long gill slits. Most other shark families have five gill slits. The mouth is very long. Cow sharks have only one dorsal fin, which is located close to the caudal fin. The caudal fin is long and low. Cow Sharks occur circumglobally in shallow to deep marine waters from bays to submarine canyons in temperate to boreal waters. Two species in the area. Page 74.

Order ECHINORHINIFORMES - Bramble Sharks

Family Echinorhinidae - Bramble Sharks

Bramble Sharks are relatively large and stout. Body and fins are covered with irregular, thorn-like denticles. The dorsal fins are small and similar in size. Anal fin is absent. Caudal fin is low and pointed. Bramble Sharks are bottom-dwelling over continental shelves and slopes. They feed on other sharks, fishes, and crabs. Litters contain 15 to 24 young. One species in the area. Page 74.

Order SQUALIFORMES - Dogfish Sharks

Family Squalidae - Dogfish Sharks

Dogfish Sharks are moderately slender, with a conical or depressed head. The spiracles are relatively large and close to the eyes. Gill slits are small and low on the body. Dorsal fins are small, with concave margins, and have a small or prominent grooveless spine. The anal fin is absent. The caudal fin is low and asymmetrical. Dogfish Sharks occur worldwide in arctic to tropical seas. They have strong jaws and prey mainly on fishes and invertebrates. Some form large schools. One species in the area. Page 74.

Family Somniosidae - Sleeper Sharks

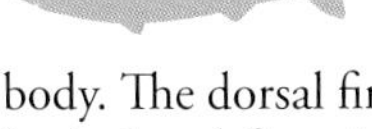

The Sleeper Shark family includes small to giant species. Spiracles are large and set just behind the eyes. The gill slits are low on the body. The dorsal fins usually lack small, grooved spines. When present, spines occur on both dorsal fins. Anal fin is absent. Caudal fin is strongly asymmetrical and notched. Most possess luminous organs. Sleeper Sharks occur circumglobally and primarily close to the bottom over continental shelves and slopes. Some are oceanic or near shore. Prey is diverse. Females bear live young. One species in the area. Page 76.

Order SQUALIFORMES - Dogfish Sharks, *cont.*

Family Dalatiidae - Kitefin Sharks

Kitefin Sharks are very small to moderately large. The head is short, the eyes are large. Some have modified jaws to take cookie-like bites out of prey. The gill slits are small and anterior to the pectoral fins. Dorsal fins either lack spines, or the first dorsal fin may have a small, grooved spine. Luminous organs are present on ventral surface. Kitefin Sharks are typically demersal in deep water over circumglobal continental slopes. One species in the area. Page 76.

Order SQUATINIFORMES - Angel Sharks

Family Squatinidae - Angel Sharks

Angel Sharks have flattened heads, bodies, and fins. The eyes and spiracles are on top of the head. The nostrils and mouth are at the front of a rounded snout. Pectoral fins are separate and triangular. Dorsal fins are near the end of the tail. Angel Sharks occur worldwide at or near bottom in tropical to temperate seas, from shore to about 4,200 ft. One species in the area. Page 76.

Order TORPEDINIFORMES - Electric Rays

Family Torpedinidae - Torpedo Electric Rays

Torpedo Electric Rays are flattened with head, body, and pectoral fins forming an almost-circular disk. The eyes and spiracles are on top of the head. Pelvic fins are rounded. First dorsal fin is larger than second. Caudal fin is triangular. Kidney-shaped electric organs are visible from above. They occur circumglobally in tropical to temperate seas. Can discharge up to 45 volts to stun prey and defend against predators. One species in the area. Page 76.

Order RAJIFORMES - Skates

Family Rhinobatidae - Guitarfishes

Guitarfishes are moderately flattened, with a wedge-shaped snout. The pectoral fins are moderately broad and rounded. Eyes and spiracles are on top of the head. Small, rounded teeth are used to crush hard-shelled prey. Dorsal fins are located on top of an elongated caudal peduncle. Guitarfishes are demersal over shallow sandy and muddy bottoms of coastal areas in tropical to warm temperate seas. Two species in the area. Page 78.

Family Rajidae - Skates

Skates are flattened, with broadly expanded pectoral fins. The snout may be elongate or blunt. The pelvic fins are typically bilobed and the tail is moderately slender, with two small dorsal fins at the end. Eyes and spiracles are on top of the head. Most have denticles and thorns dorsally. Skates are bottom-dwelling in worldwide polar to tropical seas. Males use modified pelvic fin claspers to internally fertilize females. Thirteen species in the area. Page 78.

Order MYLIOBATIFORMES - Stingrays

Platyrhinidae - Thornbacks

Thornbacks are round to heart-shaped with moderately broad pectoral fins. The tail is long and stout and has two large dorsal fins located posteriorly. A barbed spine is absent. Eyes and spiracles are on top of the head. There are one or three rows of strong thorns along the back and on dorsal surface of the tail. Thornbacks are bottom-dwelling over continental shelves in tropical to cool-temperate waters of the north Pacific. Females give birth to live young. They feed on bottom-dwelling invertebrates and fishes. Only three species in this family. One species in the area. Page 84.

Family Urotrygonidae - American Round Stingrays

American Round Stingrays have broadly rounded pectoral fins that form a nearly circular shape. The tail is about equal to the disk length and has one to two barbed spines and a small but well-developed caudal fin. A dorsal fin is present in some species. Eyes and spiracles are on top of the head. Round Stingrays occur over shallow continental shelves and upper slopes in tropical and warm-temperate waters of the eastern Pacific. Females give birth to live young. Members of this family are also called "Stingarees." One species in the area. Page 84.

Family Dasyatidae - Whiptail Stingrays

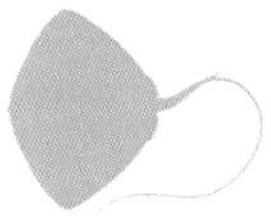

Whiptail Stingrays are moderate to very large in size. The body is flattened, expanded, and rhomboid or oval in shape. The pectoral fins are very broad and extend to the tip of the snout, which may be pointed or blunt. Eyes and spiracles are on top of head. Pelvic fins are single-lobed. Caudal fin is absent. The tail is long and whiplike and possesses one or more serrated, venomous spines. The spine is used in self-defense. Whiptail Stingrays occur worldwide in tropical to warm temperate seas. Most are demersal. They prey on fishes and invertebrates. Two species in the area. Page 84.

Family Gymnuridae - Butterfly Rays

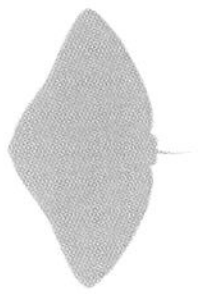

Butterfly Rays are very flattened and laterally expanded. The pectoral fins and body form a diamond-shaped disk. Eyes and spiracles are on top of the head. The tail is short and pointed and may have a serrated spine. Caudal fin is absent. They occur worldwide in tropical to temperate seas. Demersal over shallow coastal sandy and muddy bottoms, also in estuaries and river mouths. Feed on benthic invertebrates and fishes. One species in the area. Page 86.

Family Myliobatidae - Eagle Rays and Mantas

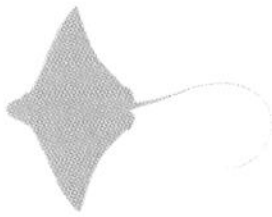

Eagle Rays and Mantas are moderate to large in size, flattened, and rhomboid to diamond-shaped. The pectoral fins are broad, pointed, and attached to the elevated head. Eyes and spiracles are on the sides of the head. Tail may be long and whiplike and may have a serrated spine. Caudal fin is absent. Eagle rays occur worldwide in tropical to temperate seas. Some feed on mollusks and crustaceans, others on plankton. Eagle Rays swim in a flapping motion and may leap from water in pursuit of prey. Three species in the area. Page 86.

Order ACIPENSERIFORMES - Sturgeons

Family Acipenseridae - Sturgeons

Sturgeons are elongate and stout with five rows of keeled, bony scutes on body. The head is covered in bony plates. The mouth is inferior and preceded by four barbels. Sturgeons occur in Northern Hemisphere fresh water or are anadromous. Demersal. They still resemble primitive fishes. Females scatter thousands of demersal eggs. Two species in the area. Page 88.

Order ELOPIFORMES - Tenpounders

Family Elopidae - Tenpounders

Tenpounders are elongate and cylindrical in shape. The mouth is large, with upper jaw extending past the rear margin of large eyes. There is a single dorsal fin, and the caudal fin is deeply forked. Tenpounders occur primarily in coastal waters of tropical to subtropical oceans. Some enter brackish or fresh water. Larvae are transparent and ribbon-like. One species in the area. Page 88.

Order ALBULIFORMES - Bonefishes

Family Albulidae - Bonefishes

Bonefishes are elongate and cylindrical in shape with a single dorsal fin and a deeply forked caudal fin. The snout is conical and the mouth is subterminal. The body is semi-translucent. Bonefishes occur near shore in worldwide tropical to warm temperate seas. They are active fishes that forage over sandy and muddy bottoms for invertebrates and small fishes. Bonefish larvae are transparent and ribbon-like. One species in the area. Page 88.

Order ANGUILLIFORMES - Eels

Family Muraenidae - Moray Eels

Moray eels are elongate and somewhat laterally compressed. The dorsal and anal fins merge with the caudal fin. The snout is short to relatively long. Teeth are well developed. Pectoral fins are absent. The body is scaleless. Moray eels occur worldwide in primarily tropical to subtropical seas. Most are marine, inhabiting rocky or coralline holes and crevices. Some live around mangroves and tidal creeks. They are predators and scavengers. The flesh of some is toxic. One species in the area. Page 88.

Family Ophichthidae - Snake Eels and Worm Eels

Snake and Worm Eels are elongate and serpentine or worm-like. The caudal fin is usually absent, with the tail ending as a hardened tip. When present, dorsal, anal, and caudal fins are confluent. Snout is short to moderately elongate. Body is scaleless. Pectoral fins are present or absent. They occur in a variety of worldwide tropical to warm temperate marine and brackish habitats. Three species in the area. Page 90.

Order CLUPEIFORMES - Herrings

Family Engraulidae - Anchovies

Anchovies are relatively small, with a blunt, rounded snout and a single dorsal fin. The snout extends beyond the jaws. Jaws are long and slender and extend almost to end of gill cover. Eyes are large. Lateral line is absent. Many have a thin to wide silvery stripe on each side. Scales are delicate and easily shed. Anchovies occur worldwide in tropical to temperate marine and brackish waters. Most are schooling and filter-feed on plankton. Many are important prey for larger fishes, marine birds, and mammals. Four species in the area. Page 90.

Family Clupeidae - Herrings

Herrings are cylindrical in shape or laterally compressed. The body is typically silvery. The mouth is usually upturned and may have a notch at the upper jaw tip. A row of scutes is usually present along the abdomen. The dorsal fin is single and the caudal fin is deeply forked. Herrings occur worldwide in tropical to polar seas. They are usually marine, coastal, and schooling. Some tolerate low salinities; others are anadromous. Most are plankton filter-feeders with numerous gill rakers. Many are important prey for larger fishes, marine birds, and mammals. Eight species in the area. Page 92.

Order ARGENTINIFORMES - Marine Smelts

Family Argentinidae - Argentines or Herring Smelts

Argentines are elongate, with small mouths and large eyes. The single dorsal fin is located at midbody and originates in front of pelvic fins. An adipose fin is always present. Scales are easily shed. They are pelagic or demersal over outer shelves and upper slopes of worldwide tropical to warm temperate seas. Prey on planktonic invertebrates and small fishes. One species in the area. Page 94.

Family Microstomatidae - Pencilsmelts

Pencilsmelts are small, elongate, and slender, with a small mouth and large eyes. The single dorsal fin is located at about midbody and is above or behind pelvic fins. An adipose fin is present in some species. Pencilsmelts are marine and pelagic in subpolar to tropical seas. Family account only.

Order OSMERIFORMES - Freshwater Smelts

Family Osmeridae - Smelts

Smelts are small and elongate, with large mouths and eyes. The single dorsal fin originates posterior to the midbody line. An adipose fin is always present. Anal fin may be expanded in males. Smelts occur in the Northern Hemisphere in coastal marine and fresh waters. Many species are anadromous. May form large schools. Smelts are an important commercial and environmental food source. Seven species in the area. Page 94.

Order SALMONIFORMES - Trouts and Salmons

Family Salmonidae - Salmonids

Salmonids are relatively long and fusiform. They have a single, spineless, midbody dorsal fin and an adipose fin. The mouth is moderately large; the eyes are small. Their coloring changes with age and environment. Salmonids are native to Northern Hemisphere waters but have been successfully introduced into other areas. Some are landlocked in fresh water, and many are anadromous and undergo extensive spawning migrations. Historically and commercially important. Thirteen species in the area. Page 98.

Order STOMIIFORMES - Dragonfishes

Family Gonostomatidae - Bristlemouths

Bristlemouths are elongate, moderately compressed, and typically small. Jaws are large and gaping and contain rows of numerous, fine teeth. Eyes are small to large. Adipose fin is present or absent. Photophores are present on the body, with those on abdomen in nearly evenly spaced rows. Scales, when present, are large and easily shed. Bristlemouths occur in deep midwaters of all tropical and temperate seas. One species in the area. Page 104.

Family Sternoptychidae - Marine Hatchetfishes

Marine Hatchetfishes are small and compressed, with many "hatchet-shaped" in profile. The mouth is large and vertical. Eyes are tubular in some. Some with a dorsal blade preceding dorsal fin. An adipose fin is usually present and often long-based. Photophores are often large and arranged in a series or in rows on sides and along ventral profile. The back is usually blackish, the sides silvery. Marine Hatchetfishes occur in deep open waters of worldwide tropical to temperate seas. Six species in the area. Page 104.

Family Stomiidae - Barbeled Dragonfishes

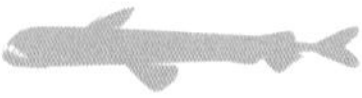

Barbeled Dragonfishes are elongate, with large toothy jaws that may extend well past the eyes. Gill rakers are absent in adults. A very long to minute chin barbel is present in most. Pectoral fins, when present, are minute to fan-like. An adipose fin and scales may be present or absent. Some with an elongate first dorsal-fin ray. Photophores may be present on the head or arranged in a series or in rows on body. Barbeled Dragonfishes occur in deep water of worldwide tropical to temperate seas. One species in the area. Page 104.

Order AULOPIFORMES - Lizardfishes

Family Aulopidae - Flagfins

Flagfins are elongate and oval in cross-section. The mouth is wide and toothy. The single dorsal fin is expanded and originates on the anterior third of the body. Adipose fin is present. Flagfins are demersal over continental shelves and slopes of worldwide tropical to warm temperate seas. One species in the area. Page 104.

Order AULOPIFORMES - Lizardfishes, *cont.*

Family Synodontidae - Lizardfishes

Lizardfishes are elongate and cylindrical in shape. The mouth is wide and toothy. The single dorsal fin is located over the midbody line. Adipose fin is present. Pelvic fins are relatively large. Lizardfishes are demersal over a variety of soft to hard bottoms in the Atlantic, Pacific, and Indian oceans. Some occur in brackish waters. They are swift, voracious predators that bury in bottom sediment. One species in the area. Page 106.

Family Alepisauridae - Lancetfishes

Lancetfishes are elongate, slender, and laterally compressed. The mouth is large and toothy and extends past the large eyes. The dorsal fin is long-based and sail-like. A small adipose fin is present. The caudal peduncle has a lateral keel. The body is scaleless and flaccid. Lancetfishes are found worldwide in warm temperate to tropical seas from surface to about 3,300 ft. One species in the area. Page 106.

Order MYCTOPHIFORMES - Lanternfishes

Family Myctophidae - Lanternfishes

Lanternfishes are small and slender to somewhat deep-bodied, with a large head and eyes. The snout is short and pointed to blunt. The mouth is very large. Adipose fin is present. Most possess a series of photophores along the abdomen and on the sides. Lanternfishes are silvery to blackish, with easily shed scales. They occur in deep water of worldwide tropical to polar seas. Many vertically migrate to surface waters at night. Seven species in the area. Page 106.

Order LAMPRIFORMES - Opahs

Family Lampridae - Opahs

Opahs are deep-bodied, compressed, and oval in profile. The mouth is small and protrusible. The pectoral fins are vertically oriented. Dorsal and anal fins are long-based. Scales are minute and reflective. Opahs are pelagic and found worldwide in tropical to temperate seas. They may wander north during the summer. They swim by flapping their pectoral fins. One species in the area. Page 108.

Family Lophotidae - Crestfishes

Crestfishes are long to very long and laterally compressed. A fleshy crest or horn extends from the head. The crest or horn bears an elongate dorsal-fin spine that supports multiple dorsal-fin rays. The remainder of the dorsal fin extends to the caudal peduncle. Pelvic fins are small and rudimentary or absent. Caudal fin is small. Crestfishes occur worldwide in tropical to warm temperate seas from surface to about 3,300 ft. May secrete an ink-like fluid when threatened. One species in the area. Page 108.

Order LAMPRIFORMES - Opahs, *cont.*

Family Trachipteridae - Ribbonfishes

Ribbonfishes are elongate and laterally compressed. The body is deep anteriorly and tapers to a very narrow caudal peduncle. The eyes are large and the mouth is protrusible. The dorsal fin extends from the top of the head to the tail, and the first several rays are elongate in some species. Caudal fin is upturned and fan-like in some. Anal fin is absent. Pelvic fins are well developed to rudimentary. Ribbonfishes occur worldwide in warm temperate to tropical seas from the surface to about 4,000 ft. Four species in the area. Page 108.

Family Regalecidae - Oarfishes

Oarfishes are very long and laterally compressed. The head is angular, the eyes are large, and the mouth is highly protrusible. The dorsal fin runs from the top of the head to the tail, with anterior rays that are long and trailing. The caudal fin is usually absent in large specimens. Anal fin is always absent. Pelvic fins are long and trailing and consist of a single ray with paddle-like projections. Oarfishes occur worldwide in warm temperate to tropical seas from surface to about 3,300 ft. They swim vertically in water column by undulating the dorsal fin. One species in the area. Page 108.

Order GADIFORMES - Cods

Family Macrouridae - Grenadiers

Grenadiers have a large head, a pointed to blunt snout, and a body that tapers to a pointed tail. The eyes are very small to large. A small chin barbel is usually present. The first dorsal fin is short-based and relatively tall. The second dorsal fin is long-based, low, and continuous with the anal fin. Caudal fin is absent. Some possess a light organ. They occur worldwide in tropical to polar seas. Most live near the bottom from about 330 to 13,000 ft. One species in the area. Page 110.

Family Moridae - Deepsea Cods

Deepsea Cods are relatively elongate with one, two, or rarely, three dorsal fins. They have one or two anal fins. The pelvic fins are small or filamentous. Caudal peduncle is narrow and caudal fin is small. Chin barbel is present or absent. Some possess a light organ. Deepsea Cods are found near the bottom of worldwide, deep continental shelves and slopes to about 8,200 ft. Six relatively shallow-water and rare species in the area. Family account only.

Family Merlucciidae - Merlucciid Hakes

Merlucciid Hakes are relatively elongate and laterally compressed posteriorly. The snout is long and depressed. The jaws are large and have strong, pointed teeth. The head has a V-shaped ridge. The first dorsal fin is short-based; the second is long-based and notched at midlength. Chin barbel is absent. Merlucciid Hakes occur in the northern and southern Atlantic, eastern Pacific, and western South Pacific oceans. Most are found near the bottom over continental shelves and upper slopes. Some occur inshore. One species in the area. Page 110.

Order GADIFORMES - Cods, *cont.*

Family Gadidae - Cods

Cods are small to large with moderately elongate bodies and one, two, or three dorsal fins. The mouth is large and a chin barbel is usually present. V-shaped ridge on head is absent. The swim bladder is not connected to the skull. Cods are typically cold-water marine fishes found in the Northern Hemisphere. Most are demersal. Seasonal migrations are common. Cods are commercially important and have been overfished. Five species in the area. Page 110.

Order OPHIDIIFORMES - Cusk-eels

Family Ophidiidae - Cusk-eels

Cusk-eels are elongate, with dorsal and anal fins that are continuous with the caudal fin and have rays of similar height. Chin barbel and opercular spine may be present or absent. Pelvic fins are usually present and are located on the throat. Cusk-eels are bottom-dwelling in shallow to deep waters of worldwide tropical to temperate seas. Two species in the area. Page 112.

Family Bythitidae - Viviparous Brotulas

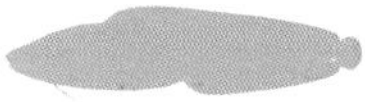

Viviparous Brotulas are moderately elongate. Dorsal and anal fins are either continuous with or free of caudal fin. Scales are present in most. Chin barbel is absent. Opercle has a well-developed spine. Pelvic fins, when present, have one or two rays. Viviparous Brotulas are bottom-dwelling in worldwide tropical to temperate seas. Some occur in freshwater caves. Females give birth to live young. One species in the area. Page 112.

Order BATRACHOIDIFORMES - Toadfishes

Family Batrachoididae - Toadfishes

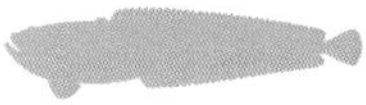

Toadfishes are small to medium in size. The head is broad and flattened, with eyes on top. Barbels or fleshy tabs are sometimes around mouth and head. First dorsal fin has two or three spines. Pelvic fins are on the throat. Toadfishes are demersal and occur worldwide in tropical to temperate waters. Most are coastal; some live in fresh water or on continental shelves. They are sluggish but voracious predators. Two species in the area. Page 114.

Order LOPHIIFORMES - Anglerfishes

Family Lophiidae - Goosefishes

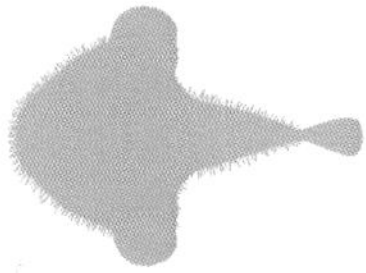

Goosefishes are moderately to greatly flattened, with a broad head and body that tapers to the tail. The mouth is broad, toothy, and usually bordered by fleshy tabs. Gill openings are behind the pectoral fins. The first dorsal-fin spine may be isolated on the snout and act as a lure. Goosefishes are bottom-dwelling in worldwide tropical to temperate seas. Two species in the area. Page 114.

Order LOPHIIFORMES - Anglerfishes, *cont.*

Family Antennariidae - Frogfishes

Frogfishes are very small to medium-sized, with rounded bodies. The mouth is large, oblique, and toothy. Gill openings are located behind limb-like pectoral fins. First dorsal-fin spine is separate and bears a fleshy lure. Frogfishes occur worldwide in tropical to subtropical seas. They use their pectoral fins to cling to or "walk" across substrate. Usually well camouflaged, often blending with surroundings. One species in the area. Page 114.

Family Ogcocephalidae - Batfishes

Batfishes are flattened, with head and body forming a circular to triangular disk. Rostrum is short to long. There is a short lure under the snout. Gill openings are behind limb-like pectoral fins. Tubercles and/or bucklers cover body. Batfishes are demersal in worldwide tropical to subtropical seas from shore to about 3,000 ft. Use pelvic and pectoral fins to "walk" across the bottom. One species in the area. Page 114.

Order MUGILIFORMES - Mullets

Family Mugilidae - Mullets

Mullets are medium to large in size. The head is typically broad and flattened. The eyes are usually partly covered by adipose lids. The snout is short and the mouth is small or moderate in size. Pectoral fins are high on the body. Mullets occur worldwide in tropical to temperate salt, brackish, and fresh water. Most tolerate varying salinities. They feed by filtering detritus or plankton. One species in the area. Page 116.

Order ATHERINIFORMES - Silversides

Family Atherinopsidae - New World Silversides

New World Silversides are typically small, translucent fishes with a silvery stripe on each side. The upper jaw is protractile. Pectoral fins are high on the body. New World Silversides occur in marine to fresh water of North, Central, and South America. They form small to large schools near the surface. Usually omnivores or planktivores. Four species in the area. Page 116.

Order BELONIFORMES - Needlefishes

Family Exocoetidae - Flyingfishes

Flyingfishes are small to medium-sized, with very long pectoral fins that are set high on the body. Pelvic fins are usually expanded. The caudal fin is deeply forked, with a long lower lobe. The lateral line is along the lower margin of the body. Flyingfishes live at the surface from inshore to well offshore in all tropical to subtropical oceans. Often leap and glide in the air for long distances. Feed on zooplankton. Three species in the area. Page 118.

Order BELONIFORMES - Needlefishes, *cont.*

Family Hemiramphidae - Halfbeaks

Halfbeaks are elongate and slender. The upper jaw is short; the lower jaw is usually very long, with a fleshy tip. Single dorsal and anal fins are near the tail. Pectoral fins are short to long and set high on the body. Scales are easily shed. Lateral line runs near lower margin of body. Halfbeaks occur near the surface of all tropical to warm temperate seas. They are typically omnivorous and feed on a variety of grasses, invertebrates, and small fishes. Some are actively sought as baitfish for game fishing. Three species in the area. Page 118.

Family Belonidae - Needlefishes

Needlefishes are long and slender, with long, pointed jaws. The mouth has many sharp teeth. Single dorsal and anal fins are near the tail. Caudal fin may be emarginate or asymmetrical. Body is either oval or round in cross-section. Needlefishes occur worldwide in tropical to warm temperate seas. Some enter brackish and fresh water. They are usually found near the surface and feed primarily on small, schooling fishes. One species in the area. Page 120.

Family Scomberesocidae - Sauries

Sauries are elongate and slender, with long, pointed jaws. Dorsal and anal fins are near the tail and are followed by five to six finlets. Caudal fin is deeply forked and symmetrical. The body is covered in small scales. Sauries are found in open surface waters of tropical to temperate seas. They are schooling and feed on a wide variety of small prey items. Sauries are an important prey item for many marine animals, including squids, billfishes, sharks, tunas, marine mammals, and sea birds. One species in the area. Page 120.

Order CYPRINODONTIFORMES - Killifishes

Family Fundulidae - Topminnows

Topminnows are small, with elongate to moderately deep bodies. The head is usually flattened and the mouth is small. The single dorsal fin is located posterior to midlength. Caudal peduncle is long and moderately deep. Males have larger anal fins than females. Females are larger than males. Topminnows occur in tropical to temperate fresh to coastal marine waters of North America to Yucatán, Cuba, and Bermuda. Many adapt to a wide range of salinities and temperatures. Two species in the area, one of which was introduced. Page 120.

Family Poeciliidae - Livebearers

Livebearers are small, with elongate to moderately deep bodies. The head is flattened and the mouth is small. The single dorsal fin is located posteriorly and may be fan-like. Caudal peduncle is moderately long and slender to stout. Livebearers occur in quiet fresh, brackish, and marine coastal waters of the Americas. Most live near the surface. Most males have a modified anal fin used for internal fertilization of females. Females may carry more than one brood and give birth to live young. One introduced species in the area. Page 120.

Order BERYCIFORMES - Alfonso Squirrelfishes

Family Anoplogastridae - Fangtooths

Fangtooths are deep-bodied and compressed. The mouth is wide and gaping, with numerous fang-like teeth. Eyes are small. The head has grooves, ridges, and spines. Scales are small or minute. Lateral line is an open groove partially covered by scales. Fangtooths are found in deep midwaters of the Atlantic, Indian, and Pacific oceans. There are only two species in this family. One species in the area. Page 122.

Order ZEIFORMES - Dories

Family Zeidae - Dories

Dories are deep-bodied, laterally compressed, and oval in shape. The mouth is large and highly protrusible. Body scales are minute or absent. May have large bucklers along base of dorsal and anal fins. Scutes may be present along abdomen. Pelvic fins may be greatly expanded. Dories occur in relatively deep marine waters over continental shelves and upper slopes of the Atlantic, Indian and Pacific oceans. One species in the area. Page 122.

Order GASTEROSTEIFORMES - Sticklebacks

Family Aulorhynchidae - Tubesnouts

Tubesnouts are elongate and tubular in shape, with bony plates on the sides. The mouth is small and the snout is long. There are 24–26 isolated spines preceding the second dorsal fin. Anal fin is located under the second dorsal fin. Tubesnouts occur coastally in the north Pacific Ocean. They are typically schooling and may gather in great numbers. One species in the area. Page 122.

Family Gasterosteidae - Sticklebacks

Sticklebacks are small, typically elongate, and scaleless. Some have bony plates on sides. Dorsal-fin spines are isolated and usually stout and sharp. Pelvic fins usually with a stout spine and one or two soft spines. Sticklebacks are found in the Northern Hemisphere in marine, brackish, and fresh water. Males build nests, court females, and defend young. Some are entirely marine; one is entirely freshwater; others are anadromous. Two species in the area. Page 122.

Family Syngnathidae - Pipefishes and Seahorses

Pipefishes and Seahorses are included in the same family despite the outward difference in appearance. All are elongate, with a body covered in bony rings. The mouth is small, the snout is tube-like. Dorsal fin is single. Anal fin, when present, is very small. Pelvic fins are absent in all. Pipefishes and Seahorses occur worldwide in warm temperate to tropical marine and brackish waters. Some are in fresh water. Females deposit eggs into males' pouch or attach eggs to males' abdomen. Males then care for eggs until hatching. Seven species in the area. Page 124.

Order GASTEROSTEIFORMES - Sticklebacks, *cont.*

Family Fistulariidae - Cornetfishes

Cornetfishes are slender, elongate, and slightly depressed in shape. The mouth is small and the snout is tube-like. First dorsal fin is absent. Second dorsal and anal fins are similar in shape. Caudal fin is forked, with a long filament trailing from the middle rays. Cornetfishes are found worldwide in tropical to subtropical seas. Occur along coastal areas, over seagrass beds, rubble, and soft bottoms, and around reefs. One species in the area. Page 126.

Family Macroramphosidae - Snipefishes

Snipefishes are small, laterally compressed, and moderately deep-bodied, with a long tube-like snout. Teeth are absent. First dorsal fin has five to eight spines. The second spine is enlarged and serrated along the rear margin. Body has bony plates above pectoral fins and/or along ventral midline. Snipefishes occur worldwide in tropical to subtropical seas. Juveniles are pelagic, adults live near bottom. One species in the area. Page 126.

Order SCORPAENIFORMES - Mail-cheeked Fishes

Family Scorpaenidae - Scorpionfishes

Scorpionfishes are small to moderate in size, with a relatively large head often bearing spines and ridges. Spines may be very strong to almost absent. The pectoral fins are rounded to fan-like. The dorsal fin is continuous and notched, with membranes on spiny portion moderately to deeply incised. Venom glands are usually present at base of dorsal, pelvic, and anal fin spines. Venom may be slightly to highly toxic. All members have an elongated bone under the eyes and across the cheeks called a "suborbital stay." They are demersal or pelagic and occur circumglobally in tropical to temperate seas. Most are well camouflaged and are adept at ambushing prey. *Sebastes* species practice internal fertilization and give birth to live young. Seventy-four species in the area. Page 126.

Family Triglidae - Searobins

Searobins are moderately elongate with a large, bony head armored with plates, ridges, and spines. Pectoral fins are small to very broad. First three pectoral-fin rays are free and fleshy. Searobins are demersal and occur circumglobablly from near shore to continental shelves of tropical to warm-temperate brackish and marine waters. The free pectoral-fin rays are used for locomotion and to detect prey items. One species in the area. Page 156.

Family Anoplopomatidae - Sablefishes

Sablefishes are elongate to moderately robust. The head lacks spines, ridges, and cirri. Dorsal fins are widely separated or set close together. Pectoral fins are moderately long and broad. Anal-fin spines small and weak. Scales on body small. Sablefishes occur near bottom in deep marine waters along the coast of the North Pacific from Japan to California. Only two species in the family; both occur in the area. Page 156.

Family Hexagrammidae - Greenlings

Greenlings are small to moderately large and moderately elongate. Head spines and ridges are absent. The dorsal fin is single, usually notched, and with a nearly equal number of spiny and soft rays. Most have more than one lateral line and one or two fleshy cirri over each eye. Scales are minute to moderate and cover the head. Greenlings are found demersally in the cold waters of the North Pacific. *Note:* Some authors place Lingcod in subfamily Ophiodontinae and Combfishes in family Zaniolepididae. Nine species in the area. Page 156.

Family Rhamphocottidae - Grunt Sculpin

The Grunt Sculpin is small and deep-bodied. The snout is long and the head is very large, with tall ridges ending with blunt spines. The preopercular spine is single and sharp. The lower pectoral-fin rays are elongate and free. The head and body are covered with multi-pointed, spiny scales. Grunt Sculpin is found demersally from the Gulf of Alaska to Santa Monica Bay, CA. Also to Japan. They occupy high-energy rocky coasts and tidepools and are also found over sandy bottoms. One species in this family. Page 162.

Family Cottidae - Sculpins

Sculpins are moderately elongate, with fan-like pectoral fins and large eyes set high on the head. The head is broad, with prominent preopercular spines. Some have spines on top of the head. Cirri may be present or absent. Dorsal fins are usually separate. The body may be scaleless or partially or entirely covered with scales or prickles. Scales are often modified into bony plates. Males of some genera have a long, penis-like appendage used to internally fertilize females. Swim bladder is absent. Most are cryptically colored. Sculpins occur in cold, northern marine and fresh waters of North America, Europe, and Asia. A few are found in the Southern Hemisphere. All are bottom-dwelling. 84 species in the area. Page 162.

Family Hemitripteridae - Searavens

Searavens are moderately elongate to robust. The head is knobby. The preopercle has blunt, skin-covered spines. Fins are typically broad. The body is covered in tiny prickles. Swim bladder is absent. Most Searavens occur in the North Pacific. One species occurs in the northwestern Atlantic. All live on or near bottom. Searavens prey primarily on invertebrates. This is a small family of fishes that is closely related to Poachers. Six species in the area. Page 194.

Family Agonidae - Poachers

Poachers are typically elongate, with overlapping, bony plates on the body. Most have two dorsal fins; some have one. Swim bladder is absent. Poachers occur in shallow to deep northern waters of the Atlantic and Pacific and also off southern South America. They are sluggish, and females scatter their eggs. Poachers are closely related to Sculpins and Searavens despite the differences in external appearance. Twenty-six species in the area. Page 198.

Order SCORPAENIFORMES - Mail-cheeked Fishes, *cont.*

Family Psychrolutidae - Fathead Sculpins

Fathead Sculpins have a large head and forebody that tapers to a narrow caudal peduncle. Bony arches on the head may bear spines. Dorsal fins are usually continuous but may be separate. Body is scaleless or has prickly plates. Females are larger than males. Fathead Sculpins occur in the Atlantic, Indian and Pacific oceans from shallow inshore waters to about 9,000 ft. Most are demersal. Six species in the area. Page 208.

Family Cyclopteridae - Lumpfishes

Lumpfishes have rounded bodies that are usually covered with tubercles. Two short dorsal fins are usually present; the first is sometimes covered with skin. Pelvic fins, when present, form a suction disk used to adhere to hard surfaces. Pectoral fins extend ventrally and in front of suction disk. Lumpfishes occur in the cold, northern waters of the Pacific and Atlantic oceans. Most are demersal; some are pelagic. Some migrate hundreds of miles to spawn. After spawning, males defend egg clusters. Some are sought for their eggs, which are sold as caviar. Three to ten species in the area. Page 210.

Family Liparidae - Snailfishes

Snailfishes are small, often tadpole-shaped, and soft-bodied. Dorsal and anal fins are confluent, or nearly so, with the caudal fin. Pectoral fins are usually bilobed. Pelvic fins, when present, form a suction disk used to adhere to hard surfaces. Snailfishes are a large family. They are widely distributed and are found in all oceans from the intertidal zone to about 23,000 ft. Some are commensal. Thiry-two to thirty-three species in the area. Page 212.

Order PERCIFORMES - Perches

Family Moronidae - Temperate Basses

Temperate Basses are moderately elongate and somewhat laterally compressed. The opercle has two rounded to pointed spines on the posterior margin. The two dorsal fins are completely to nearly separate. The anal fin has three strong spines and the lateral line extends onto the caudal fin. Temperate Basses are native in fresh, brackish, and inshore marine waters of North America, Europe, and North Africa. They have been successfully introduced to other areas. Some species can tolerate changes in salinity. Some are important game and food fish. One species in the area. Page 220.

Family Polyprionidae - Wreckfishes

Wreckfishes are large, oblong, and moderately compressed. The head has a low to raised bony knob between the eyes and one at the nape. The opercle has two spines, with the lower spine at the end of a long, horizontal ridge. Wreckfishes occur in the Atlantic, Indian, and Pacific oceans and in the Mediterranean. Adults are found over continental shelves and seamounts. Some juveniles are pelagic. One species in the area. Page 220.

Order PERCIFORMES - Perches, *cont.*

Family Serranidae - Sea Basses

Sea Basses are usually robust with large mouths. The opercle has three spines with the middle spine usually most conspicuous. The dorsal fin is usually continuous; spiny portion usually deeply notched. Anal fin has three spines. Lateral line extends only to end of caudal peduncle. Sea Basses occur circumglobally from near shore to moderately deep tropical to temperate seas. A few live in fresh water. The majority are hermaphrodites that mature first as females and then change into males. Some function simultaneously as both sexes. Eight species in the area. *Note:* this family was recently separated from Epinephelidae. Page 222.

Family Epinephelidae - Groupers

Groupers are usually robust, comparatively deep-bodied, with large mouths. The opercle has three spines, with the middle spine usually most conspicuous. The dorsal fin is usually continuous, and the anal fin has three spines. Lateral line runs comparatively low and extends only to end of caudal peduncle. Groupers occur circumglobally from near shore to moderately deep tropical to temperate seas. A few live in fresh water. The majority are hermaphrodites that mature first as females and then change into males. Eight species in the area. *Note:* this family was recently separated from Serranidae. Page 224.

Family Priacanthidae - Bigeyes

Bigeyes are moderately to very deep-bodied and laterally compressed with oblique jaws and very large, reflective eyes. The preopercle has a single spine at the lower corner. The dorsal fin is continuous. Pelvic fins are small to greatly expanded, with inner rays attached to the abdomen by a membrane. Color of head, body, and irises are shades of red. Bigeyes occur circumglobally in tropical to warm temperate seas. Most Bigeyes are nocturnal. One species in the area. Page 228.

Family Apogonidae - Cardinalfishes

Cardinalfishes are small, with a short snout, large mouth and eyes, and a long peduncle. Dorsal fins are separate. Coloration is variable and may pale at night. They occur circumglobally in shallow tropical to warm temperate seas, usually around coral reefs and seaweed beds. Some are in brackish or fresh water. They are generally secretive and nocturnal. Some are commensal. Many species mouth-brood eggs. Three species in the area. Page 228.

Family Malacanthidae - Tilefishes

Tilefishes are oblong to elongate, with a single low dorsal fin. The head profile is gently to steeply sloping. The mouth is moderately large and fleshy. A prominent to reduced predorsal ridge is present in most. The opercle has a single blunt or notched spine. Tilefishes are demersal and occur worldwide in tropical to warm temperate seas from shore to about 1,600 ft. They inhabit caves and crevices or may construct mounds or burrows. Some authors divide Tilefishes into two families. Two species in the area. Page 230.

Family Nematistiidae - Roosterfish

The Roosterfish is large-bodied and somewhat laterally compressed with elongate and trailing dorsal-fin spines. Soft dorsal fin is long-based, caudal fin is forked, and the pectoral fins are long and pointed. Scales are small and iridescent. The body is marked with four to five distinct black bands. Roosterfishes occur only from southern CA to Peru, including the Gulf of California and offshore islands. It is frequently found along sandy beaches and sometimes in estuaries from nearshore and surface to about 66 ft. A popular sportfish. One species in this family. Page 230.

Family Coryphaenidae - Dolphinfishes

Dolphinfishes are elongate and laterally compressed. The head profile is rounded in females, steeply sloping in adult males. The spineless dorsal fin originates on the head and reaches to the caudal peduncle. The caudal fin is deeply forked. Dolphinfishes occur circumglobally in tropical to warm temperate seas. They live in open ocean waters over continental shelves and slopes from surface to about 660 ft. They are often associated with *Sargassum* seaweed, flotsam, and oil rigs. Sought as food fish. One species in the area. Page 230.

Family Echeneidae - Remoras

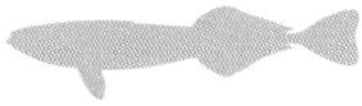

Remoras are elongate and round in cross-section. The head is broad and flattened. The first dorsal fin is modified into an oval-shaped cephalic disk that the Remora uses to attach itself to a host. Remoras occur worldwide in tropical to warm temperate seas. They are associated with a variety of hosts, including sharks, rays, tarpons, seabasses, cobias, jacks, parrotfishes, billfishes, molas, dolphins, and whales. Some are host-specific, some attach to a range of hosts, and others are free-swimming. Seven species in the area. Page 232.

Family Carangidae - Jacks and Pompanos

Jacks and Pompanos are small to large with body shapes varying from deep and compressed to elongate and fusiform. The eyes have an adipose lid that can be strongly developed in some species. First dorsal fin may be well developed or a series of spines. Most have a forked caudal fin. First one or two anal-fin spines are separate and may be embedded. Many have bony scutes along lateral line. Body shape and coloring change dramatically with age. Jacks and Pompanos occur circumglobally in tropical to warm temperate brackish to oceanic habitats. Most are schooling and pursue pelagic prey. Fifteen or sixteen species in the area. Page 234.

Family Bramidae - Pomfrets

Pomfrets are medium to large, laterally compressed, and round to teardrop-shaped. The eyes are large; the lower jaw protrudes. Dorsal and anal fins are long-based and may be either low, lobed, or fan-like. Scales covering head and body are large and often keeled. Pomfrets occur worldwide in temperate to warm-temperate seas. Most are pelagic and schooling. Four species in the area. Page 240.

Family Lutjanidae - Snappers

Snappers are elongate and moderately deep-bodied. The upper jaw is somewhat protrusible. Well-developed canine-like teeth are usually present. The cheeks are scaled, but the snout is scaleless. Maxilla are usually scaleless. Dorsal fin is unnotched to deeply notched. Color is often reddish but may be purplish to grayish with stripes or bars. Snappers occur circumglobally in tropical to warm temperate seas from shore to about 1,800 ft. Most are demersal; some live in fresh water. Most are nocturnal. Females spawn several times per season. Five species in the area. Page 242.

Family Lobotidae - Tripletails

Tripletails are oval to oblong, deep-bodied, and laterally compressed. The head profile is steeply sloping. The snout is short and blunt. Second dorsal, anal, and caudal fins are broad and rounded. Tripletails occur circumglobally in tropical to warm temperate seas from coastal waters to well offshore. Some live in brackish or fresh water. They are sluggish fishes, often floating on their sides with flotsam or *Sargassum* seaweed. Juveniles are believed to mimic mangrove leaves. One species in the area. Page 244.

Family Gerreidae - Mojarras

Mojarras are moderately slender to deep-bodied and laterally compressed. The upper jaw is highly protrusible. The dorsal fin is continuous with a slight to deep notch. The bases of the dorsal and anal fins are scaled. Caudal fin is deeply forked. Scales are typically large and usually silvery. Mojarras occur circumglobally in tropical to warm temperate seas. Most are found along the coast, many enter brackish water, and some live in fresh water. Two species in the area. Page 244.

Family Haemulidae - Grunts

Grunts are oblong, moderately deep-bodied, and laterally compressed. The dorsal head profile is almost straight to convex. The snout is moderately short to long. Mouth is small to moderate with thick lips. Dorsal fin is continuous and either notched or unnotched. Scales are absent on snout and lips. Grunts occur circumglobally in tropical to warm temperate seas in shallow coastal, brackish, and occasionally, fresh water. All are capable of sound production. Five species in the area. Page 244.

Family Sparidae - Porgies

Porgies are small to medium in size and oblong to oval in profile. The body is usually deep and moderately to strongly compressed. The head profile is usually steep. The mouth is small with a slightly protrusible upper jaw. Teeth in sides of jaws are molar-like. Dorsal fin is continuous and weakly to slightly notched. Porgies are found circumglobally in tropical to warm temperate seas. Most occur over continental shelves; some enter estuaries or live in fresh water. Some are hermaphroditic. One species in the area. Page 246.

Family Polynemidae - Threadfins

Threadfins are moderately elongate, with widely separated dorsal fins and a deeply forked caudal fin. The snout is short and projects beyond the large, inferior mouth. The pectoral fin is low on the body, with lower rays that are separate and filamentous. Threadfins occur circumglobally in shallow tropical to warm temperate seas. Some enter estuaries and rivers. Some are hermaphroditic. Two species in the area. Page 248.

Family Sciaenidae - Drums and Croakers

Drums (aka Croakers) are small to large, with a gently to steeply sloping dorsal head profile. The body may be short and deep to moderately elongate. All have pores on the snout and chin, a bony flap above the gill opening, and lateral-line scales that extend onto the caudal fin. The first dorsal fin is short-based and usually continuous with the long-based second dorsal fin. A gas bladder is always present and produces drumming sounds. Drums are found in a wide range of habitats in worldwide tropical to warm temperate seas. Some live in estuaries or fresh water. Many are migratory and use estuaries as nursery grounds. Many are sought as sport and food fish. Eight species in the area. Page 248.

Family Mullidae - Goatfishes

Goatfishes are moderately elongate, with a convex dorsal profile and nearly straight ventral profile. The dorsal fins are separate; the caudal fin is forked. Two well-developed barbels extend from chin. The upper jaw is slightly protrusible. Goatfishes are found near bottom over continental shelves and slopes of circumglobal tropical to warm temperate seas. Some enter estuaries. Use barbels to locate prey. One species in the area. Page 252.

Family Kyphosidae - Sea Chubs

Sea Chubs are oval to oblong in profile and moderately compressed. The head is short; the snout is blunt. The mouth is small and horizontal. The dorsal fin is continuous, with the spiny portion depressible into a groove. Sea Chubs are found worldwide in tropical to subtropical seas. They are associated with coral reefs, rocky bottoms, hard structures, and *Sargassum* seaweed. Five species in the area. Page 252.

Family Chaetodontidae - Butterflyfishes

Butterflyfishes are deep-bodied, compressed, and oval, round, or rhomboid in profile. The mouth is small and the snout is blunt to long and pointed. Preopercular margin may be serrated, but a prolonged spine is absent at the lower angle. Dorsal fin is continuous to slightly notched. Spiny portion may be tall and deeply incised. Butterflyfishes occur circumglobally in shallow tropical to subtropical seas. Most are reef associated. Some are found in brackish or fresh water. They graze on small invertebrates and plankton. Butterflyfishes usually occur singly or in pairs. Popular in the aquarium trade. Two species in the area. Page 254.

Family Pomacanthidae - Angelfishes

Angelfishes are deep-bodied, compressed, and oval to round in profile. The mouth is small and the snout is blunt. Preopercular margin is often serrated and a prolonged spine is always present at the lower angle. Dorsal fin is continuous and may have trailing middle soft rays. Angelfishes occur circumglobally in shallow tropical to subtropical seas. They are generally associated with coral reefs, where they are often seen singly or in pairs. One species in the area. Page 254.

Family Embiotocidae - Surfperches

Surfperches are oval, round, or oblong in profile. The mouth is small and often fleshy. The dorsal fin is continuous with tallest spines that are shorter, taller, or equal to tallest soft rays. Scales are smooth and the caudal fin is moderately to deeply forked. Males of some species develop a thickened area on the anterior anal-fin lobe. Surfperches reproduce by internal fertilization, with females giving birth to live young. Some young are nearly mature at birth. The majority of Surfperches occur in shallow nearshore waters of the northeastern Pacific. Eighteen species in the area. Page 256.

Family Pomacentridae - Damselfishes

Damselfishes are small, oval to oblong in profile, and laterally compressed. The mouth is small and toothless. The lateral line is either incomplete or interrupted. The dorsal fin is continuous and the anal fin has only two spines. Caudal fin is shallowly to deeply forked. Damselfishes are found circumglobally in tropical to warm temperate seas. Most are in the Indo-Pacific. They occur over reefs, sandy, rubble, and seagrass bottoms. Most are territorial and defend eggs and feeding areas. Five species in the area. Page 262.

Family Labridae - Wrasses and Parrotfishes

Wrasses and Parrotfishes are oblong to elongate and slightly to strongly compressed. Head profile is gently to steeply sloping. Upper jaw is protrusible in most species. Teeth are separate and canine-like or fused. Dorsal fin is continuous. Scales are relatively large to very large. Wrasses and Parrotfishes occur circumglobally in shallow tropical to warm temperate seas. Many are reef associated; others inhabit seagrass beds or sandy to rocky bottoms. Most are hermaphroditic. Color and pattern often change with sex. Four species in the area. *Note:* Family Scaridae was recently grouped with Labridae. Page 264.

Family Bathymasteridae - Ronquils

Ronquils are small and elongate, with large mouths and broad pectoral fins. The dorsal fin is continuous but not attached to caudal fin. Anal fin is long-based and pelvic fins are thoracic. The lateral line is high on the body. Some species are capable of color change. Ronquils occur in cold, coastal waters of the northern Pacific Ocean and Bering Sea. They are demersal and usually over rocky bottoms and primarily near shore. Six species in the area. Page 266.

Family Zoarcidae - Eelpouts

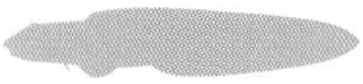

Eelpouts are small to large and elongate, with long-based dorsal and anal fins that are confluent with a pointed caudal fin. The head is relatively large. The mouth is large and often big and with fleshy lips. The upper jaw usually projects beyond the lower. Pelvic fins (when present) are minute and thoracic. Scales (when present) are small and usually embedded. Lateral line is single or double and complete or incomplete. Eelpouts occur worldwide in cold waters. Most are demersal but some are found in midwater. Nineteen species in the area. Page 270.

Family Stichaeidae - Pricklebacks

Pricklebacks are elongate, rounded to oval in cross-section anteriorly, and compressed posteriorly. The dorsal fin is long-based and entirely spiny and separated from caudal fin in most species. Pectoral fins are small to large. Pelvic fins, when present, are small and thoracic. Scales are minute and often buried or absent on part of the body. Some with a fleshy crest; others with few to many cirri above eyes and head. All have one to four lateral lines that may or may not be complete. Pricklebacks occur in cold northern waters of the Atlantic and Pacific oceans from shore to about 820 ft. Thirty-two species in the area. Page 276.

Family Cryptacanthodidae - Wrymouths

Wrymouths are slender and eel-like. The mouth is large and strongly oblique. Head is broad and flat on top. Eyes are set high on the head. Dorsal fin is entirely spiny and confluent with caudal fin. Pelvic fins are absent. Wrymouths occur demersally in shallow inshore waters of the northwest Atlantic and northern Pacific oceans. Some are found in mud burrows. This is a small family of fishes that is closely related to Pricklebacks. Two species in the area. Page 284.

Family Pholidae - Gunnels

Gunnels are small, elongate, and laterally compressed. The mouth is relatively small and oblique. The dorsal fin is entirely spiny. Anal fin is about half as long as dorsal fin. Pectoral fins are comparatively small to minute or absent. Pelvic fins are minute in most and absent in others. Body is covered in minute, embedded scales. Lateral line is absent. Gunnels occur in northern Atlantic and Pacific coastal waters. Often found around rocks and in tide pools. Nine species in the area. Page 286.

Family Anarhichadidae - Wolffishes

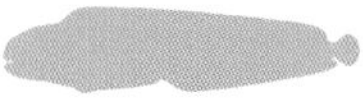

Wolffishes are large and robust anteriorly, slender to tapered posteriorly. The jaws are large and strong, with large conical teeth in front and molar-like teeth in the sides. The dorsal fin is tall, flexible, and entirely spiny. Caudal fin is either separate or confluent with dorsal and anal fins. Pectoral fins are large and fan-like. Pelvic fins are absent. Body is naked or with minute scales. Lateral lines are faint. Wolffishes are demersal in coastal northern waters of the Atlantic and Pacific oceans. Females produce large eggs that are possibly internally fertilized. Some species exhibit brooding behavior. Wolffishes feed on a variety of invertebrates. They occur singly or in small groups. Two species in the area. Page 290.

Family Ptilichthyidae - Quillfish

The Quillfish is extremely long and eel-like. The lower jaw protrudes and has a fleshy tip. The dorsal fin consists of 81 to 90 isolated spines anteriorly and 137 to 157 soft rays posteriorly. Anal fin is long-based, with 185 to 196 rays. Dorsal and anal fins are confluent with the reduced caudal fin that tapers to a filament. Pelvic fins and lateral line are absent. Quillfishes occur in the northern Pacific Ocean from surface to about 1,200 ft. They burrow in soft sediments. One species in this family. Page 290.

Family Zaproridae - Prowfish

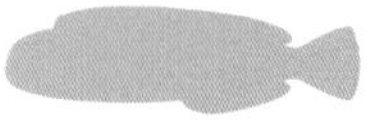

The Prowfish is stout and laterally compressed, with a small oblique mouth and a blunt snout. The head is rounded in profile, with large white pores around eyes, along chin, and on cheeks. The pectoral fins are large and broad. Dorsal and anal fins are long-based and evenly tall. The caudal fin is large. Pelvic fins are absent. There may be as many as three lateral lines that consist of widely spaced pores. The flesh is flaccid. Prowfishes occur in the Bering Sea and northern Pacific Ocean. Adults are found near bottom; juveniles are pelagic and often found near the surface over deep waters. One species in this family. Page 290.

Family Scytalinidae - Graveldiver

The Graveldiver is elongate, round in cross-section anteriorly, and laterally compressed posteriorly. The mouth is large and upturned. The eyes are very small and set high on the head. Pectoral fins are very small. Dorsal and anal fins begin near midbody, are buried in flesh, and are confluent with the caudal fin. Pelvic fins, lateral line, and scales are absent. Graveldivers occur along shallow open coasts in the northeastern Pacific from the Aleutian Islands to San Luis Obispo County, CA. They burrow into soft and mixed bottoms. One species in this family. Page 292.

Family Trichodontidae - Sandfishes

Sandfishes are relatively small, deep bodied, and laterally compressed with a nearly vertical mouth that has fringed lips. The preopercle has five sharp spines. Pectoral fins are long and broad. The dorsal fins are separate. Anal fin is long-based. Scales are absent. Sandfishes occur demersally over soft bottoms from near shore to deep offshore waters of the northern Pacific from California to Korea. They commonly bury in sand, with only the head protruding. One species in the area. Page 292.

Family Ammodytidae - Sand Lances

Sand Lances are slender and elongate. The head is small; the snout is pointed. The lower jaw protrudes. Teeth are absent. Dorsal and anal fins lack spines. Caudal fin is separate and forked. Pelvic fins are usually absent. Lateral line is high on the body. Sand Lances occur from shore to offshore in the cold to tropical marine waters of the Atlantic, Arctic, Pacific, and Indian oceans. Some burrow rather quickly into bottom sediment. They are an important prey for larger fishes, marine mammals, and birds. Sought commercially. One species in the area. Page 292.

Family Uranoscopidae - Stargazers

Stargazers have a large head and a robust body. The mouth is large and oblique to nearly vertical. Lips are fringed. Eyes are either on or near top of head. First dorsal fin can be absent. Pectoral fins are fan-like. A blunt or sharp spine is present over the pectoral fins and is venomous in most species. The body is naked or covered in small, smooth scales. Stargazers are demersal and occur circumglobally in tropical to temperate seas from shore to about 3,000 ft. Some species have a lure-like filament in the mouth. One genus has an electric organ used to stun prey. Two species in the area. Page 292.

Family Blenniidae - Combtooth Blennies

Combtooth Blennies are small and scaleless. The head is usually very blunt. Nostrils sometimes with fleshy flaps. Eyes often with cirri, fleshy flaps, or tentacles. Mouth is usually small, with fixed or movable comb-like teeth that are closely packed in a single row. Some have canine teeth. Pectoral-fin rays are unbranched. Caudal-fin rays are branched or unbranched. Combtooth Blennies occur worldwide in tropical to warm temperate seas. All are demersal; most are in shallow water. Four species in the area. Page 294.

Family Clinidae - Kelp Blennies

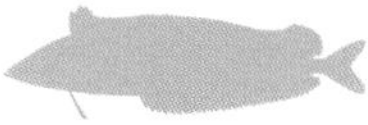

Kelp Blennies are small to relatively large and elongate. The dorsal fin is long-based, sometimes highest anteriorly, and with more spines than rays. Dorsal- and anal-fin rays are usually unbranched. Cirri may be present on head but are absent from nape. The body is covered in small, usually embedded scales in all but one species. Kelp Blennies are typically demersal and are found circumglobally in temperate seas. Four species in the area. Page 296.

Family Labrisomidae - Labrisomid Blennies

Labrisomid Blennies are small, with a body that is usually scaled. The head is blunt to pointed. Cirri are often present on nostrils and eyes. Nape may have a single cirrus, multiple cirri, or fleshy tabs. The mouth is small to moderate, with fleshy lips. Teeth are variable in size and arrangement. All fin rays, including caudal, are unbranched. Labrisomid Blennies are demersal in the tropical to warm-temperate western Atlantic and eastern Pacific oceans. They are generally reef associated in shallow water. Males and females are often colored and patterned differently. Three species in the area. Page 298.

Family Chaenopsidae - Tube Blennies

Tube Blennies are small, elongate, and scaleless. Cirri on nostrils and over eyes vary in complexity or may be absent. Lateral line is absent. Caudal fin may be continuous or separate from dorsal and anal fins. Males are generally darker or more brightly colored than females. Some male species have a much higher anterior dorsal-fin lobe than females. Tube Blennies are demersal in the tropical to subtropical western Atlantic and eastern Pacific oceans. Many live in abandoned worm tubes and invertebrate shells. Four species in the area. Page 298.

Family Icosteidae - Ragfish

The Ragfish is compressed and limp, with long-based dorsal and anal fins. The head and eyes are small. Adults lack pelvic fins. Caudal-fin margin is crescent-shaped to forked in adults, rounded in juveniles. Body is elongate in adults, deep and rounded in juveniles. Scales absent or embedded in adults. Caudal pedulncle is narrow in all. Ragfishes occur in marine waters from the Bering Sea to southern California. Also to Hokkaido, Japan. Young are in shallow water to near surface in offshore waters. Adults occur near bottom in deep water. One species in this family. Page 300.

Family Gobiesocidae - Clingfishes

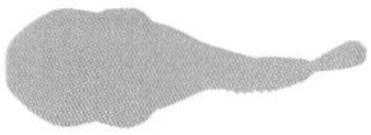

Clingfishes are small and scaleless. The head and anterior portion of the body are generally rounded and flattened. Eyes are on top of the head. Dorsal fin is single. Pelvic and pectoral fins are modified into an adhesive sucking disk. Clingfishes occur circumglobally in shallow tropical to warm temperate seas and in brackish and fresh water. Most are bottom-dwelling. Use adhesive disk to attach to hard substrates and plants. Eight species in the area. Page 300.

Family Callionymidae - Dragonets

Dragonets are small, elongate, and variably flattened anteriorly. The head may be moderately to greatly expanded. Gill openings are reduced to small pores on upper head. Preopercle has a strong, variably shaped spine. First dorsal fin is often tall and sometimes filamentous. Pelvic fins are attached by a membrane to the pectoral-fin base. Color may be bright to dull and cryptic. Dragonets occur circumglobally in shallow to deep tropical seas. A few enter fresh water. One rare species in the area. Family account only.

Family Eleotridae - Sleepers

Sleepers are small to moderately sized and elongate to somewhat stout. The head is usually flattened on top. The eyes are widely separated. Two dorsal fins are present. Second dorsal-fin base is shorter than the distance between its rear origin and the caudal-fin origin. Pelvic fin bases close together or united, but fins are separate. Sleepers are found in most shallow tropical to subtropical fresh, brackish, and marine waters. One species in the area. Page 302.

Family Gobiidae - Gobies

Gobies are small to very small and elongate. The head is short and broad. First dorsal fin, when present, is separate from second dorsal fin. Second dorsal-fin base is usually longer than the distance between its rear origin and caudal-fin origin. Pelvic fins separate, somewhat connected, or completely united to form a disk. Lateral line is absent. Body is scaleless, partially scaled, or entirely scaled. Gobies occur circumglobally in a wide variety of tropical to warm temperate waters. They are most diverse in reef habitats. Most are marine and demersal. Some are free-swimming. Gobies are the most diverse family of marine fishes, with more than 1,500 known species. Sixteen species in the area. Page 304.

Family Ephippidae - Spadefishes

Spadefishes are deep-bodied and circular to oblong in profile. The head is short and the body is deeply compressed. Jaws have slender, movable, brush-like teeth. First dorsal fin is distinct from second and is usually notched in most. Preopercle is smooth to serrate and lacks a prominent spine at corner. Spadefishes occur circumglobally in tropical to warm temperate seas and rarely in brackish waters. They form schools over coral, rocky, and artificial reefs. Juveniles are pelagic; some mimic floating leaves or debris. One species in the area. Page 310.

Family Luvaridae - Louvar

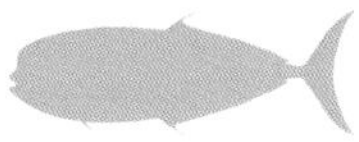

The Louvar is oblong in profile, with a slender, keeled, caudal peduncle. The snout is blunt; the forehead is compressed and keel-like. The mouth and eyes are small. Dorsal fin is long-based and originates just posterior to eyes in juveniles and posterior to body midpoint in adults. First dorsal- and anal-fin spines are long in juveniles and shorten with age. Louvars are oceanic and pelagic in relatively deep water in worldwide tropical to temperate seas. Females can produce up to 47 million eggs. One species in this family. Page 310.

Family Sphyraenidae - Barracudas

Barracudas are elongate and small to moderately large. The head is long, the snout is pointed, and the lower jaw protrudes. The large jaws and roof of the mouth have numerous sharp conical or flattened teeth. Dorsal fins are short-based and widely separated. Barracudas occur worldwide in tropical to warm temperate seas. They occur in a wide variety of coastal habitats from surface to near bottom. Small species and juveniles of large species form schools. Barracudas are swift, voracious predators. Two species in the area. Page 310.

Family Gempylidae - Snake Mackerels

Snake Mackerels are very elongate to moderately deep-bodied. The mouth is large, with a protruding lower jaw. Teeth are strong; some are fang-like. First dorsal fin is long-based. Finlets follow second dorsal fin in some species; finlets absent in others. Pelvic fins are small, rudimentary, or absent. Lateral line is single and straight to wavy, branched, or double. Scales are minute to moderately large and modified in some. One genus with keels on caudal peduncle. Snake Mackerels occur worldwide in tropical to subtropical seas. Four shallow water species in the area. Page 312.

Family Trichiuridae - Cutlassfishes

Cutlassfishes are elongate and strongly compressed. The mouth is large, with a protruding lower jaw. Fang-like teeth are usually present. Dorsal fin is extremely long-based and is notched in some. Pelvic fins are very small or absent. Anal fin reduced or embedded. Caudal fin is small and forked or absent. Scales absent. Cutlassfishes occur worldwide in tropical to warm temperate seas from surface to about 6,500 ft. Two shallow-water species in the area. Page 312.

Family Scombridae - Mackerels and Tunas

Mackerels and Tunas are medium to large and elongate to very robust. The body is somewhat compressed to rounded in cross-section. The spiny dorsal fin is short- to long-based and inserts into a groove. A series of finlets follow second dorsal and anal fins, and two oblique keels at caudal-fin base are present in all. Many also with a lateral keel on the caudal peduncle. Some have a well-developed corselet of thickened scales anteriorly. Mackerels and Tunas occur worldwide in tropical to temperate seas. Most are marine and pelagic, some are coastal. Some are highly migratory. Important food fish. Fourteen species in the area. Page 314.

Family Xiphiidae - Swordfish

The Swordfish is large, robust anteriorly, and rounded in cross-section. The upper jaw forms a flattened, sword-like bill. First dorsal fin is short-based and widely separated from second dorsal fin in adults. The caudal peduncle is deeply notched at the caudal-fin base and has strong lateral keels. Pelvic fins are absent. Oceanic in worldwide offshore tropical to temperate seas. Highly migratory. Important food fish. One species in this family. Page 318.

Family Istiophoridae - Billfishes

Billfishes are elongate and moderately compressed. The body tapers from head to tail. The upper jaw forms a spear-like bill that is round in cross-section. First dorsal fin is long-based in all and sail-like in one adult species. The first dorsal and first anal fins insert into grooves. Pelvic fins are long and narrow. The caudal peduncle is shallowly notched in front of the caudal-fin base. Caudal fin with two keels at base. Billfishes occur worldwide in tropical to warm temperate seas and usually offshore in upper water column. Swift predators. Use bill to injure, impale, and stun prey. Important sport fishes. Five species in the area. Page 320.

Family Centrolophidae - Medusafishes

Medusafishes are medium to large and elongate to robust. The mouth is relatively large, with a single row of small conical teeth. The dorsal fin is continuous and the caudal peduncle is thick. Pelvic fins insert into a groove. Medusafishes occur worldwide in tropical to temperate seas. Absent from mid-Pacific. Adults occur over deep shelves, canyons, and submerged islands. Juveniles are associated with sea jellies and flotsam. One species in the area. Page 322.

Family Nomeidae - Driftfishes

Driftfishes are moderate to large, slender to deep-bodied, and laterally compressed. The mouth is small; the snout is moderately blunt to very blunt. Dorsal fins are separate. First dorsal fin and pelvic fins insert into a groove. Second dorsal and anal fins are about the same shape and are scaled at the base. Caudal keels are absent. Driftfishes occur circumglobally in tropical to subtropical oceanic waters. Adults inhabit midwaters or are demersal over continental slopes. Juveniles occur near surface and are associated with sea jellies. Two species in the area. Page 322.

Order PERCIFORMES - Perches, *cont.*

Family Tetragonuridae - Squaretails

Squaretails are medium-sized, elongate, and rounded in cross-section. The snout is blunt. The lower jaw fits completely within the upper jaw when closed. First dorsal fin is long-based and inserts into a groove. Second dorsal and anal fins are similar in shape. Modified scales form two low keels on caudal-fin base. Body scales are keeled and geometrically arranged. Squaretails occur in deep water of tropical to subtropical oceanic seas. One species in the area. Page 322.

Family Stromateidae - Butterfishes

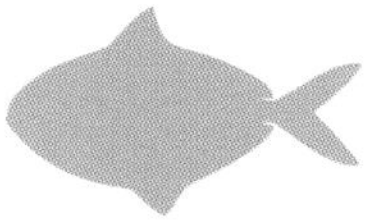

Butterfishes are deep-bodied and laterally compressed. Snout is short; mouth is small. Eyes are covered by adipose eyelids. Dorsal and anal fins are single, long-based, scaled, and similar in shape. Anterior fin lobes may be elongate. Pelvic fins are absent. Butterfishes are found worldwide in tropical to warm temperate seas. Absent from southwestern Pacific and Indian oceans. They are pelagic along the coast, over continental shelves, and sometimes in estuaries. One species in the area. Page 322.

Order PLEURONECTIFORMES - Flatfishes

Family Paralichthyidae - Sand Flounders

Sand Flounders are deep-bodied and laterally compressed. The mouth is large and protrusible. Eyes are large to relatively small and usually left-facing. Anterior dorsal-fin rays may be long and mostly free of membrane. Pelvic fins are symmetrical in shape and short-based. Sand flounders occur worldwide in tropical to temperate seas. Some in fresh water. All are demersal and usually over soft bottoms. Usually burrow into sediment to await prey. Many are quick color changers. Six species in the area. Page 324.

Family Pleuronectidae - Righteye Flounders

Righteye Flounders are deep-bodied and laterally compressed. The mouth is very small to large. Eyes almost always face right. Pelvic fins are symmetrical in shape and short-based. Lateral line is well developed on both sides of body. Righteye Flounders primarily occur in temperate to cold marine waters. Some in deep, tropical, or brackish waters. Demersal but may leave bottom to find prey. Many are quick color changers. Twenty-eight species in the area. Page 326.

Family Bothidae - Lefteye Flounders

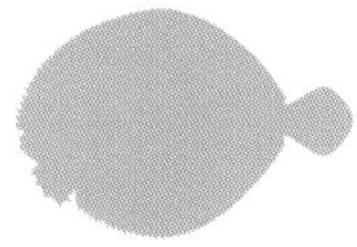

Lefteye Flounders are somewhat to very deep-bodied and laterally compressed. The mouth is moderate to large and protrusible. Eyes are left-facing (rarely right-facing) and are close-set or widely separated. Pelvic fin on eyed side is larger and with longer base than blind-side fin. Lefteye Flounders occur worldwide in tropical to warm temperate seas. Demersal, usually over soft bottoms. In many species males have separated eyes and elongate pectoral fin. One rare species in the area. Family account only.

Order PLEURONECTIFORMES - Flatfishes, *cont.*

Family Cynoglossidae - Tonguefishes

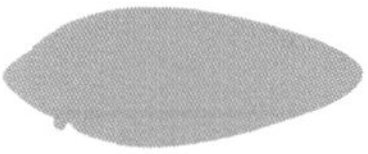

Tonguefishes are lance- or tongue-shaped, with a blunt to pointed tail. The mouth is small. Eyes are small and left-facing. Dorsal and anal fins are confluent with caudal fin. Pectoral fins are absent. Eyed-side pelvic fin, when present, is confluent with anal fin. Lateral line is absent. Tonguefishes are demersal over a variety of bottoms from shore to about 5,000 ft. in worldwide tropical to temperate seas. One species in the area. Page 338.

Order TETRAODONTIFORMES - Plectognaths

Family Balistidae - Triggerfishes

Triggerfishes are deep-bodied and moderately compressed. The mouth is small, with a single row of strong teeth. Gill openings are short and slit-like. A patch of enlarged scales above the pectoral fins is usually present. First dorsal fin with three stout spines; second locks first upright. Second dorsal- and anal-fin rays branched. Scales plate-like and geometric. Triggerfishes occur circumglobally in tropical to warm temperate seas from surface to bottom in a variety of habitats. Three species in the area. Page 338.

Family Ostraciidae - Boxfishes

Boxfishes are triangular or square to pentangular in cross-section. The mouth is small, with fleshy lips and moderately strong teeth. Gill openings are slit-like. Spiny dorsal fin and pelvic skeleton are absent. The body is encased in close-set hexagonal to polygonal plates. Plates are either visible or concealed by skin. Plates may form "horns" over eyes in some. Boxfishes occur circumglobally in tropical to warm temperate seas. Demersal and pelagic over rocky and coral reefs and sandy and seagrass bottoms. One species in the area. Page 340.

Family Tetraodontidae - Puffers

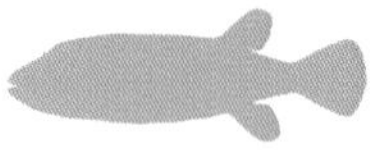

Puffers are slender to somewhat robust, becoming rounded when inflated. The mouth is small, with four fused teeth forming a "beak." Gill openings are slit-like. Spiny dorsal fin and pelvic skeleton are absent. Body is covered in tough, elastic skin. Most with small, spine-like prickles on abdomen, sides, and back. Puffers occur circumglobally in tropical to warm temperate seas. Most in shallow, nearshore waters. Some in brackish or fresh water. Many have highly toxic organs and flesh. Three species in the area. Page 340.

Family Diodontidae - Porcupinefishes

Porcupinefishes are somewhat robust, becoming round when inflated. The mouth is small, with fused teeth forming a strong "beak." Gill openings are slit-like. Spiny dorsal fin and pelvic skeleton are absent. Body is covered in tough, elastic skin and short to long spine-like scales. Some with spines always erect. Porcupinefishes occur circumglobally in tropical to warm temperate seas. Most are demersal and solitary; one is pelagic. Some are known to have toxic organs and flesh. Three species in the area. Page 340.

Family Molidae - Molas

Molas are large, deep-bodied, and laterally compressed. Body profile is rounded or oval to oblong. Mouth and gill openings are small. Dorsal and anal fins are set back on the body and are similar in shape. A caudal peduncle and true caudal fin are absent. Caudal fin is replaced by a leathery structure that is supported by modified dorsal- and anal-fin rays. Pelvic fins are absent. Molas are pelagic in worldwide tropical to warm temperate seas. Often seen swimming on their sides at the surface. Juveniles are schooling. Two species in the area. Page 342.

Notes

SPECIES

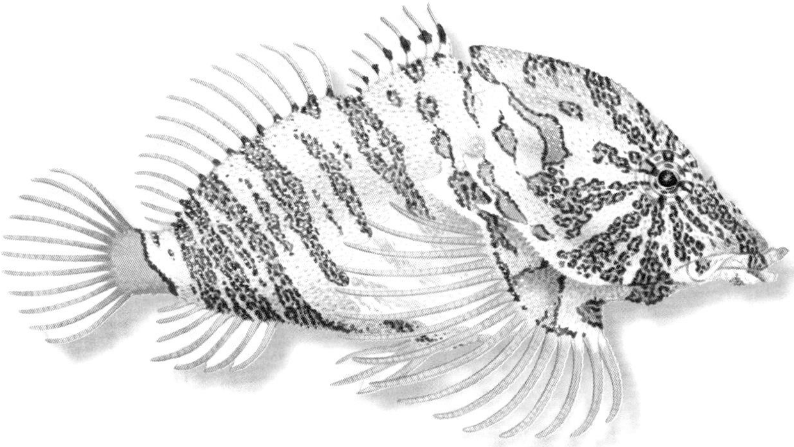

Myxinidae - Hagfishes

Black Hagfish - *Eptatretus deani* (Evermann & Goldsborough, 1907)

FEATURES: Body uniformly blackish, purplish black, or dark brown; occasionally with pale spots and blotches. Pale patches on head above rudimentary eyes. Four barbels around mouth and nostril. Usually 11 gill openings, with the first about ⅕ to ⅙ of total length from snout tip. Last opening may be enlarged. HABITAT: SE AK to Baja California. Demersal from about 350 to 9,000 ft. BIOLOGY: Scavenge on variety of prey.

Pacific Hagfish - *Eptatretus stoutii* (Lockington, 1878)

FEATURES: Body shades of tan, brown, or gray; may have pale spots and blotches. Pale patches may be above rudimentary eyes. Four barbels around mouth and nostril. Usually 12 gill openings, with the first about ⅓ to ¼ of total length from snout tip. HABITAT: Vancouver, BC, to central Baja California. Demersal from about 50 to 3,000 ft.

ALSO IN THE AREA: Shorthead Hagfish, *Eptatretus mcconnaugheyi*, page 347.

Petromyzontidae - Northern Lampreys

Pacific Lamprey - *Entosphenus tridentatus* (Richardson, 1837)

FEATURES: Dark brown to bluish black dorsally. Pale to silvery below. Fins dusky; blotches absent. Mouth with four lateral tooth plates; central two pairs with three cusps. HABITAT: Eastern AK to northern Baja California. Also to Japan. In rivers, streams, and coastal waters to about 1,500 ft. BIOLOGY: Anadromous and parasitic.

Western River Lamprey - *Lampetra ayresii* (Günther, 1870)

FEATURES: Bluish black to silvery gray or yellowish to brown dorsally. Pale below. Caudal fin with dark gray blotch. Mouth with three lateral tooth plates; central pair with three cusps. HABITAT: Southeastern AK to central CA drainage. In rivers, estuaries, and shallow coastal waters. BIOLOGY: Anadromous and migratory. Parasitic.

Arctic Lamprey - *Lethenteron camtschaticum* (Tilesius, 1811)

FEATURES: Bluish black to brownish dorsally, yellowish to pale brown below. Dark blotch on second dorsal fin and caudal fin. Mouth with three lateral tooth plates; central pair with two cusps. HABITAT: Nearly circumpolar. Northern to southern AK. In rivers, lakes to coastal waters to about 160 ft. BIOLOGY: Anadromous and parasitic.

Chimaeridae - Shortnose Chimaeras

Spotted Ratfish - *Hydrolagus colliei* (Lay & Bennett, 1839)

FEATURES: Brownish to silvery dorsally, with iridescent highlights. Numerous pale spots on dorsal surface. Males with club-like clasper on head. First dorsal fin with erect spine. Pectoral fins broad, caudal fin long, tapered. HABITAT: SE AK to southern Baja California. Usually demersal to about 3,000 ft. Also at surface or in intertidal areas. BIOLOGY: Feed on a variety of invertebrates. Perform elaborate courtship patterns.

Hagfishes

Black Hagfish
to 20 in.

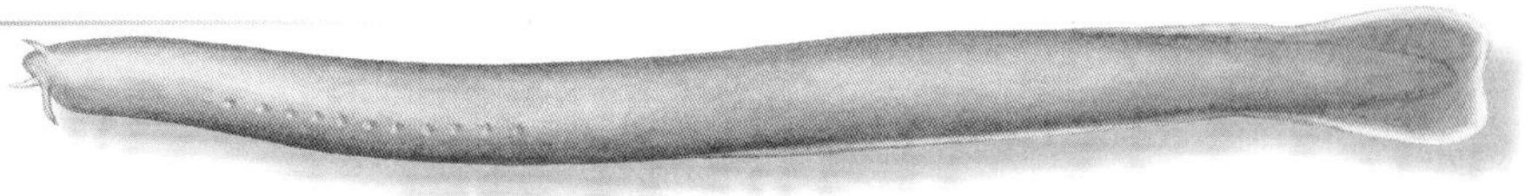

Pacific Hagfish
to 25 in.

Northern Lampreys

Pacific Lamprey
to 27 in.

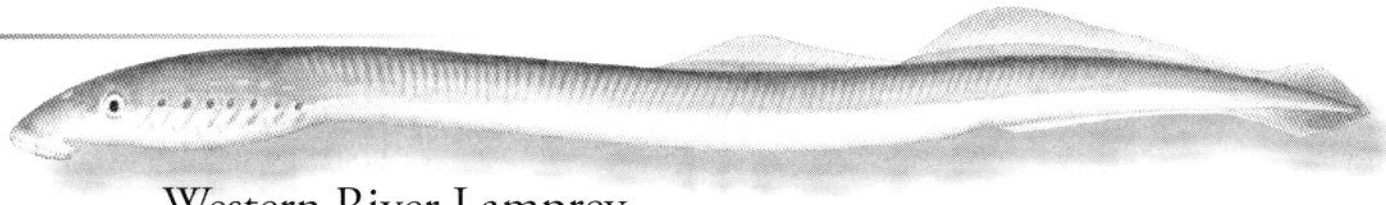

Western River Lamprey
to 12 in.

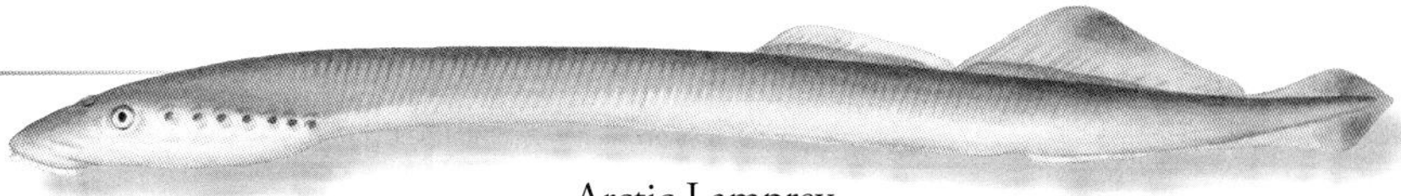

Arctic Lamprey
to 24 in.

Shortnose Chimaeras

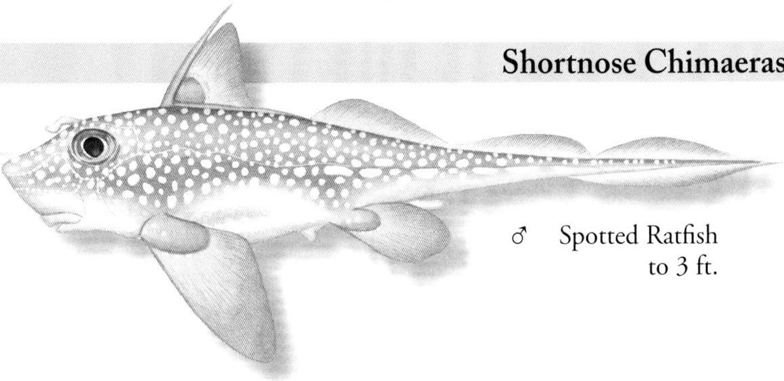

♂ Spotted Ratfish
to 3 ft.

Heterodontidae - Bullhead Sharks

Horn Shark - *Heterodontus francisci* (Girard, 1855)

FEATURES: Body and fins shades of brown to gray with darker mottling. Usually with dark brown to blackish spots that overlie dark mottling. Juveniles may also have pale spots. Head broad and elevated, with a pig-like snout. Raised ridges over eyes. Dorsal fins with single, stout spine. HABITAT: Northern CA to Gulf of California. Over rocky areas, in crevices and caves. Also over sandy areas and around kelp beds. Occur from intertidal zone to about 655 ft. BIOLOGY: Horn Shark are solitary, sluggish, and more active at night. May share caves and crevices with Spiny Lobsters or California Morays. Feed on crustaceans and small fishes.

Rhincodontidae - Whale Shark

Whale Shark - *Rhincodon typus* Smith, 1828

FEATURES: Dark gray, greenish gray, or reddish dorsally. Pale below. Pattern of white or yellow spots and bars covers upper body and fins. Snout short; mouth broad. Largest fish in the world. HABITAT: Circumglobal in tropical and warm temperate seas. Santa Cruz, CA, to Northern Chile and Galápagos Islands. Pelagic, at or near surface, near shore or in the open sea. BIOLOGY: Whale shark are harmless filter feeders. Large quantities of planktonic crustaceans, squids, and fishes are captured as water flows over large, fibrous gill plates. Often seen near the surface. Placid and approachable by divers. Females carry up to 300 young. Currently listed as Vulnerable.

Odontaspididae - Sand Tiger Sharks

Ragged-tooth Shark - *Odontaspis ferox* (Risso, 1810)

FEATURES: Brownish to grayish brown dorsally. Pale below. Often with dark spots and blotches on back, sides, and fins. Snout broadly pointed, upturned. Teeth protruding. Dorsal fins almost equal in size. First dorsal-fin origin over rear pectoral-fin margin. Caudal keels absent. Body robust. HABITAT: Scattered circumglobally in temperate waters. Newport Beach and Santa Barbara Island, CA, to Gulf of California. Found near surface to about 1,740 ft. Rare to uncommon. BIOLOGY: Poorly understood. Feed on bony fishes, squids, and shrimps. Caught in trawl nets on longlines. Reportedly inquisitive and docile around divers. Currently listed as Vulnerable.

Mitsukurinidae - Goblin Shark

Goblin Shark - *Mitsukurina owstoni* Jordan, 1898

FEATURES: Shades of white to pale pinkish or gray. Snout long, flattened, and bluntly pointed. Jaws highly protrusible, with narrow pointed teeth. Dorsal fins small, low, and similar in shape. Body soft and flabby. HABITAT: Scattered circumglobally in temperate and tropical seas. Possibly semi-oceanic. Recorded from about 130 to 4,265 ft. Uncommon. BIOLOGY: Poorly known. Prey on small fishes, squids, and crustaceans. Caught as bycatch in seines and by hook-and-line.

Bullhead Sharks

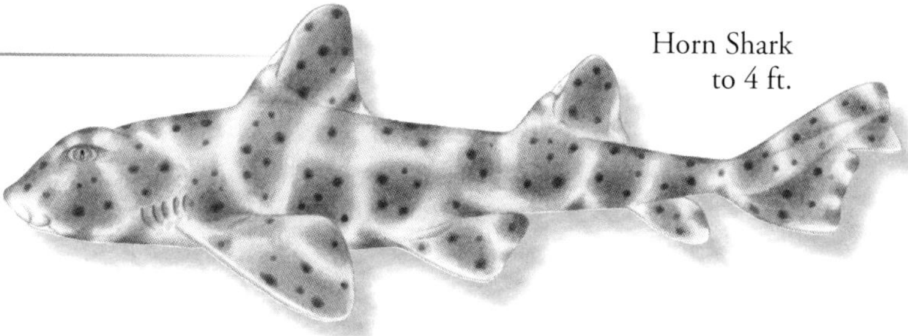

Horn Shark
to 4 ft.

Whale Shark

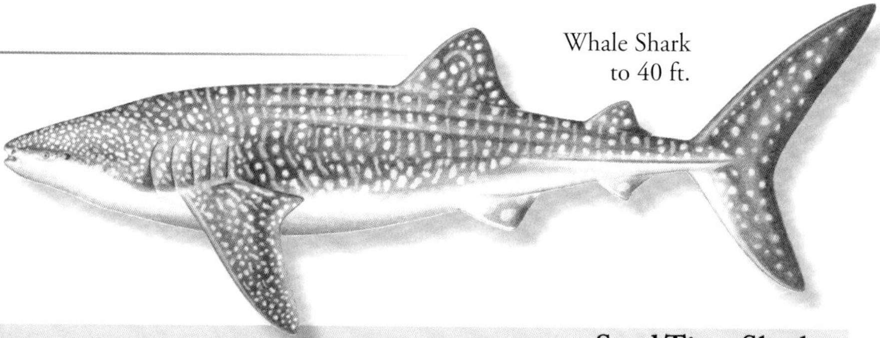

Whale Shark
to 40 ft.

Sand Tiger Sharks

Ragged-tooth Shark
to 5.5 ft.

Goblin Shark

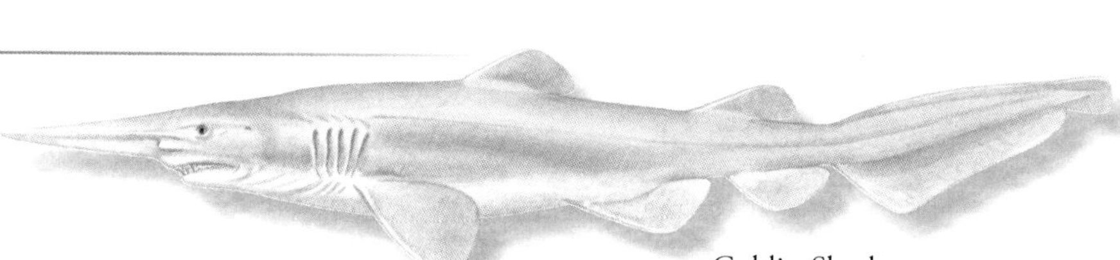

Goblin Shark
to 16.4 ft.

Megachasmidae - Megamouth Shark

Megamouth Shark - *Megachasma pelagios* Taylor, Compagno & Struhsaker, 1983

FEATURES: Grayish brown dorsally and on sides. Pale below. Lower jaw may be spotted. Head broadly rounded, with small eyes. Mouth wide and gaping, with numerous rows of very small, hook-like teeth. Lining of mouth dark and mottled. Pectoral fins long and pointed. Caudal fin long and low. HABITAT: Probably scattered in circumtropical seas. Oceanic. Recorded from southern CA. Occur from surface to about 545 ft. Rare. BIOLOGY: Known from very few records. Feed on plankton.

Alopiidae - Thresher Sharks

Pelagic Thresher - *Alopias pelagicus* Nakamura, 1935

FEATURES: Dark blue to purplish black dorsally. White below. White patch above pectoral fins absent. Labial furrows absent. Pectoral fins blunt at tips. Tail as long or longer than body. HABITAT: Circumtropical. Southern CA to Panama. Oceanic. Occur from surface to about 2,300 ft. BIOLOGY: Poorly known. Active and strong-swimming. May leap out of the water. Use tail to wound prey. Listed as Vulnerable.

Bigeye Thresher - *Alopias superciliosus* (Lowe, 1841)

FEATURES: Bluish brown dorsally. Pale below. Skin has iridescent highlights. Forehead prominently indented over very large eyes. Tail almost as long as body. HABITAT: Circumglobal. Southern CA to Gulf of California. Possibly off Peru and northern Chile. Found coastally over the continental shelf to deep water. BIOLOGY: Feed on pelagic and bottom fishes and squids. Use tail to stun prey. Litters usually contain only two young. Sought commercially. Currently listed as Vulnerable.

Thresher Shark - *Alopias vulpinus* (Bonnaterre, 1788)

FEATURES: Brown, gray, blackish, or blue gray dorsally and under snout. Two distinct white patches ventrally. Tail almost as long as body. HABITAT: Circumglobal in tropical to temperate seas. SE Alaska and British Columbia to Chile. Possibly absent from Gulf of California. Coastal and pelagic from surface to about 1,200 ft. BIOLOGY: Active and strong-swimming, often leaping out of the water. Use tail to herd and stun prey. Sometimes hunt in groups. Taken by longlines and nets. Listed as Vulnerable.

Cetorhinidae - Basking Shark

Basking Shark - *Cetorhinus maximus* (Gunnerus, 1765)

FEATURES: Brownish, grayish, or blackish; often with obscure, irregular stripes. Mouth large; snout long and fleshy. Gill slits nearly circle throat. A very large shark. HABITAT: Circumglobal in temperate seas. In the eastern Pacific from Aleutian Islands to Baja California. Pelagic. Also off Chile. Near shore and offshore. BIOLOGY: Sluggish and nonaggressive. Gill rakers possess numerous bristle-like combs that filter large quantities of plankton. Often seen feeding at the surface. Migrate long distances. Occur singly or in groups up to 100. Currently listed as Vulnerable.

Megamouth Shark

Megamouth Shark
to 18 ft.

Thresher Sharks

Pelagic Thresher
to 13 ft.

Bigeye Thresher
to 16 ft.

Thresher Shark
to 21 ft.

Basking Shark

Basking Shark
to 32 ft.

Lamnidae - Mackerel Sharks

White Shark - *Carcharodon carcharias* (Linnaeus, 1758)

FEATURES: Pale to dark gray, blue gray, or blackish dorsally. White ventrally. Snout blunt, pointed. First dorsal-fin origin located behind pectoral-fin base. Prominent caudal keels. Large-bodied. HABITAT: Circumglobal in tropical to temperate seas, less common in tropical waters. Northern Gulf of Alaska to Gulf of California. Also Panama to Chile. Nearshore. BIOLOGY: White Shark are strong and predatory. Inquisitive. They feed on a wide variety of marine life, from sea birds to seals. Slow-growing. Females produce small litters. Well documented but often maligned. Threatened by overfishing and exploitation. Currently listed as Vulnerable.

Shortfin Mako - *Isurus oxyrinchus* Rafinesque, 1810

FEATURES: Gray blue, purplish, or dark blue dorsally. White ventrally. Snout pointed. Dorsal-fin origin located behind pectoral fins. Pectoral fins shorter than head. Prominent caudal keels. HABITAT: Circumglobal in tropical to temperate seas. British Columbia to Chile, and Galápagos Islands. Pelagic. BIOLOGY: Shortfin Mako are active and strong-swimming. Acrobatic and bold. They feed on schooling fishes and other sharks. Populations are threatened by overfishing. Currently listed as Vulnerable.

Salmon Shark - *Lamna ditropis* Hubbs & Follett, 1947

FEATURES: Bluish gray to blackish dorsally. White ventrally. May have few to many dark blotches on lower head and abdomen. Snout blunt, pointed. First dorsal-fin origin located above pectoral-fin base. Prominent caudal keels. Stout, round-bodied. HABITAT: Gulf of Alaska to central Baja California. Also western Pacific to Korea. Pelagic and coastal. BIOLOGY: Highly migratory. Females larger than males. May live up to 30 years. As their name implies, Salmon Shark feed on salmons and other fishes.

Scyliorhinidae - Catsharks

Brown Catshark - *Apristurus brunneus* (Gilbert, 1892)

FEATURES: Uniformly dark brown, with darker margins on fins. Others described with pale margins on fins. Head long and laterally expanded. Dorsal fins low, with first dorsal-fin base over pelvic-fin base. HABITAT: Icy Point, AK, to northern Baja California. Also Panama to Chile. Found near bottom and in the water column from about 110 to 4,285 ft. BIOLOGY: Females carry 1 to 16 eggs that may take up to 27 months to hatch. Feed on crustaceans, fishes, squids, and shrimps.

Swell Shark - *Cephaloscyllium ventriosum* (Garman, 1880)

FEATURES: Shades of pale brown with darker bars and saddles that are interspersed with darker and paler spots. Head is broad and flattened. First dorsal-fin base located over pelvic-fin base. HABITAT: Monterey Bay to Acapulco, Mexico, including Gulf of California. Also Chile. Occur near bottom around nearshore reefs and algae beds. BIOLOGY: Slow-moving and usually solitary, but may be found in small groups. Occupy caves and crevices during the day; forage at night.

Mackerel Sharks

White Shark
to 19 ft.

Shortfin Mako
to 8 ft.

Salmon Shark
to 10 ft.

Catsharks

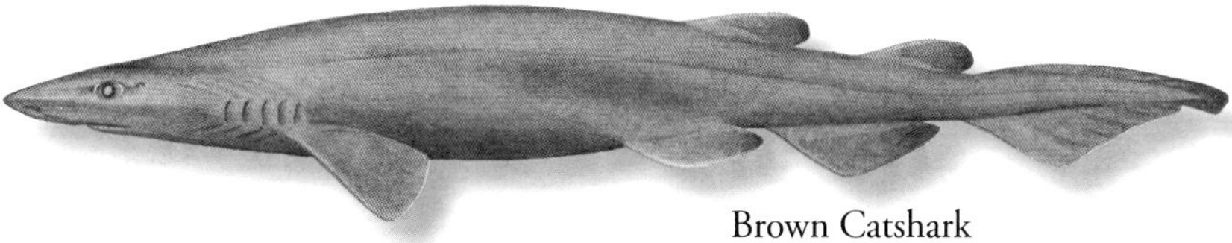

Brown Catshark
to 2.3 ft.

Swell Shark
to 3.3 ft.

Scyliorhinidae - Catsharks, *cont.*

Filetail Catshark - *Parmaturus xaniurus* (Gilbert, 1892)

FEATURES: Brownish, with irregular pale spots and blotches. May also be uniformly brown. Snout moderately long and blunt. Dorsal fins similarly shaped. Enlarged, saw-like denticles along upper edge of caudal fin. HABITAT: OR to CA. Usually near bottom from about 300 to 4,100 ft. Juveniles may occur in midwater. BIOLOGY: Enlarged gills allow Filetail Catshark to live in low-oxygen conditions.

Triakidae - Hound Sharks

Tope - *Galeorhinus galeus* (Linnaeus, 1758)

FEATURES: Uniformly brownish to grayish dorsally. Pale below. Second dorsal fin much smaller than first and similar in size to anal fin. Caudal fin with expanded upper terminal lobe. HABITAT: Occur nearly worldwide in cold to temperate seas. N British Columbia to Ecuador. Pelagic along the coast from surf zone to about 3,610 ft. BIOLOGY: Highly migratory, active. Populations declining. Listed as Vulnerable.

Gray Smoothhound - *Mustelus californicus* Gill, 1864

FEATURES: Uniformly gray dorsally. Pale below. Snout comparatively short, rounded. First dorsal fin closer to pelvic fins than pectoral fins. HABITAT: In the eastern Pacific from northern CA to Mazatlán, Mexico, including Gulf of California. Occur from surf zone to about 310 ft. BIOLOGY: Migrate north in summer; residential in southern waters. Litters with two to five pups.

Brown Smoothhound - *Mustelus henlei* (Gill, 1863)

FEATURES: Usually uniformly bronze-brown with iridescent highlights. Some may be grayish. Dorsal and caudal fins with dark inner margins; outer margins frayed, scaleless. Snout moderately long, bluntly pointed. HABITAT: In the eastern Pacific from N WA to Gulf of California, and Ecuador to Peru. Found from surf zone to about 920 ft. Usually demersal but also occur in midwater and at surface.

Sicklefin Smoothhound - *Mustelus lunulatus* Jordan & Gilbert, 1882

FEATURES: Uniformly grayish brown to grayish dorsally. Pale below. Snout comparatively short, pointed. Rear margins of dorsal fins falcate. Second dorsal fin comparatively low. HABITAT: In the eastern Pacific from southern CA to Peru, including Gulf of California. Found inshore to about 300 ft. offshore.

Leopard Shark - *Triakis semifasciata* Girard, 1855

FEATURES: Tannish to pale gray on upper body and fins, with dark saddles and spots. Saddles and spots develop pale centers with age. Snout rounded. HABITAT: In eastern Pacific from OR to Mazatlán, Mexico, including Gulf of California. Occur from surf zone to about 515 ft. Prefer soft bottoms and rocky areas near reefs. BIOLOGY: Active and strong-swimming. May form migratory schools with other sharks. Observed forming complex mating clusters.

Catsharks

Filetail Catshark
to 22 in.

Hound Sharks

Tope
to 6.4 ft.

Gray Smoothhound
to 4 ft.

Brown Smoothhound
to 3 ft.

Sicklefin Smoothhound
to 5.6 ft.

Leopard Shark
to 6 ft.

Carcharhinidae - Requiem Sharks

Narrowtooth Shark - *Carcharhinus brachyurus* (Günther, 1870)

FEATURES: Bronze, coppery, or grayish brown, with greenish highlights dorsally. Pale below. Prominent white streak on sides. Snout broad, bluntly pointed. Second dorsal fin low, nearly triangular, and positioned over anal fin. HABITAT: Scattered circumglobally in warm temperate seas. Southern CA to central Peru, including Gulf of California. Occur inshore to offshore over continental shelves. BIOLOGY: Active and seasonally migratory. Other names: Bronze Whaler, Copper Shark. Near Threatened.

Bull Shark - *Carcharhinus leucas* (Müller & Henle, 1839)

FEATURES: Gray dorsally. White ventrally. Fins with dusky tips. Eyes well forward on head. Snout short, wide, blunt. Body stout. Broad triangular to falcate dorsal fins. HABITAT: Worldwide in warm temperate to tropical seas. Southern CA (rare) to Peru, including Gulf of California. Generally near shore. May enter bays, estuaries, and rivers. Rare to uncommon in the area. Reports from CA may be erroneous. BIOLOGY: Bull Shark are large and powerful and may be aggressive. They have the uncommon ability to thrive in fresh water. Their diet is wide and varied. Listed as Near Threatened.

Oceanic Whitetip Shark - *Carcharhinus longimanus* (Poey, 1861)

FEATURES: Gray or brown dorsally. White below. Tips of first dorsal, pectoral, and caudal fins mottled white. Black blotches present on fin tips of immature specimens. Dorsal fin very tall and rounded at top. Pectoral fins large and paddle-like. Body large and stout. HABITAT: Circumtropical from edges of continental shelves to mid-ocean. Southern CA to Peru. Reported from central CA. Found at or near surface to about 655 ft. BIOLOGY: Oceanic Whitetip Shark feed primarily on fishes but also invertebrates, sea birds, and carrion. Inquisitive and competitive. Currently listed as Vulnerable.

Dusky Shark - *Carcharhinus obscurus* (Lesueur, 1818)

FEATURES: Shades of gray to bronze dorsally. White ventrally. Most fins dusky-tipped. Pectoral fins long and pointed. Low interdorsal ridge present. A large-bodied shark. HABITAT: Worldwide in tropical to warm temperate seas. Occur from shoreline to outer continental shelves in warm waters. Southern CA to Ecuador, including Gulf of California. BIOLOGY: Dusky Shark are active and migratory. They are slow to mature and have small litters. Prey include fishes, other sharks, rays, carrion, and a variety of invertebrates. Protected in US. Currently listed as Vulnerable.

Tiger Shark - *Galeocerdo cuvier* (Péron & Lesueur, 1822)

FEATURES: Shades of gray, with darker bars and spots dorsally. Markings muted in older sharks. Snout short, blunt. Body large and stout. Trailing fin tips pointed. Small caudal keel present. HABITAT: Circumglobal in tropical to temperate seas. Southern CA to Peru. Also reported from Prince William Sound, AK. Coastal and offshore, at surface or bottom. BIOLOGY: Tiger Shark are voracious predators with an indiscriminate and varied diet. Prey include: invertebrates, fishes, sharks, marine reptiles, birds, and mammals. Litters with 10–82 young. Listed as Near Threatened.

Narrowtooth Shark
to 9.6 ft.

Bull Shark
to 11 ft.

Oceanic Whitetip Shark
to 13 ft.

Dusky Shark
to 12 ft.

Tiger Shark
to 18 ft.

Blue shark - *Prionace glauca* (Linnaeus, 1758)

FEATURES: Shades of blue dorsally. White ventrally. Snout is long and pointed. Body is slender. Pectoral fins long and pointed. Caudal fin long and tall. HABITAT: Circumglobal in all tropical to temperate seas. Western Gulf of Alaska to Chile. Widely distributed. Oceanic, pelagic. BIOLOGY: Blue Shark are inquisitive and brazen. They are most active in early evening and night. Migrate long distances with prey and during reproductive seasons, sometimes making oceanic crossings. A heavily fished shark. Listed as Near Threatened.

Pacific Sharpnose Shark - *Rhizoprionodon longurio* (Jordan & Gilbert, 1882)

FEATURES: Grayish brown to grayish dorsally. White below. Snout long and broadly pointed. Eyes relatively large. First dorsal-fin origin over rear margin of pectoral fins. Second dorsal-fin origin over center of anal-fin origin. Anal fin preceded by long preanal ridge. HABITAT: Southern CA to Peru, including Gulf of California. Occur inshore from estuaries to about 330 ft. BIOLOGY: Feed on small fishes and crustaceans. Usually two to five pups per litter.

Sphyrnidae - Hammerhead Sharks

Scalloped Hammerhead - *Sphyrna lewini* (Griffith & Smith, 1834)

FEATURES: Brownish gray or gray dorsally. White ventrally. Head flattened; anterior edge arched, with three notches near middle. Fins usually tipped dusky or black. HABITAT: Circumtropical in warm seas over continental shelves. Southern CA to Peru. From surface to about 900 ft. BIOLOGY: Scalloped Hammerhead occur singly, in pairs, or in huge congregations. Migratory. Prey on fishes, crustaceans, and squids. Adults placid and popular with tourists. Vulnerable to overfishing. Listed as Endangered.

Bonnethead - *Sphyrna tiburo* (Linnaeus, 1758)

FEATURES: Gray or grayish brown dorsally; may have small dark spots. Pale to white ventrally. Head flattened, convex; front edge lacks notches. Second dorsal and anal fins similar in size. HABITAT: Southern CA to Peru, including Gulf of California. Also western Atlantic. Occur in warm, shallow, coastal waters over sand and mud, in river mouths, and around reefs. BIOLOGY: Bonnethead are sluggish. Preyed upon by other, larger sharks. They prey primarily on crabs and shrimps. Common bycatch of shrimp fisheries. Popular in public aquaria.

Smooth Hammerhead - *Sphyrna zygaena* (Linnaeus, 1758)

FEATURES: Gray or brownish gray dorsally. Pale or white ventrally. Fins unmarked or tipped dusky or black. Head flattened, arched, with no notch at center. HABITAT: Circumglobal in tropical to temperate seas. Central CA to Gulf of California and Chile. Occur close inshore to offshore. Near surface in deep water. BIOLOGY: Smooth Hammerhead are strong and seasonally migratory. Young form very large schools. Prey include fishes, other sharks, rays, crustaceans, and squids. Listed as Vulnerable.

Requiem Sharks

Blue Shark
to 12.5 ft.

Pacific Sharpnose Shark
to 3.6 ft.

Hammerhead Sharks

Scalloped Hammerhead
to 13.8 ft.

Bonnethead
to 5 ft.

Smooth Hammerhead
to 13 ft.

Chlamydoselachidae - Frill Sharks

Frill Shark - *Chlamydoselachus anguineus* Garman, 1844

FEATURES: Dark brownish to grayish brown. Mouth gaping, with rows of three-cusped teeth. Six pairs of gill slits; first gill slits connected under throat. Single dorsal fin is low and much smaller than anal fin. Body long, serpent-like. HABITAT: Scattered worldwide in tropical to temperate seas. Central CA to Chile. Demersal to pelagic or at surface. Usually in deep water; occasionally at surface. Rare. BIOLOGY: Feed on squids and fishes. Litters with 6–12 pups. Caught as bycatch. Listed as Near Threatened.

Hexanchidae - Cow Sharks

Bluntnose Sixgill Shark - *Hexanchus griseus* (Bonnaterre, 1788)

FEATURES: Brownish to grayish dorsally, paler below. Body usually unmarked, but may have dark spots on sides. Lateral line usually pale. Head broad, snout blunt. Six long gill slits. HABITAT: Scattered circumglobally in temperate seas. Aleutian Islands to southern Baja California, and to Chile. Found from surface to about 8,200 ft. BIOLOGY: Feed on squids, fishes, and small sharks and rays. Larger specimens feed on cetaceans and seals. Occur singly or in groups. Listed as Near Threatened.

Broadnose Sevengill Shark - *Notorynchus cepedianus* (Péron, 1807)

FEATURES: Grayish to grayish brown, with numerous small dark spots on body and fins. Occasionally uniformly colored with pale spots. Head broad, snout blunt. Seven long gill slits. HABITAT: Scattered worldwide in temperate inshore waters over continental shelves. Northern BC to Chile. Occur from shore to about 1,870 ft. BIOLOGY: Active, strong, and migratory. May form aggregations.

Echinorhinidae - Bramble Sharks

Prickly Shark - *Echinorhinus cookei* Pietschmann, 1928

FEATURES: Shades of dark brown to gray or blackish. Underside of head pale. Body covered with bristle-like denticles. Dorsal fins located posteriorly; similarly shaped. Anal fin absent. HABITAT: Scattered in Pacific Ocean; OR to Gulf of California, and Costa Rica to Chile. Over continental shelves, slopes, and seamounts. Found near bottom from about 15 to 3,600 ft. BIOLOGY: Sluggish. Alone or in groups. Near Threatened.

Squalidae - Dogfish Sharks

Pacific Spiny Dogfish - *Squalus suckleyi* (Girard, 1855)

FEATURES: Shades of brown to gray dorsally. Pale ventrally. White spots usually on upper sides. Pectoral fins with pale margins. Snout pointed. Body slender. Dorsal fins with round spines. HABITAT: In North Pacific from Yellow Sea to Gulf of California. Also western Atlantic. Occur coastally from intertidal zone to continental shelves. BIOLOGY: Unusually slow-growing and long-lived. Seasonally migratory or residential; schooling. May form large feeding packs. Populations declining from overfishing.

Frill Sharks

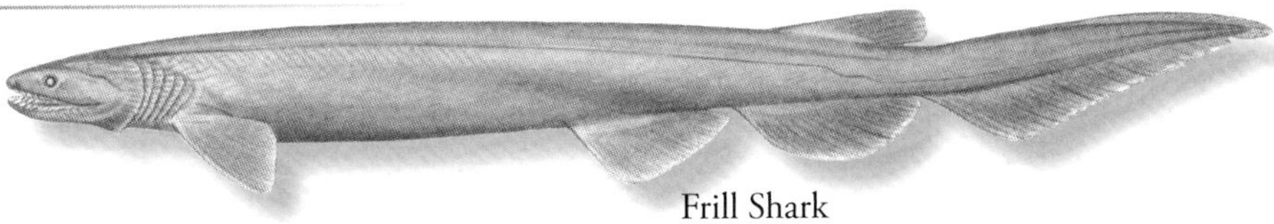

Frill Shark
to 6.4 ft.

Cow Sharks

Bluntnose Sixgill Shark
to 15.8 ft.

Broadnose Sevengill Shark
to 9.5 ft.

Bramble Sharks

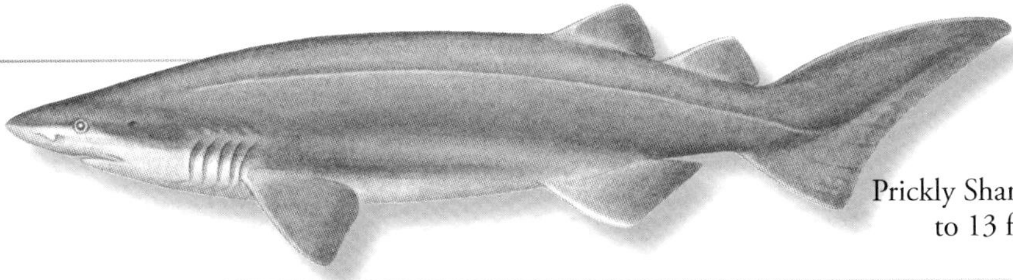

Prickly Shark
to 13 ft.

Dogfish Sharks

Pacific Spiny Dogfish
to 6.5 ft.

Somniosidae - Sleeper Sharks

Pacific Sleeper Shark - *Somniosus pacificus* Bigelow & Schroeder, 1944

FEATURES: Brown, dark gray, blackish, or greenish. Snout short and rounded. Mouth small. Gill slits short and low on body. Dorsal fins small and similar in shape. Anal fin absent. Body robust and covered with bristle-like denticles. HABITAT: Found in northern Pacific. Northern AK to southern Baja California. Occur on or near bottom over continental shelves and slopes to about 6,625 ft. Also occasionally at surface in intertidal zone. BIOLOGY: Diet is wide and varied and includes fishes, marine mammals, invertebrates, and carrion. Believed to eat by sucking in prey. Usually sluggish but may make long, vertical migrations in the water column.

Dalatiidae - Kitefin Sharks

Cookiecutter Shark - *Isistius brasiliensis* (Quoy & Gaimard, 1824)

FEATURES: Shades of brown. Abdomen paler, with luminescent organs. Broad, dark band behind head. Fins with dark inner margins, pale outer margins. Snout very short and rounded. Eyes very large. Mouth equipped with sharp, triangular lower teeth. Dorsal fins set close together. Body cylindrical. HABITAT: Scattered circumglobally. Northern Pacific to northern Baja California. BIOLOGY: Migrate vertically in the water column at night. Takes swift, deep, twisting bites from abdomen and sides of prey that leave round, cookie-shaped wounds. Prey on fishes, marine mammals, and squids.

Squatinidae - Angel Sharks

Pacific Angel Shark - *Squatina californica* Ayres, 1859

FEATURES: Brownish, grayish, or reddish to blackish dorsally. Usually with irregular spots and mottling. Spots may form rosettes. Pale ventrally. Mouth broad, with pointed teeth. Nostrils with several barbels. Enlarged denticles form middorsal row. Flattened head, body, and fins form rhomboid shape. Tail long, caudal fin expanded. HABITAT: Puget Sound, WA, to southern Baja and Gulf of California to southern Chile. Occur from surf zone to about 600 ft. Found over soft bottoms in bays and around rocks and canyons. BIOLOGY: An ambush predator that buries in bottom sediment and swiftly engulfs prey with protrusible jaws. More active at night. Listed as Near Threatened.

Torpedinidae - Torpedo Electric Rays

Pacific Electric Ray - *Torpedo californica* Ayres, 1855

FEATURES: Dark bluish black, purplish black to gray, or brownish dorsally, with irregular dark spots. Spots sometimes absent. Ventral surface white to slate gray. Eyes very small. Disk nearly circular in shape. First dorsal fin over pelvic fins. Caudal fin expanded, triangular. Skin smooth. HABITAT: In eastern Pacific from northern British Columbia to central Baja California. Found from surf zone to about 2,970 ft over soft bottoms in bays and near kelp beds, rocks, or canyons. Also around rocks and kelp beds. BIOLOGY: Feed nocturnally by electroshocking a variety of fishes.

Sleeper Sharks

Pacific Sleeper Shark
to 14 ft.

Kitefin Sharks

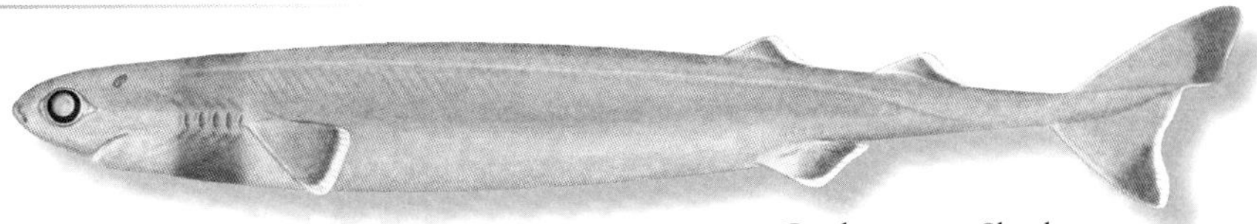

Cookiecutter Shark
to 22 in.

Angel Sharks

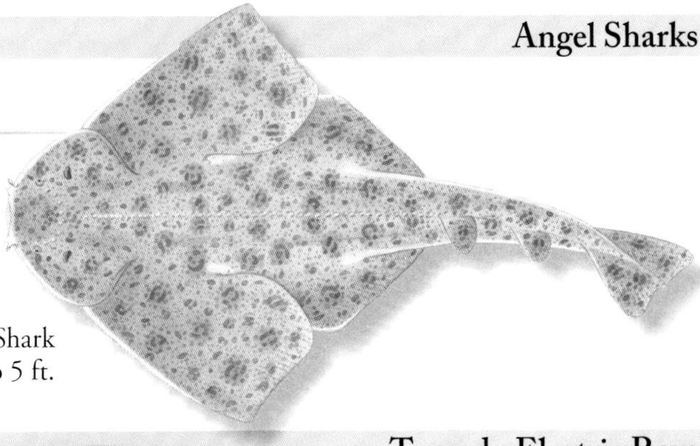

Pacific Angel Shark
to 5 ft.

Torpedo Electric Rays

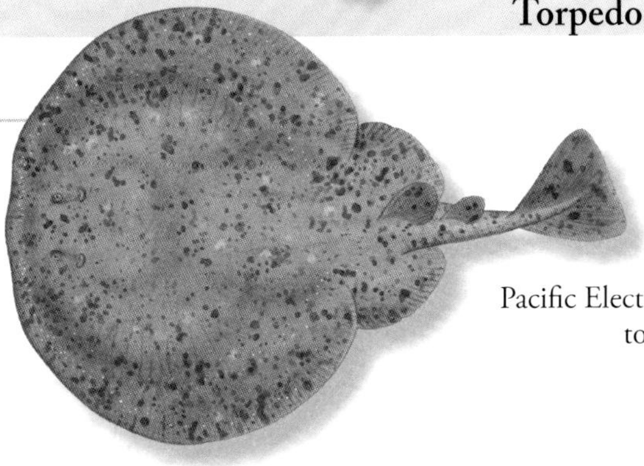

Pacific Electric Ray
to 4.5 ft.

Rhinobatidae - Guitarfishes

Shovelnose Guitarfish - *Rhinobatos productus* Ayres, 1854

FEATURES: Uniformly pale brown to grayish brown dorsally. White below. Some may display obscure, pale bands. Head and pectoral fins form a spade-shaped disk that is longer than it is wide. Back and tail with a single row of spines. The tail is long, with two widely separated dorsal fins. HABITAT: San Francisco, CA, to southern Mexico, including Gulf of California. Demersal over soft bottoms from near shore to about 300 ft. BIOLOGY: Form large aggregations and make inshore migrations to mate and give birth. Often found buried in soft sediment during the day. Near Threatened.

Banded Guitarfish - *Zapteryx exasperata* (Jordan & Gilbert, 1880)

FEATURES: Shades of pale brown, with wavy dark bands on dorsal surface. Ventral surface is pale with dark speckles. Head and pectoral fins form wedge-shaped disk that is about as long as it is wide. There is a single row of spines along back and tail. Tail is moderately long, with two separated dorsal fins. HABITAT: Central CA to Gulf of Mexico. Also reported from Peru. Demersal over rocky and soft bottoms from shallow tidepools to about 655 ft. BIOLOGY: Enter shallow waters to mate and give birth.

Rajidae - Skates

Deepsea Skate - *Bathyraja abyssicola* (Gilbert, 1896)

FEATURES: Uniformly purplish gray to grayish brown, tan, or dark brown dorsally. Large males may have irregular pale blotches and dark spots. Pale blotches are small to absent in females. Ventral surface is dark, with areas of white. Snout long, pointed, and flexible. Scapular thorns absent. One to five nuchal thorns are separated from row of thorns on trunk and tail. Dorsal and ventral surfaces covered with tiny denticles. HABITAT: AK to northern Baja California. Also to Sea of Japan. Usually found over continental slopes from about 1,195 to 9,530 ft.

Aleutian Skate - *Bathyraja aleutica* (Gilbert, 1896)

FEATURES: Dorsal surface grayish brown to dusky olive, with obscure darker patches. Some described with pale areas dorsally. Ventral surface white with grayish margins. Snout long, pointed, and flexible. One or two strong scapular thorns present. Continuous or broken row of thorns along back to dorsal fins. Dorsal surface with fine denticles; ventral surface smooth except around head. HABITAT: AK to northern CA. Also to Sea of Japan. Demersal from about 50 to 5,255 ft.

Sandpaper Skate - *Bathyraja interrupta* (Gill & Townsend, 1897)

FEATURES: Shades of brown to blackish dorsally; may have pale mottling and a white spot on each side of tail. Ventral surface white; underside of tail primarily dark. Snout short, flexible. One or two scapular thorns present. Rows of thorns along back continuous or interrupted. Tail comparatively long. Dense denticles cover dorsal surface. Ventral surface smooth except around snout tip. HABITAT: Bering Sea and Aleutian Islands to southern CA. Found demersally from about 655 to 4,500 ft.

Guitarfishes

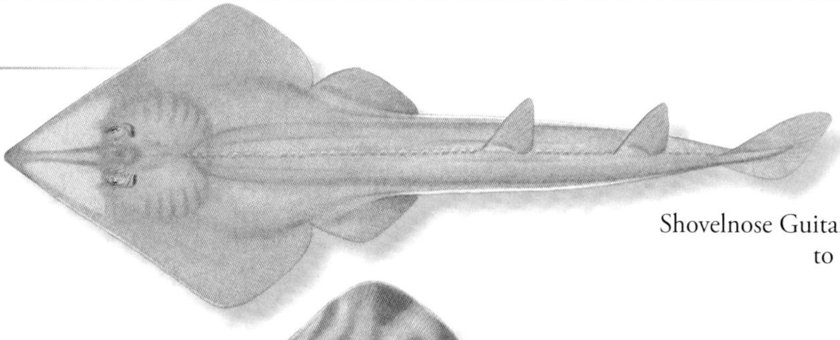

Shovelnose Guitarfish
to 5 ft.

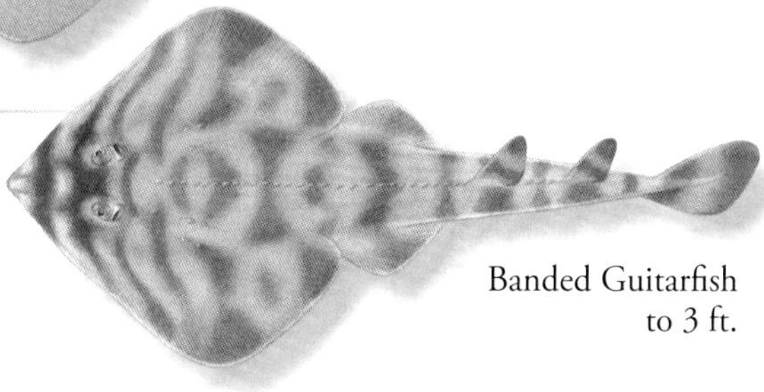

Banded Guitarfish
to 3 ft.

Skates

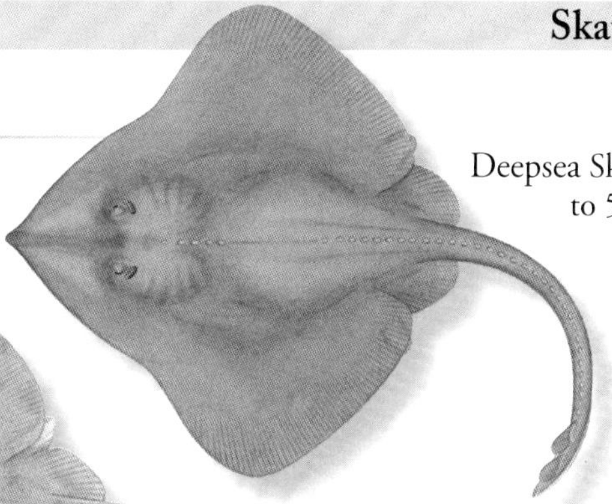

Deepsea Skate
to 5 ft.

Aleutian Skate
to 5.3 ft.

Sandpaper Skate
to 5.3 ft.

Commander Skate - *Bathyraja lindbergi* Ishiyama & Ishihara, 1977

FEATURES: Dorsal surface usually dark grayish brown; rarely with pale mottling. Ventral surface usually dark grayish brown, with pale area around mouth; sometimes pale brown. Snout short and flexible. Scapular thorns absent. Row of thorns usually continuous along back to dorsal fins. Thorns well developed along tail. HABITAT: Okhotsk Sea to Bering Sea, and Aleutian Islands. Found near bottom from about 390 to 3,940 ft; possibly to 6,560 ft. Rarely encountered in the area.

Whiteblotched Skate - *Bathyraja maculata* Ishiyama & Ishihara, 1977

FEATURES: Described as gray dorsally, with scattered white to yellowish blotches. Observed specimen dark purplish brown, with paler spots and blotches on dorsal surface. Ventral surface described as gray, with darker margins on underside of tail. Scapular thorns absent. Nuchal thorns number one to about four. Thorns along tail well developed. Dorsal surface covered with denticles. Ventral surface smooth. HABITAT: Sea of Okhotsk and northern Sea of Japan to Bering Sea and Aleutian Islands. Demersal from about 240 to 3,915 ft.

Butterfly Skate - *Bathyraja mariposa* Stevenson, Orr, Hoff & McEachran, 2004

FEATURES: Brown, with symmetrically arranged yellowish blotches and darker spots dorsally. Tail distinctly banded. Ventral area whitish with dark margins along underside of tail. Snout moderately long. Scapular thorns and thorns along back absent. Thorns along tail small or absent. Dorsal surface densely covered with denticles. HABITAT: Aleutian Islands from Petrel Bank to the Islands of Four Mountains. Found near bottom from about 295 to 1,470 ft.

Whitebrow Skate - *Bathyraja minispinosa* Ishiyama & Ishihara, 1977

FEATURES: Dorsal and ventral surfaces dark brown to dark grayish brown. Area around and between eyes usually white, but may be dark. Snout short, flexible. Eyes comparatively close. Scapular thorns absent. Nuchal thorns usually absent; when present, usually fewer than three. Thorns along back absent. Thorns along tail strongly developed. Fine denticles cover dorsal surface. Ventral surface smooth. HABITAT: Sea of Okhotsk off Hokkaido, off Kuril Islands to Commander Islands and Bering Sea to Aleutian Islands. Bottom dwelling from about 525 to 4,685 ft. Rare in the area.

Alaska Skate - *Bathyraja parmifera* (Bean, 1881)

FEATURES: Grayish brown with darker spots and blotches dorsally. May have a pair of variable pale spots posteriorly. Dark blotches on tail. Snout moderately long, blunt, and flexible. Thorns around eyes present. Strong scapular thorns present; usually two on each side. Thorns along back to dorsal fins continuous or interrupted on back. Tail comparatively short. Dorsal surface with coarse denticles. Ventral surface mostly smooth. HABITAT: Sea of Okhotsk and northern Sea of Japan to Bering Sea and the eastern Gulf of Alaska. Found demersally from about 65 to 4,700 ft.

Commander Skate
to 3.5 ft.

Whiteblotched Skate
to 4.8 ft.

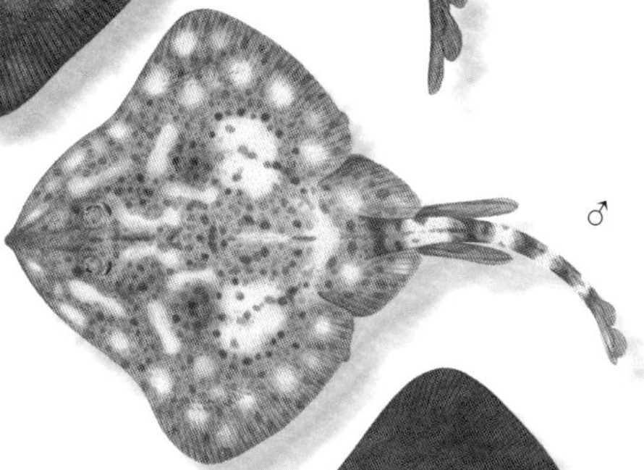

♂ Butterfly Skate
to 2.5 ft.

Whitebrow Skate
to 2.7 ft.

dark form

Alaska Skate
to 4.3 ft.

Mud Skate - *Bathyraja taranetzi* (Dolganov, 1983)

FEATURES: Shades of reddish brown to grayish brown dorsally, with darker spots and blotches. A pair of pale blotches on pectoral-fin base usually present. Ventral surface white, with brown on tail. Snout short and blunt. Nuchal thorns usually absent but weak and number fewer than three when present. Scapular thorns absent. Thorns along tail may be unevenly spaced and very small distally. Dorsal surface covered with denticles. Ventral surface smooth. HABITAT: Pacific coast of Kamchatka and northern Kuril Islands to Bering Sea and Aleutian Islands. Demersal from about 190 to 3,460 ft.

Big Skate - *Raja binoculata* Girard, 1855

FEATURES: Shades of brown dorsally with a pair of variably ocellated spots at pectoral-fin bases. Pale spots form irregular wavy lines and circles. Ventral surface white to sooty; may have dark blotches. Snout long, stiff, and pointed. One nuchal thorn present. Scapular thorns absent. Tail with row of thorns. Pelvic fins shallowly notched. Dorsal surface covered with small prickles. Ventral surface with prickles on snout, abdomen, and between gills. HABITAT: E Bering Sea and Aleutian Islands to Baja California; rarely south of Point Conception. Demersal from about 7 to 2,625 ft. NOTE: Near Threatened.

California Skate - *Raja inornata* Jordan & Gilbert, 1881

FEATURES: Shades of brown, with scattered pale blotches and small darker spots dorsally. May also have pairs of faint ocellated spots on pectoral fins. Ventral surface tan. Snout moderately long, pointed, and stiff. Zero to seven nuchal thorns. Scapular thorns absent. Row of thorns along tail flanked with rows of smaller thorns. Pelvic fins deeply notched. Dorsal surface with small, scattered prickles. Ventral surface smooth. HABITAT: Strait of Juan de Fuca to southern Baja California, including Gulf of California. Occur over bottom from about 40 to 5,250 ft.

Longnose Skate - *Raja rhina* Jordan & Gilbert, 1880

FEATURES: Pale to dark brown dorsally. May have scattered pale and dark spots and a pair of dark ocellated spots on pectoral-fin bases. Ventral surface mottled pale to dark grayish. Snout long, stiff, and sharply pointed. One or two nuchal thorns present. Scapular thorns absent. Rows of thorns along tail present in large specimens. Pelvic fins deeply notched. Dorsal surface with sharp prickles in adults. Ventral surface prickly in females, smooth in males. HABITAT: Southeastern Bering Sea to Cedros Is., Baja California. Also in Gulf of California. Demersal from about 30 to 3,500 ft.

Starry Skate - *Raja stellulata* Jordan & Gilbert, 1880

FEATURES: Tan to brownish dorsally, with scattered darker spots and rosettes. Pair of dark, ocellated blotches and white patches on pectoral-fin bases. Ventral surface whitish with dark margins. Snout short, blunt, and stiff. Thorns along back continuous with tail thorns. Dorsal surface with numerous star-shaped prickles. HABITAT: OR to central Baja California. Demersal from shore to about 2,400 ft.

ALSO IN THE AREA: Okhotsk Skate, *Bathyraja violacea*, page 347.

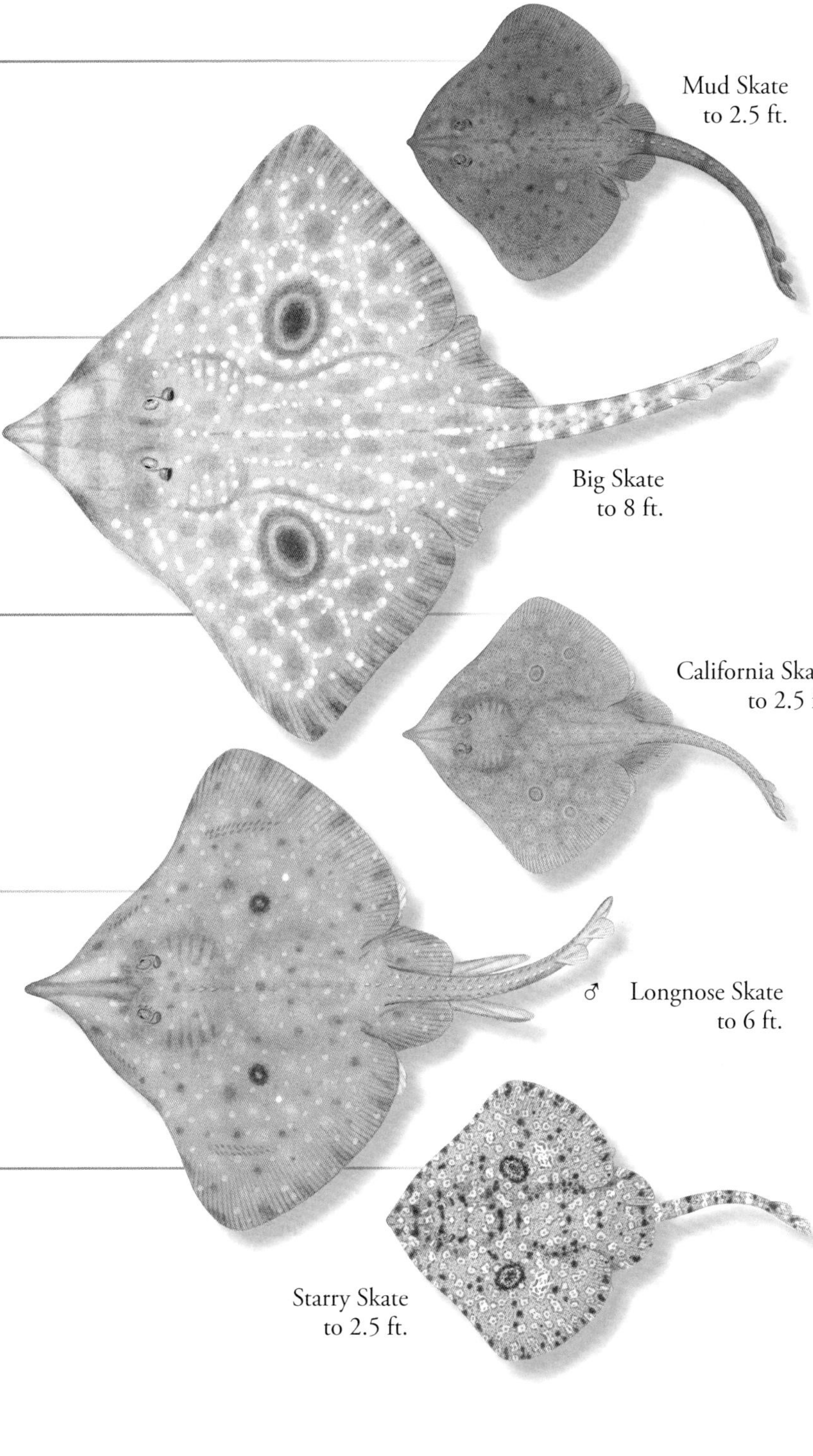

Mud Skate
to 2.5 ft.

Big Skate
to 8 ft.

California Skate
to 2.5 ft.

♂ Longnose Skate
to 6 ft.

Starry Skate
to 2.5 ft.

Platyrhinidae - Thornbacks

Thornback - *Platyrhinoidis triseriata* (Jordan & Gilbert, 1880)

FEATURES: Uniformly grayish brown to olivaceous dorsally. Ventral surface creamy to yellowish. Head and pectoral fins form rounded heart shape that is somewhat wider than long. Usually two pairs of scapular thorns present. Three rows of spines extend down tail. Tail moderately long with two dorsal fins posteriorly. HABITAT: Northern CA to Gulf of California. Also reported from Ecuador. Found over muddy to sandy bottoms of bays, sloughs and beaches and around kelp beds from surf zone to about 450 ft. BIOLOGY: Form large aggregations in bays and sloughs, to mate and give birth.

Urotrygonidae - American Round Stingrays

Round Stingray - *Urobatis halleri* (Cooper, 1863)

FEATURES: Dorsal coloration variable. Shades of tan, brown, gray, or blackish with pale or dark spots that form rosettes and vermiculations. Sometimes uniformly grayish to blackish. Ventral surface white to yellowish. Snout somewhat angular. Head and pectoral fins form nearly round disk. Tail relatively short and stout with a long barb at midlength and a rounded caudal fin. Dorsal fin absent. HABITAT: Northern CA to Ecuador, including Gulf of California. Found from shoreline to about 70 ft over sandy to muddy or rocky bottoms and around reefs. BIOLOGY: Migrate inshore to mate and give birth. Excavate round depressions in the bottom while searching for bottom-dwelling invertebrates.

Dasyatidae - Whiptail Stingrays

Diamond Stingray - *Dasyatis dipterura* (Jordan & Gilbert, 1880)

FEATURES: Dorsal surface brownish to grayish or blackish. May be reddish near disk margins. White ventrally. Snout is bluntly pointed. Head and pectoral fins form a diamond shape that is slightly wider than long. Tail is long, slender, and whip-like and is less than one and a half times as long as disk. One or more spines are located at about one-third length of tail. Narrow dorsal and ventral tail folds present. HABITAT: Central CA to northern Chile, Galápagos Islands, and Gulf of California. Found over soft bottoms from surf zone to about 230 ft. BIOLOGY: Often found buried in bottom sediment. Feed on a variety of bottom-dwelling crustaceans and fishes. Females may live to 28 years. Other name: Shorttail Stingray.

Pelagic Stingray - *Pteroplatytrygon violacea* (Bonaparte, 1832)

FEATURES: Dark bluish gray to purplish black dorsally. Ventral area is a paler shade of dorsal color. Snout is short and barely protrudes from rounded anterior fin margins. Tail thick at base, narrows at barb. Narrow ventral tail fold present, dorsal fold absent. HABITAT: Circumglobal in temperate to tropical seas. In eastern Pacific from British Columbia to central Chile. Pelagic and epipelagic over continental shelves and oceanic waters. BIOLOGY: Feed on a variety of fishes, crustaceans, squids, and sea jellies. Females may live to about 10 years. Litters contain three to nine young.

Thornbacks

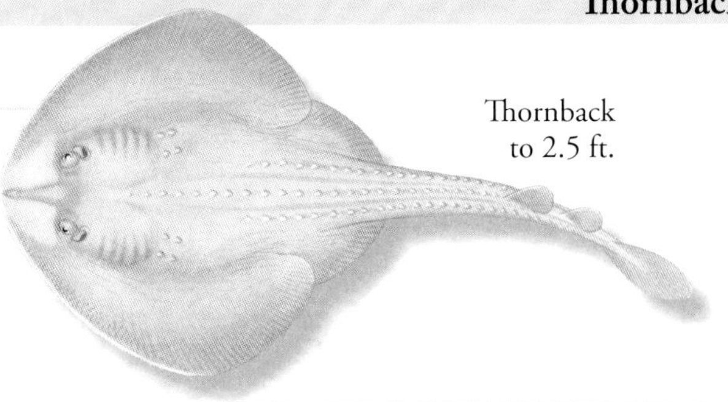

Thornback
to 2.5 ft.

American Round Stingrays

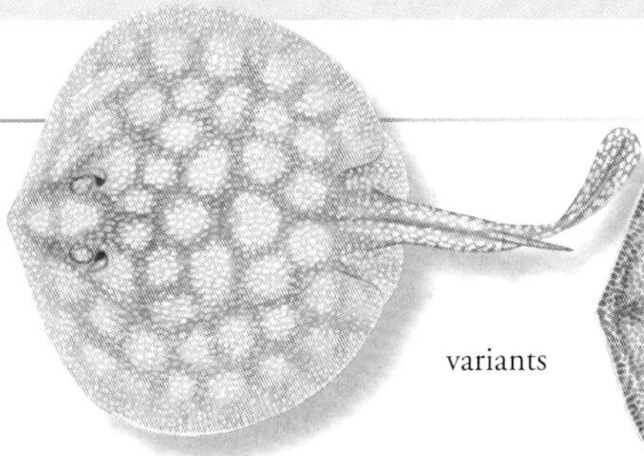

variants

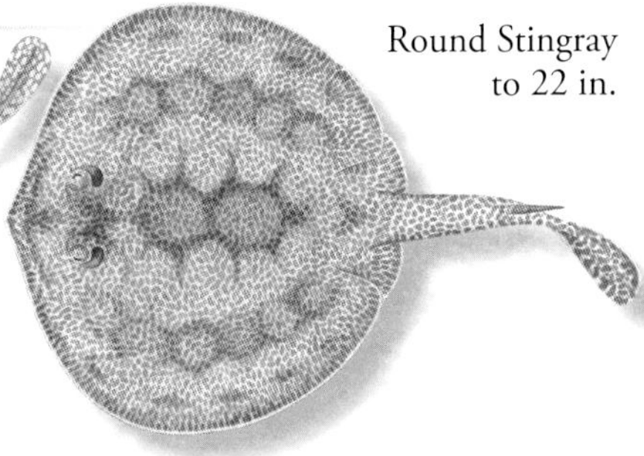

Round Stingray
to 22 in.

Whiptail Stingrays

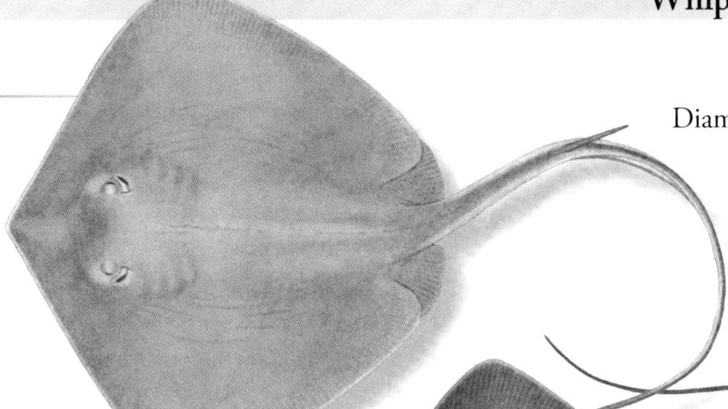

Diamond Stingray
width to 4 ft.

Pelagic Stingray
width to 2.6 ft.

Gymnuridae - Butterfly Rays

California Butterfly Ray - *Gymnura marmorata* (Cooper, 1864)

FEATURES: Shades of brown to grayish brown dorsally. May be densely mottled with small dark spots or almost uniformly colored. White ventrally. Body and pectoral fins form a compressed, nearly triangular disk that is almost twice as wide as it is long. Eyes very small. Tail short, with a barbed spine near base. Caudal fin absent. HABITAT: Southern CA to Peru, including the Gulf of California. Demersal over soft bottoms along beaches and in shallow bays from surf zone to about 310 ft. BIOLOGY: California Butterfly Ray root bottom sediments for prey that include shrimps, bivalves, worms, gastropods, and fishes. Females give birth to live young.

Myliobatidae - Eagle Rays and Mantas

Giant Manta - *Manta birostris* (Walbaum, 1792)

FEATURES: Color varies. Blackish, reddish, or olivaceous to brown dorsally. Dorsal surface either uniform or with pale patches, spots, or chevrons. Ventral surface either white or blotched. Head very broad, with mouth at front. Pectoral fins wide, arched, and pointed. Anterior margins of pectoral fins modified into cephalic fins. Tail short. Barbed spine absent. HABITAT: Circumglobal in tropical to warm temperate seas over continental shelves. Southern CA to Peru, including Gulf of California. Occur from near surface to about 100 ft. BIOLOGY: Giant Manta use cephalic fins to direct flow of water into the mouth. Use specialized gill plates to strain planktonic food. May be seen leaping from the water. Currently listed as Vulnerable.

Spinetail Mobula - *Mobula japanica* (Müller & Henle, 1841)

FEATURES: Blackish dorsally, with small white patch behind each eye. May also have pale areas behind head. Tip of dorsal fin pale. Pale ventrally. Head protruding, with mouth on underside. Pectoral fins wide, arched, and pointed; anterior margins form cephalic fins. Tail long and whip-like with barbed spine near base. HABITAT: Circumglobal in tropical to temperate seas. Central CA to Peru, including Gulf of California. Found near surface to about 100 ft. BIOLOGY: Spinetail Mobula travel singly or in groups. Feed on plankton by using cephalic fins to direct water into mouth and over gill plates. May leap from the water. Caught as bycatch. Currently listed as Near Threatened.

Bat Ray - *Myliobatis californica* Gill, 1865

FEATURES: Shades of brown to blackish dorsally; tips of pectoral fins may be pale. Ventral surface pale. Head elevated and protruding. Mouth with rows of plate-like teeth. Pectoral fins wide, arched, and bluntly pointed. Tail long and whip-like, with one to three barbed spines at base. HABITAT: In eastern Pacific from OR to Gulf of California, with an isolated population at the Galápagos Islands. Found over soft bottoms in bays and sloughs and around kelp beds from intertidal zone to about 355 ft. BIOLOGY: Bat Ray use snout and disk edges to dig in bottom sediment for prey. Use strong, crushing jaws to eat hard-shelled mollusks and other invertebrates. May bury itself in sand.

Butterfly Rays

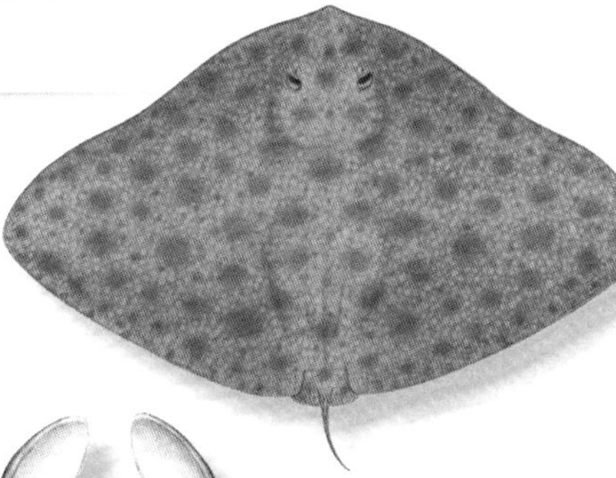

California Butterfly Ray
width to 4 ft.

Eagle Rays and Mantas

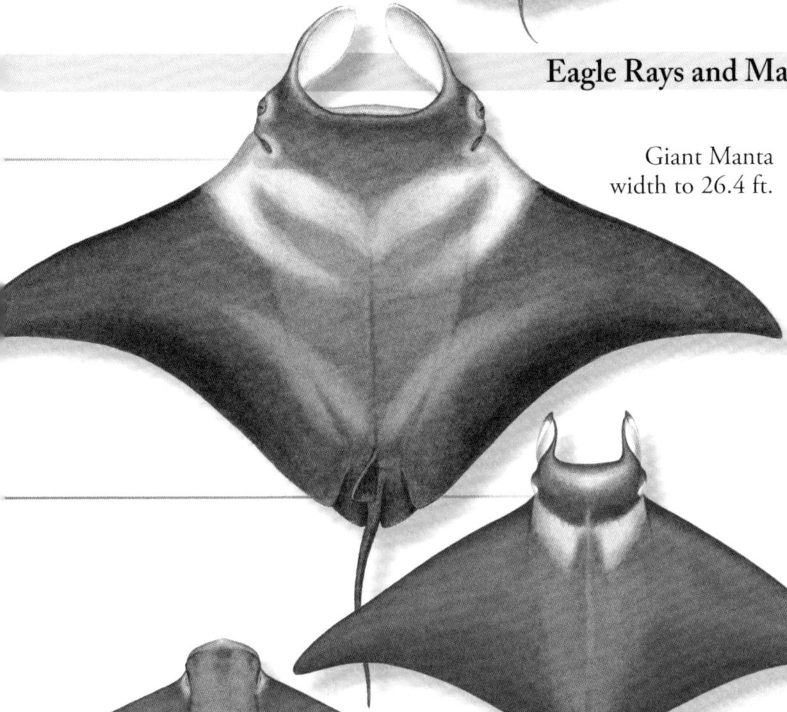

Giant Manta
width to 26.4 ft.

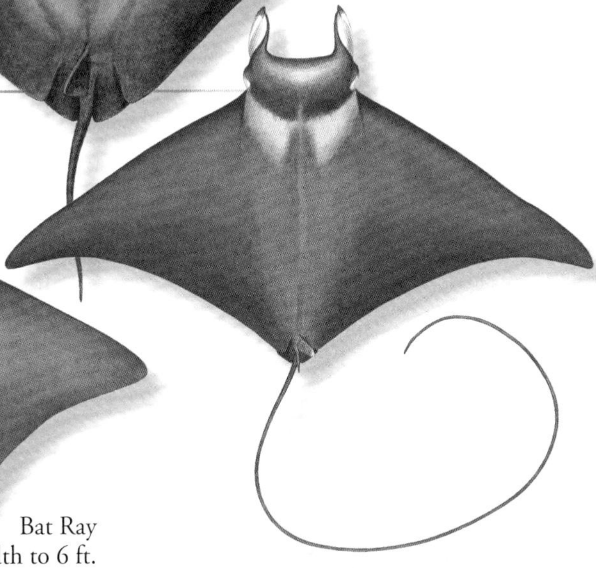

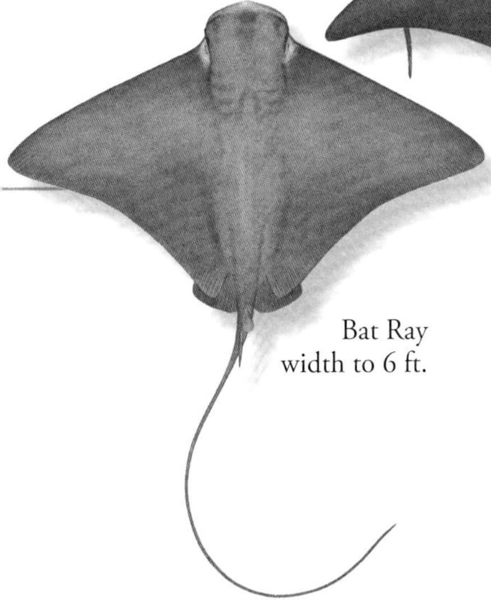

Bat Ray
width to 6 ft.

Spinetail Mobula
width to 10.2 ft.

Acipenseridae - Sturgeons

Green Sturgeon - *Acipenser medirostris* Ayres, 1854

FEATURES: Olive to grayish dorsally, pale below. May have stripes on lower sides and abdomen. Barbels closer to mouth than tip of snout. Back with 8–11 scutes; 23–30 scutes on midline; abdomen with two rows of 7–11 scutes; and 1–2 scutes on upper peduncle. HABITAT: Bering Sea and Gulf of Alaska to northern Baja California. Also to Sea of Japan. Anadromous; in rivers and along coast to about 260 ft. NOTE: Near Threatened.

White Sturgeon - *Acipenser transmontanus* Richardson, 1837

FEATURES: Shades of gray dorsally and on sides. Pale below. Barbels closer to tip of snout than to mouth. Back with 11–14 scutes; 38–48 scutes on midline; abdomen with two rows of 9–12 scutes. Scutes on upper peduncle absent. Scutes may be barely visible in large specimens. HABITAT: Northern Gulf of Alaska to northern Baja California. Anadromous; in rivers and along coast to about 400 ft.

Elopidae - Tenpounders

Machete - *Elops affinis* Regan, 1909

FEATURES: Bluish green to brownish dorsally. Silvery on sides and below. Head yellowish. Margin of lower caudal-fin lobe white. Snout moderately short and pointed. Mouth and eyes large. Body elongate and somewhat compressed. Dorsal-fin origin over pelvic-fin origin or slightly posterior. Caudal fin deeply forked. HABITAT: Southern CA to Peru, including Gulf of California. Occur in shallow inshore waters along coast, including bays, lagoons, and mangrove areas. Frequently enter fresh water.

Albulidae - Bonefishes

Cortez Bonefish - *Albula gilberti* Pfeiler & van der Heiden, 2011

FEATURES: Silvery grayish to greenish dorsally. May have faint darker saddles. Silvery on sides and below. Pectoral fins pale to dark yellowish. Base of pelvic fins pale to dark yellowish. May have some yellow on lower parts of head. Dorsal fin with 18–20 rays. Pelvic fins with 9–12 rays; tips may reach posterior edge of anus. HABITAT: Central CA (rare) to Gulf of California. Occur in shallow waters along coast. BIOLOGY: Forage singly or in groups for invertebrates and fishes. Prized game fish.

Muraenidae - Moray Eels

California Moray - *Gymnothorax mordax* (Ayres, 1859)

FEATURES: Head and body densely mottled in shades of dark brown to olive. Gill opening blackish. Dorsal and anal fins continuous around tail. Pectoral fins absent. HABITAT: Pt. Conception, CA, to southern Baja California. Occur around reefs and rocky areas along coast from intertidal zone to about 130 ft. BIOLOGY: Often seen with head protruding from holes and crevices. Feed on fishes, crustaceans, and octopi.

Sturgeons

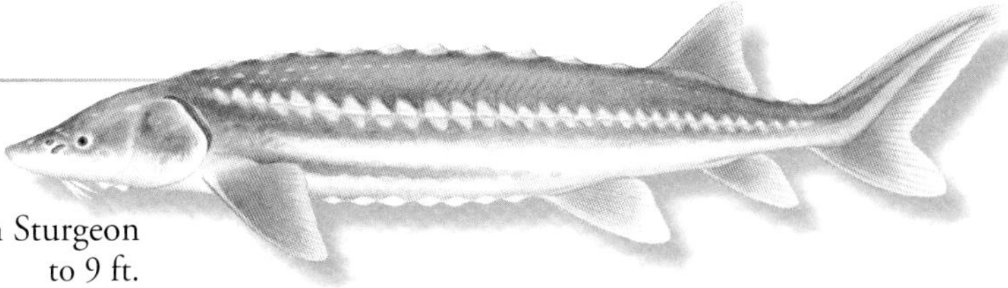

Green Sturgeon
to 9 ft.

White Sturgeon
to 20 ft.

Tenpounders

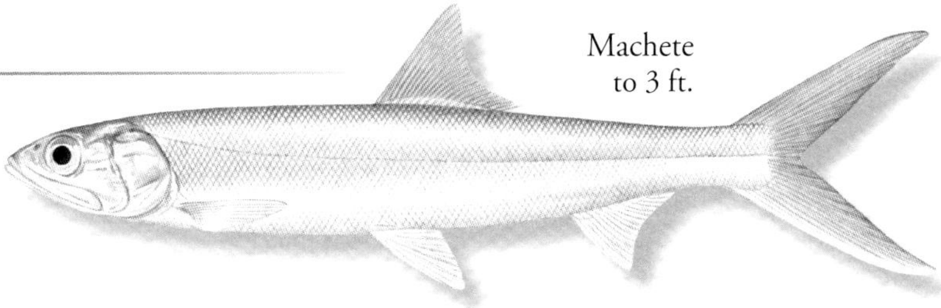

Machete
to 3 ft.

Bonefishes

Cortez Bonefish
to 18 in.

Moray Eels

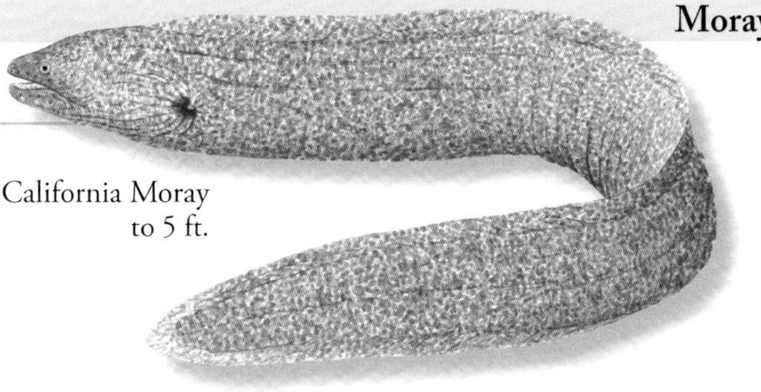

California Moray
to 5 ft.

Ophichthidae - Snake Eels and Worm Eels

Pacific Snake Eel - *Ophichthus triserialis* (Kaup, 1856)

FEATURES: Tannish with rows of small and large dark brown spots on head and body. Dorsal and anal fins banded. Anterior nostrils tubular and located under snout. Barbel midway between nostrils. Tip of tail finless. HABITAT: Northern CA to Peru, including Gulf of California and Galápagos Islands. Occur over sandy and muddy bottoms from intertidal zone to about 510 ft. Also at surface. BIOLOGY: Burrow in soft bottoms.

Yellow Snake Eel - *Ophichthus zophochir* Jordan & Gilbert, 1882

FEATURES: Pale purplish to brownish, with black around gill opening. Pectoral fins blackish. Outer margins of dorsal and anal fins blackish. Barbels present between and behind nostrils. Tip of tail finless. HABITAT: Northern CA to Peru, including Gulf of California. Occur over rocky and sandy bottoms from intertidal zone to about 360 ft. BIOLOGY: Reclusive. Often found buried in sand or mud. Feed on crustaceans.

ALSO IN THE AREA: Pacific Worm Eel, *Myrophis vafer*, page 347.

Engraulidae - Anchovies

Deepbody Anchovy - *Anchoa compressa* (Girard, 1858)

FEATURES: Brownish to greenish dorsally, with silvery stripe on sides. Snout short, somewhat blunt. Mouth large, with rather blunt maxilla tip that does not reach to preopercular margin. Anal fin with 29–33 rays. HABITAT: Central CA to southern Baja California. Found in embayments and along coast from surf zone to about 60 ft.

Slough Anchovy - *Anchoa delicatissima* (Girard, 1854)

FEATURES: Greenish dorsally, pale below, with silvery stripe on sides. Snout short, blunt. Mouth large, with pointed maxilla reaching to interoperculum but not to gill margin. Anal fin with 23–26 rays. HABITAT: Southern CA to southern Baja California. Found in embayments and along coast from shore to about 50 ft. BIOLOGY: Schooling.

Anchoveta - *Cetengraulis mysticetus* (Günther, 1867)

FEATURES: Bluish to greenish dorsally, silvery below. Juveniles with silvery stripe on sides. Snout short, relatively pointed. Mouth large, with bluntly rounded maxilla tip that does not reach rear margin of preopercle. Pelvic-fin origin slightly anterior to dorsal-fin origin. Body compressed. HABITAT: Southern CA to Peru, including Gulf of California. Occur inshore and along coast from surface to about 105 ft.

Northern Anchovy - *Engraulis mordax* Girard, 1854

FEATURES: Iridescent greenish to bluish dorsally. Silvery below. Juveniles with silvery stripe on sides. Snout short, pointed. Mouth large, with pointed maxilla tip that reaches onto preopercle. Anal-fin origin slightly behind or just below last dorsal-fin ray. Body round in cross-section. HABITAT: Eastern Gulf of Alaska (rare) to southern Baja California and Gulf of California. Coastal from near surface to about 1,015 ft.

Snake Eels and Worm Eels

Pacific Snake Eel
to 4 ft.

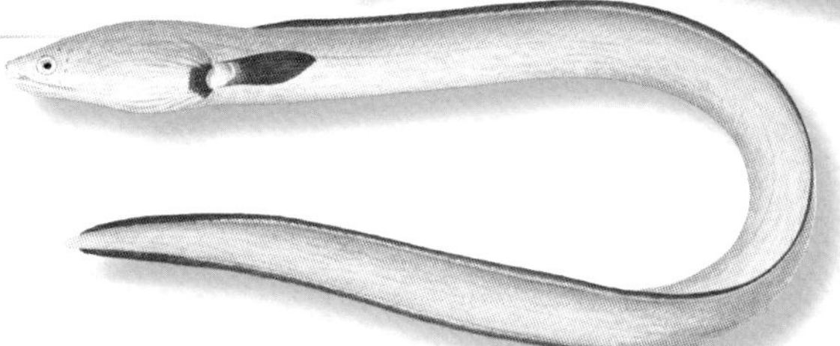

Yellow Snake Eel
to 3.2 ft.

Anchovies

Deepbody Anchovy
to 6.5 in.

Slough Anchovy
to 4.7 in.

Anchoveta
to 8.7 in.

Northern Anchovy
to 9.7 in.

Clupeidae - Herrings

American Shad - *Alosa sapidissima* (Wilson, 1811)

FEATURES: Dark blue or blue green dorsally with metallic hues. Silvery below. Large, dark spot behind operculum, followed by one or two rows of fainter spots. Upper jaw has median notch. Anal fin with 19–23 rays. Body deep and compressed, becoming deeper with age. HABITAT: Along coast from Kamchatka to SE Bering Sea and Gulf of Alaska to northern Baja California. BIOLOGY: Introduced into the area. Anadromous.

Pacific Herring - *Clupea pallasii* Valenciennes, 1847

FEATURES: Bluish to greenish dorsally. Silvery below. Spots on sides absent. Upper jaw lacks median notch. Pelvic fins located under dorsal fin. Anal fin with 13–20 rays. Body elongate and laterally compressed. HABITAT: Arctic Alaska to northern Baja California. Also to Japan and Korea. Pelagic. Found coastally and offshore to about 820 ft. BIOLOGY: Schooling. Enter large estuaries during spawning.

Threadfin Shad - *Dorosoma petenense* (Günther, 1867)

FEATURES: Olivaceous to bluish black dorsally. Silvery below. Large, dark spot behind operculum. Upper jaw notched. Dorsal midline scaleless. Dorsal fin with single long, trailing ray. Anal fin with 17–27 rays. Body relatively deep and compressed. HABITAT: Occur from OR to CA. Pelagic. Primarily in quiet fresh water. Also found in brackish to saltwater estuaries and bays. BIOLOGY: Introduced into the area. Anadromous and schooling. Females cast up to 21,000 eggs. Primarily a filter-feeder.

Round Herring - *Etrumeus acuminatus* Gilbert, 1890

FEATURES: Greenish to bluish dorsally. Silvery below. Dorsal fin slightly forward of midbody. Pelvic fins located behind or just below rear margin of dorsal fin. Anal fin with 10–19 rays. Body elongate and cylindrical in cross-section. HABITAT: Circumglobal in warm seas. Monterey Bay to Chile. Usually occur in deep water off coast and along continental shelves and slopes. Pelagic. BIOLOGY: Migrate vertically in the water column. Spawning occurs at night.

Flatiron Herring - *Harengula thrissina* (Jordan & Gilbert, 1882)

FEATURES: Bluish dorsally. Faint yellow stripe on upper body. Silvery below. Dark spot behind operculum. Operculum lacks bony striations; rear margin with two fleshy lobes. Dorsal fin lacks elongate ray. Anal fin with 15–17 rays. Body somewhat deep and compressed. HABITAT: Occur along coast and in estuaries from southern CA to Peru, including Gulf of California. BIOLOGY: Form large, dense schools. Feed on plankton.

Deepbody Thread Herring - *Opisthonema libertate* (Günther, 1867)

FEATURES: Bluish dorsally, with faint stripes along scales. Usually with a dark spot behind operculum. Several scattered dark spots on sides often present. Last dorsal-fin ray long and trailing. Anal fin with 19–20 rays. Body relatively deep and compressed. HABITAT: Found near shore from southern CA (rare) to Chile, including Gulf of California. BIOLOGY: Form dense schools. Feed on plankton.

American Shad
to 2.5 ft.

Pacific Herring
to 18 in.

Threadfin Shad
to 13 in.

Round Herring
to 12 in.

Flatiron Herring
to 7.3 in.

Deepbody Thread Herring
to 12 in.

Clupeidae - Herrings, *cont.*

Middling Thread Herring - *Opisthonema medirastre* Berry & Barrett, 1963

FEATURES: Bluish to bluish green dorsally, with faint spot along scales. Silvery below. Dark spot behind operculum absent. Scales absent along anterior dorsal midline. Last dorsal-fin ray long and trailing. Anal fin with 19–23 rays. Body relatively deep and compressed. HABITAT: Southern CA to Peru, including Gulf of California. Pelagic. Usually occur near shore to about 12 ft. BIOLOGY: Form large schools.

Pacific Sardine - *Sardinops sagax* (Jenyns, 1842)

FEATURES: Bluish to greenish dorsally. Silvery below. Usually with two rows of irregularly shaped dark spots on sides. Operculum with oblique striations. Anal fin with 17–20 rays. Body elongate, somewhat compressed. HABITAT: SE AK to southern Baja California, including Gulf of California. Also to eastern Kamchatka and Japan. Found near shore from surface and surf zone to about 495 ft. BIOLOGY: Schooling.

Argentinidae - Argentines or Herring Smelts

Pacific Argentine - *Argentina sialis* Gilbert, 1890

FEATURES: Head and body silvery; darker along back. May display faint dark bars. Snout long and pointed. Mouth small, not reaching anterior margin of eye. Eyes large. Dorsal and anal fins tall. Anal fin with 12–15 rays. Adipose fin present. HABITAT: OR to Gulf of California, and probably northern Peru. Found on or near bottom from about 35 to 1,065 ft. Larvae are pelagic. BIOLOGY: Form small to large schools. Preyed upon by fishes, marine mammals, and squids.

Osmeridae - Smelts

Whitebait Smelt - *Allosmerus elongatus* (Ayres, 1854)

FEATURES: Greenish gray dorsally, with a silvery stripe on sides. Pale below. Mouth with a large canine tooth in vomer; sometimes with a smaller tooth on either side. Maxilla extends to below about center of eye. Diameter of eye about equal to depth of caudal peduncle. Operculum lacks striations. Males with broader anal fin than females. Lateral line incomplete. HABITAT: Vancouver Island, BC, to San Francisco, CA. Found along coast and in bays in shallow marine waters, perhaps to 340 ft. BIOLOGY: Schooling. Spawn on subtidal sandy banks. Preyed upon by larger fishes.

Surf Smelt - *Hypomesus pretiosus* (Girard, 1854)

FEATURES: Both sexes olivaceous dorsally, with a silvery stripe on sides. Pale below. Females are brighter green than males. Males are darker and browner than females. Breeding males with numerous tubercles on head, sides, and fins. Maxilla ends before middle of eye. All teeth are small. Operculum lacks striations. Lateral line incomplete. HABITAT: Izembek Bay, AK, to S CA. Occur in surf zone and along shallow coastal waters; sometimes found in brackish water, rarely in fresh water. BIOLOGY: Spawn year-round during daytime over coarse sandy beaches near high-tide line.

Herrings

Middling Thread Herring
to 10.8 in.

Pacific Sardine
to 16 in.

Argentines or Herring Smelts

Pacific Argentine
to 8.7 in.

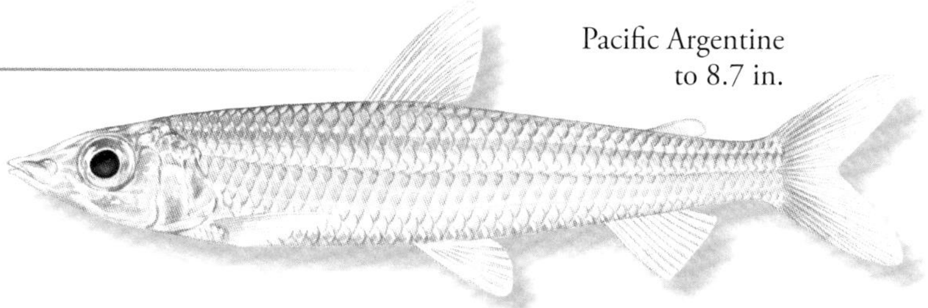

Smelts

Whitebait Smelt
to 18 in.

Surf Smelt
to 18 in.

Capelin - *Mallotus villosus* (Müller, 1776)

FEATURES: Greenish to translucent dorsally. Silvery below. Maxilla extends to or almost to mid-eye. Adipose fin rectangular and long-based. Males have an expanded anal-fin base. Lateral line complete. HABITAT: Circumpolar. In NE Pacific from Bering Sea to Strait of Juan de Fuca. In cold salt water over deep offshore banks and in coastal areas. BIOLOGY: Schooling and migratory. Spawning takes place from surf zone to relatively shallow depths. Economically and environmentally important.

Pacific Rainbow Smelt - *Osmerus dentex* Steindachner & Kner, 1870

FEATURES: Translucent greenish to olivaceous dorsally. Sides with broad, iridescent, silvery stripe. Silvery below. Small spots pepper body and fins. Maxilla extends to or past posterior margin of eye. Operculum lacks striations. Lateral line incomplete. HABITAT: Circumpolar. In NE Pacific from Bering Sea to Heceta Head, OR. In bays and estuaries. Also in northern land-locked lakes and ponds. BIOLOGY: Pacific Rainbow Smelt are schooling and anadromous. Able to tolerate subzero water temperatures.

Night Smelt - *Spirinchus starksi* (Fisk, 1913)

FEATURES: Greenish brown dorsally. Silvery on sides and below. Maxilla extends almost to or slightly beyond posterior margin of eye. Mouth with a band of small teeth in vomer; large canine teeth absent. Diameter of eye less than depth of caudal peduncle. Operculum lacks striations. Pectoral fins comparatively short, never reaching pelvic-fin origin. Anal fin comparatively low. Lateral line incomplete. HABITAT: SE Gulf of Alaska to Point Arguello, CA. Found from surf zone and at surface to about 420 ft. BIOLOGY: Spawn at night in surf on coarse sandy beaches.

Longfin Smelt - *Spirinchus thaleichthys* (Ayres, 1860)

FEATURES: Greenish brown to brownish dorsally. Breeding males darker than females with diffuse stippling dorsally and on sides. Silvery on sides and below. Mouth with small teeth in a row on vomer; canine teeth present on tongue. Maxilla extends to or almost to posterior margin of eye. Operculum lacks striations. Pectoral fins extend to or beyond pelvic-fin origin. Anal fin greatly expanded in males, moderately expanded in females. Lateral line incomplete. HABITAT: SW Gulf of Alaska to Monterey Bay, CA. Found nearshore and inshore from surf zone and at surface to about 450 ft. BIOLOGY: Anadromous. Ascend rivers to spawn. Some landlocked populations.

Eulachon - *Thaleichthys pacificus* (Richardson, 1837)

FEATURES: Bluish black to bluish or brownish, with diffuse stippling dorsally. Silvery on sides and below. Maxilla extends to or beyond posterior margin of eye. Operculum with well-defined striations. Pelvic fins expanded in males. Lateral line complete. HABITAT: Eastern Bering Sea to Point Conception, CA. Occur coastally from surface to about 990 ft. BIOLOGY: Anadromous. Spawning takes place in estuaries and freshwater rivers over a variety of substrates. Historically economically important. Populations declining.

♂ Capelin
to 10 in.

Pacific Rainbow Smelt
to 14 in.

Night Smelt
to 9 in.

♂ Longfin Smelt
to 8 in.

Eulachon
to 10 in.

♂

♀

Arctic Cisco - *Coregonus autumnalis* (Pallas, 1776)

FEATURES: Brownish to dark green dorsally, silvery below. Spots on body and fins absent. Dorsal, caudal, and adipose fins dusky, other fins almost colorless. Snout slightly longer than eye diameter. Lower jaw not protruding beyond upper. Teeth absent from jaws in adults. HABITAT: From Point Barrow, AK, east to Arctic Canada, Europe, and Siberia. Found coastally and in estuaries, lagoons, bays, river drainages, and lakes. Anadromous and landlocked. BIOLOGY: Spend six to eight years at sea before returning to rivers to spawn in summer and fall. May travel up to 932 miles upriver to spawn.

Bering Cisco - *Coregonus laurettae* Bean, 1881

FEATURES: Brownish to dark greenish dorsally, silvery below. May have black dots with faint halos on body, or white spots on fins, or both. Dorsal, caudal, and adipose fins dusky, other fins almost colorless. Snout about equal to eye diameter. Lower jaw not protruding to slightly protruding. Teeth usually absent in adults; rarely with a few teeth in lower jaw. HABITAT: From Oliktok Point, AK, southwest to northern Gulf of Alaska. Also to eastern Siberia. Primarily in fresh water and along coast, but anadromous populations are known. BIOLOGY: Most spend winters in salt or brackish water near river mouths; others may winter far upstream. Undertake extensive spawning migrations.

Humpback Whitefish - *Coregonus pidschian* (Gmelin, 1789)

FEATURES: Dark brownish to bluish dorsally, silvery below. Color of fins similar to dorsal coloration. Snout usually longer than eye diameter. Upper jaw overhangs lower. Teeth in jaws absent. Adults develop prominent nuchal hump behind head. HABITAT: From Beaufort Sea southwest to northern Gulf of Alaska. Also westward across Siberia. Occur coastally near shore. Anadromous, with some populations remaining in fresh water. BIOLOGY: May undertake extensive spawning migrations.

Least Cisco - *Coregonus sardinella* Valenciennes, 1848

FEATURES: Brownish to dark greenish dorsally, silvery below. Back and dorsal fin may have dark spots. Pelvic fins with dark membranes. Snout usually shorter than eye diameter. Lower jaw protrudes beyond upper. Teeth in jaws absent. HABITAT: From northern AK to northern Gulf of Alaska. Also Arctic coast of Canada to Cambridge Bay. Also Bering Sea and Siberia to White Sea. Occur coastally near shore, in estuaries, and in fresh water. Anadromous and landlocked populations.

Cutthroat Trout - *Oncorhynchus clarkii* (Richardson, 1837)

FEATURES: When at sea, bluish green to bluish dorsally. Silvery below. Upper head and body with numerous, variable, and irregular black spots. Dorsal, adipose, and entire caudal fin spotted. Pale to deep orange or red streak under lower jaw. Teeth present on rear of tongue. HABITAT: Gulf of Alaska to Eel Point, CA. Anadromous and nonmigratory freshwater populations. BIOLOGY: Prefer small coastal rivers and streams. Juveniles and post-spawning adults usually enter the ocean in spring and summer. While at sea may stay close to shore or migrate up to 150 miles. While at sea Cutthrout Trout prey on fishes and a variety of invertebrates.

Arctic Cisco
to 24 in.

Bering Cisco
to 19 in.

Humpback Whitefish
to 18 in.

Least Cisco
to 18.5 in.

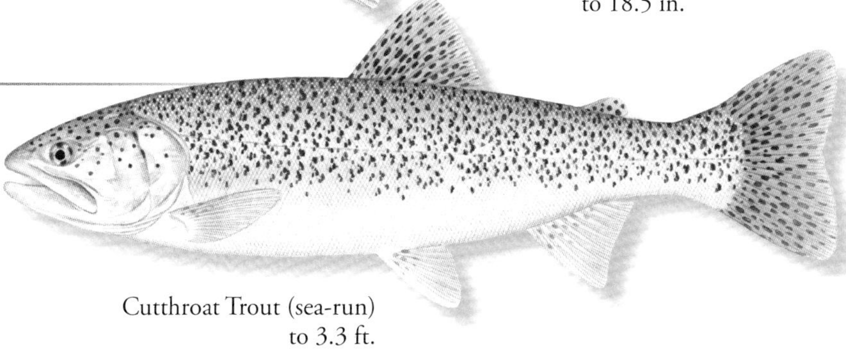

Cutthroat Trout (sea-run)
to 3.3 ft.

Pink Salmon - *Oncorhynchus gorbuscha* (Walbaum, 1792)

FEATURES: When at sea, metallic bluish green to bluish dorsally. Silvery below. Upper back with scattered blackish oval spots and semi-circles. Dorsal fin and entire caudal fin spotted. Mouth white, with black gums. Juveniles greenish dorsally, silvery below; parr marks absent. HABITAT: Northern AK to La Jolla, CA; more commonly to Puget Sound. Also to Japan, northern Siberia, and to Mackenzie River delta, Canada. Anadromous. Found throughout North Pacific and Bering Sea above 40°N. BIOLOGY: After hatching, fry move to coast, then offshore when length reaches about 0.3 inches. After about one year, adults return to intertidal zones of coastal creeks or fresh water close to shore to spawn. Some travel far upstream. Other name: Humpy.

Chum Salmon - *Oncorhynchus keta* (Walbaum, 1792)

FEATURES: When at sea, metallic blue dorsally. Silvery below. Distinct spots or speckling on body or fins absent. Tips of pelvic and anal fins whitish in mature specimens. Mouth white with white gums; tongue may be black. Parr marks on juveniles narrow and mostly above lateral line; fade with age. HABITAT: Northern AK to Del Mar, CA; more commonly to OR. Also to Japan, Arctic coast of Siberia, and Mackenzie River delta, Canada. Anadromous. Found throughout northern Pacific and Bering Sea. BIOLOGY: Soon after hatching, most juveniles enter brackish and salt water, then move offshore after several months. Adults typically spend between three to five winters at sea before returning to natal rivers and streams in summer to fall. Some populations return in winter.

Coho Salmon - *Oncorhynchus kisutch* (Walbaum, 1792)

FEATURES: When at sea, metallic bluish green on back. Silvery below. Upper back with small, scattered black spots. Dorsal fin with small dark spots at base. Caudal fin with spots on upper lobe; spots may be very faint. Mouth mostly pale, with white gum line. Parr marks on juveniles fairly narrow and midlateral and fade with age; anterior anal-fin rays elongate. HABITAT: Point Hope, AK, to northern Baja California. More commonly to Monterey Bay, CA. Anadromous. Occur mainly over the continental shelf. BIOLOGY: After hatching, juveniles either stay in fresh water or migrate to estuaries. After 1–4 (usually 1–2) winters in fresh or brackish water, move offshore. Adults return to rivers and streams usually at three years of age in fall and winter. Other name: Silver Salmon.

Rainbow Trout - *Oncorhynchus mykiss* (Walbaum, 1792)

FEATURES: When at sea, metallic bluish green dorsally. Silvery below. Upper head, back, and dorsal and caudal fins with numerous small black flecks and spots. May have faint pinkish stripe on sides. Inside of mouth and gums whitish to pale pinkish. Juveniles with small spots above, between, and below parr marks. Anal fin with 13–16 rays. HABITAT: Kuskokwim River, AK, to northern Baja California. In the sea only to central CA. Also to Sea of Okhotsk and Peace River, Canada. Anadromous and freshwater populations. BIOLOGY: Anadromous juveniles remain in fresh water from 1–4 years and usually enter the ocean in spring, quickly moving offshore. After a few months or up to three years at sea, immatures and adults return to freshwater streams in summer and winter. They feed through fall and winter, then return to sea again. Other name: Steelhead.

Pink Salmon (sea-run)
to 2.5 ft.

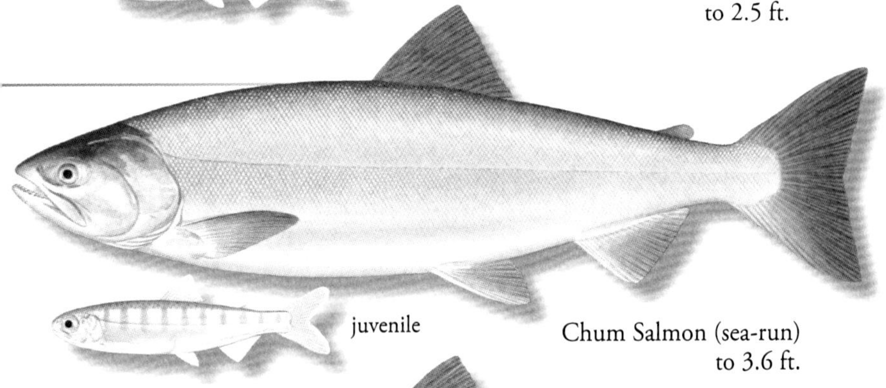

Chum Salmon (sea-run)
to 3.6 ft.

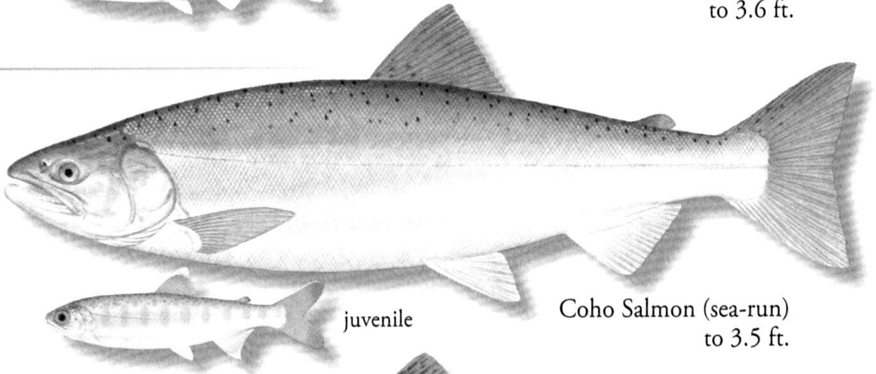

Coho Salmon (sea-run)
to 3.5 ft.

Rainbow Trout (sea-run)
to 4 ft.

Sockeye Salmon - *Oncorhynchus nerka* (Walbaum, 1792)

FEATURES: When at sea, metallic bluish dorsally. Silvery below. Head is bluish black on top. Spots typically absent from body and fins. Eyes bright gold, comparatively large. Mouth white with white gums. Parr marks on juveniles vary in size and relation to lateral line; fade with age. HABITAT: Point Hope, AK, to central CA. Also to Japan and Siberia. Widely introduced elsewhere. Anadromous and freshwater populations. BIOLOGY: After hatching, anadromous juveniles run to sea after one to three years in lakes or after a few weeks in streams close to sea. They stay close to shore, then move offshore for one to four years before migrating back to natal waters to spawn.

Chinook Salmon - *Oncorhynchus tshawytscha* (Walbaum, 1792)

FEATURES: When at sea, metalic greenish, greenish blue, grayish, or black dorsally. Silvery below. All with scattered, irregular black spots on back, dorsal fin, and entire caudal fin at sea and during spawning migration. Gums at base of teeth in lower jaw black. Parr marks on juveniles are wide, narrowly spaced, and more or less midlateral; fade with age. HABITAT: Point Hope, AK, to Ventura, CA; strays to Baja California. Also to Japan and Coppermine River, Canada. Introduced in Great Lakes and elsewhere. Anadromous. BIOLOGY: After hatching, juveniles live a few weeks to a year or more in fresh water before moving to estuaries. Reside in estuaries for a few or more months, then move offshore. Return to natal rivers after one to five years at sea. May travel up to 2,000 miles upstream to spawn. Other name: King Salmon.

Atlantic Salmon - *Salmo salar* Linnaeus, 1758

FEATURES: When at sea, brownish or greenish to bluish dorsally with bronze to silver reflections. Silvery below. Numerous small black flecks on back and sides; some X- or Y-shaped. Larger black spots on head. Caudal fin sometimes with spots. Juveniles with midlateral parr marks with red spots in between; fade with age. Head comparatively small in adults. Maxilla extends below mid-eye, beyond mid-eye only in large adult males. Anal fin with 8–11 rays. HABITAT: Native to NE Atlantic; introduced in Pacific. Eastern Bering Sea, Aleutian Islands to WA. Anadromous. BIOLOGY: Juveniles live in fresh water for 2–4 years, then move to sea. Spend 1–6 years at sea, then return to natal rivers to spawn. Natural reproduction takes place in at least a few rivers in BC.

Dolly Varden - *Salvelinus malma* (Walbaum, 1792)

FEATURES: Sea-run specimens olivaceous to bluish dorsally and on sides. Abdomen pale. Back and sides with numerous small creamy spots dorsally and reddish ocellated spots on sides. Pectoral, pelvic, and anal fins white on anterior margin and blackish on submargin. Juveniles similarly colored, patterned. Anal fin with 9–13 rays, usually 11. HABITAT: Arctic AK to northern WA. Also to Mackenzie River, Canada, and to Japan and Korea. Anadromous, freshwater, and some populations with anadromous females. BIOLOGY: Generally, anadromous juveniles live for one to five years in fresh water. Most migrate to sea after three to four years. Migrate to coastal waters in summer months and form small schools. Return to rivers and streams in late summer to early fall. Timing and behavior is complex and variable. Dolly Varden may live to 18 years.

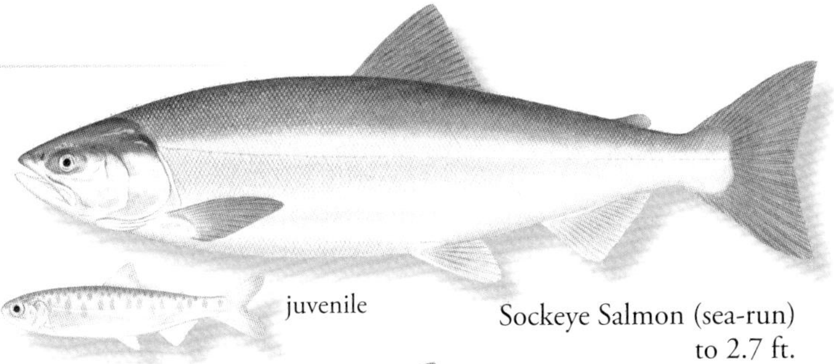

Sockeye Salmon (sea-run)
to 2.7 ft.

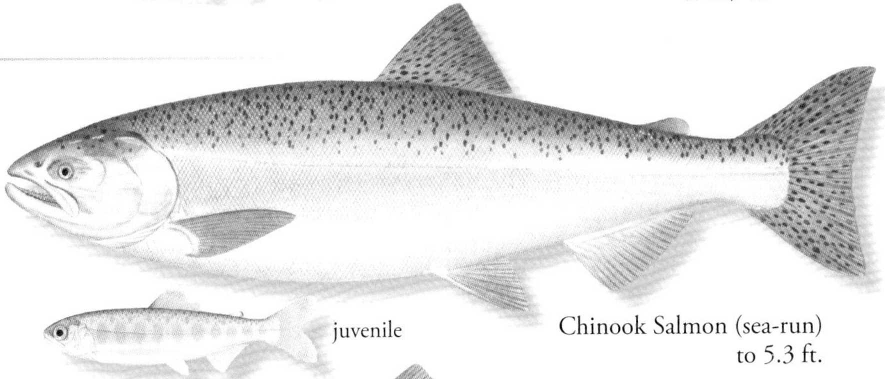

Chinook Salmon (sea-run)
to 5.3 ft.

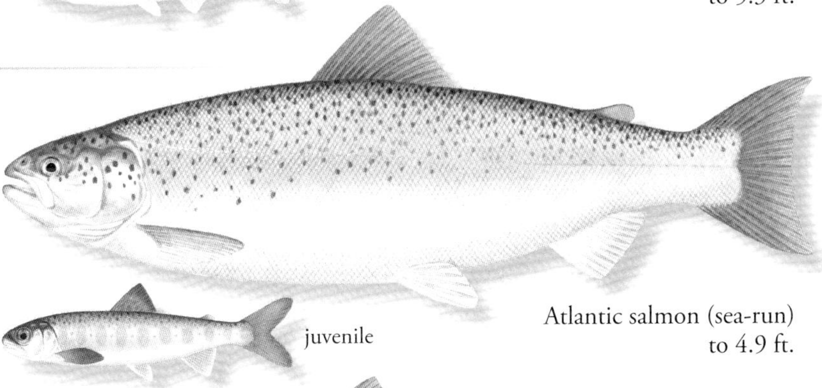

Atlantic salmon (sea-run)
to 4.9 ft.

Dolly Varden (sea-run)
to 3.3 ft.

Gonostomatidae - Bristlemouths

Benttooth Bristlemouth - *Cyclothone acclinidens* Garman, 1899

FEATURES: Head and body brownish to grayish brown. Mouth broad and gaping with numerous forward-facing, bristle-like teeth. Rows of photophores along abdomen and above anal-fin base. Well-developed photophore on upper caudal peduncle. HABITAT: Circumglobal and oceanic in warm seas. OR to central Chile from about 65 to 6,230 ft; primarily below 1,300 ft. BIOLOGY: Feed nocturnally on copepods.

Sternoptychidae - Marine Hatchetfishes

Slender Hatchetfish - *Argyropelecus affinis* Garman, 1899

FEATURES: Blackish on back, silvery on sides and below. Lateral midline unmarked. Upper preopercular spine short. Anal fin with 13 sequential rays. Ventral photophores in a nearly straight line. Body short and moderately keeled. HABITAT: Circumglobal in tropical to subtropical seas. OR to northern Chile from about 330 to 3,700 ft. BIOLOGY: Reportedly do not migrate vertically. Feed on a variety of invertebrates.

Lowcrest Hatchetfish - *Argyropelecus sladeni* Regan, 1908

FEATURES: Blackish on back, silvery on sides and below. Dark spots on posterior lateral midline. Upper preopercular spine long. Anal fin with seven anterior rays, five posterior rays. Ventral photophores unevenly arranged. Body short and deeply keeled. HABITAT: Circumglobal in tropical to subtropical seas. W Bering Sea and British Columbia to Chile from about 180 to 3,700 ft. BIOLOGY: Migrate vertically during the night.

ALSO IN THE AREA: There are four other Hatchetfishes in the area, pages 347–348.

Stomiidae - Barbeled Dragonfishes

Pacific Viperfish - *Chauliodus macouni* Bean, 1890

FEATURES: Head and body brown to blackish. Mouth large and gaping, with many long and short sharp teeth. First dorsal-fin ray long, filamentous. Body elongate, with rhomboid scales and rows of photophores on abdomen. HABITAT: Bering Sea to central Baja California and Gulf of California. Also to Japan. Pelagic from about 80 to 14,000 ft.

ALSO IN THE AREA: Longfin Dragonfish, *Tactostoma macropus*, page 348.

Aulopidae - Flagfins

Eastern Pacific Flagfin - *Aulopus bajacali* Parin & Kotlyar, 1984

FEATURES: Reddish brown dorsally, pale to silvery below. Upper back with dark saddles. Dark blotches on sides. Snout short. Mouth large, with maxilla extending to rear margin of large eyes. Males with elongate anterior dorsal-fin rays. HABITAT: Southern CA (rare), southern Baja California to Ecuador, including Gulf of California and Galápagos Islands. Demersal from about 270 to 1,000 ft.

Bristlemouths

Benttooth Bristlemouth
to 2.8 in.

Marine Hatchetfishes

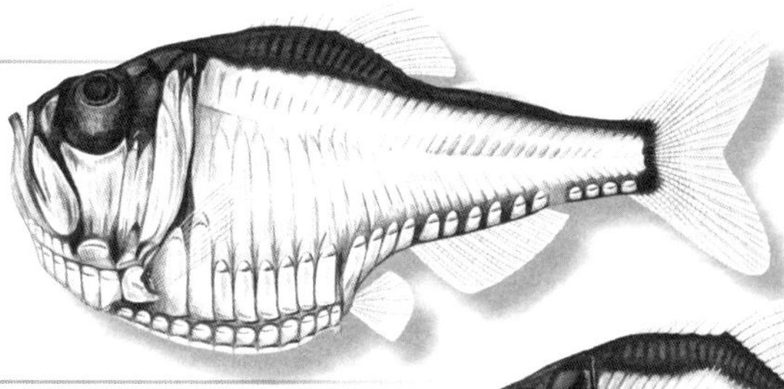

Slender Hatchetfish
to 3.7 in.

Lowcrest Hatchetfish
to 2 in.

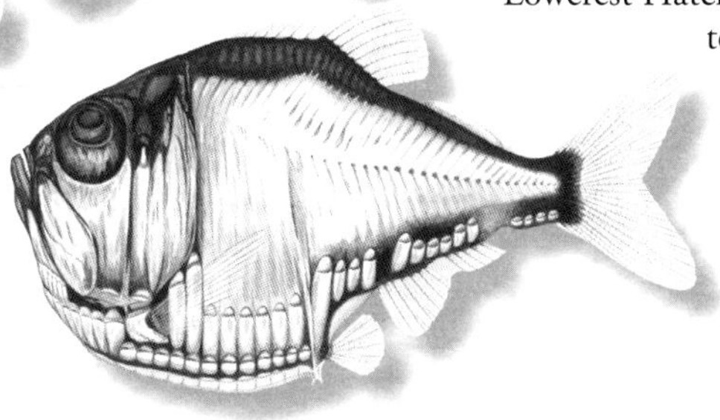

Barbeled Dragonfishes

Pacific Viperfish
to 11 in.

Flagfins

♂ Eastern Pacific Flagfin
to 9.7 in.

Synodontidae - Lizardfishes

California Lizardfish - *Synodus lucioceps* (Ayres, 1855)

FEATURES: Dusky dorsally with wavy brownish blotches and darker saddles. Oblong brownish blotches along lateral line. Silvery below. Snout comparatively long, bluntly pointed. Mouth large, gaping, with maxilla extending well beyond eyes. HABITAT: Cape Beal, British Columbia (rare) to Guyamas, Gulf of California. Demersal from about 5 to 750 ft. BIOLOGY: Perch or bury in bottom sediment; ambush prey. Feed on fishes, squids, and shrimps.

Alepisauridae - Lancetfishes

Longnose Lancetfish - *Alepisaurus ferox* Lowe, 1833

FEATURES: Back dark bluish green. Sides iridescent silvery to coppery. Jaws large and toothy. Dorsal fin long-based, generally tall, and sail-like. Body elongate, laterally compressed. HABITAT: Circumglobal in offshore cold to tropical seas; occasionally in inshore waters. Occur from surface to about 6,000 ft. BIOLOGY: Voracious predator of cephalopods, crustaceans, and fishes. Preyed upon by tunas, toothed whales, and seals.

Myctophidae - Lanternfishes

California Headlightfish - *Diaphus theta* Eigenmann & Eigenmann, 1890

FEATURES: Bluish black dorsally, iridescent silvery on sides and below. Well-developed pair of "headlight" photophores in front of eyes. Large luminous scale above pectoral-fin base. Large photophores on upper and lower caudal peduncle absent. HABITAT: Southern Bering Sea to Chile. Also to Japan. Pelagic from surface to about 5,500 ft.

Northern Lampfish - *Stenobrachius leucopsarus* (Eigenmann & Eigenmann, 1890)

FEATURES: Bluish to brownish black dorsally, iridescent silvery on sides and below. "Headlight" photophore in front of eyes absent. Eye diameter less than caudal peduncle depth. Pectoral fins very small. Well-developed photophores on upper and lower caudal peduncle present in both sexes. HABITAT: Southern Bering Sea to northern Baja California. Also to Japan. Pelagic from about 100 to 9,500 ft. BIOLOGY: Migrate vertically in the water column. May live to about eight years old.

Blue Lanternfish - *Tarletonbeania crenularis* (Jordan & Gilbert, 1880)

FEATURES: Blackish dorsally, iridescent silvery on sides and below. Snout slightly projecting. Eye diameter greater than caudal peduncle depth. Only first two or three scales with visible lateral-line pores. Well-developed photophores present on upper and lower caudal peduncle in males; upper photophore absent in females. HABITAT: Aleutian Islands and Gulf of Alaska to Baja California. Also to Japan. Found from surface to about 3,700 ft. BIOLOGY: Migrate to surface at night.

ALSO IN THE AREA: There are four other Lanternfishes in the area, page 348.

Lizardfishes

California Lizardfish
to 2 ft.

Lancetfishes

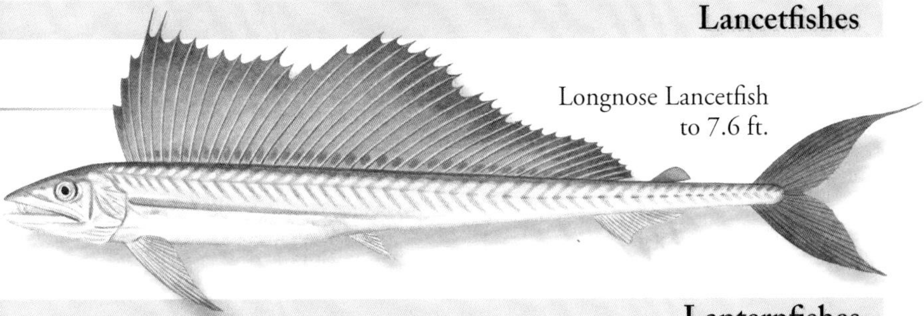

Longnose Lancetfish
to 7.6 ft.

Lanternfishes

California Headlightfish
to 4.6 in.

Northern Lampfish
to 5 in.

♂ Blue Lanternfish
to 5 in.

Lampridae - Opahs

Spotted Opah - *Lampris guttatus* (Brünnich, 1788)

FEATURES: Iridescent bluish to purplish or greenish dorsally, pinkish on sides and below. Body covered in silvery spots. All fins red. Body very deep and compressed. HABITAT: Circumglobal in tropical to temperate seas. Gulf of Alaska to Chile, including Gulf of California. Also to Japan. Pelagic in open water from surface to about 2,400 ft. BIOLOGY: Muscular and predatory. Swim by flapping pectoral fins.

Lophotidae - Crestfishes

North Pacific Crestfish - *Lophotus capellei* Temminck & Schlegel, 1845

FEATURES: Head and body silvery. Fins red. Head with low, squarish to slightly forward-projecting forehead. First dorsal-fin rays tall. Pelvic fins rudimentary. Pectoral, caudal, and anal fins small in adults. HABITAT: Occur from Point Arena to Morro Bay, CA. Found well offshore or near coast, from near surface to about 300 ft. BIOLOGY: A poorly understood species. Previously identified as *Lophotus lacepede*.

Trachipteridae - Ribbonfishes

King-of-the-Salmon - *Trachipterus altivelis* Kner, 1859

FEATURES: Head and body silvery. Young specimens with evenly spaced dark spots on sides. Fins red, except pectorals. First three to six dorsal-fin rays elongate. Anal fin absent. Caudal fin vertically oriented with lower rays greatly reduced. Caudal and pelvic fins reduce with age. Dorsal and ventral profiles evenly taper to tail. HABITAT: SE Bering Sea and Gulf of Alaska to Chile. Found near surface to about 3,900 ft.

Tapertail Ribbonfish - *Trachipterus fukuzakii* Fitch, 1964

FEATURES: Head and body silvery. Some described with two blackish spots on sides. Dorsal and caudal fins red. Pectoral fins clear to pinkish. Anal fin absent. Caudal fin vertically oriented with lower rays reduced. Body tapers noticeably at about two-thirds of length, forming a long, narrow tail. HABITAT: Purisma Point, CA (rare), to Chile, including Gulf of California. Oceanic. At or near surface at night.

ALSO IN THE AREA: Whiptail Ribbonfish, Scalloped Ribbonfish, pages 348–349.

Regalecidae - Oarfishes

Oarfish - *Regalecus glesne* Ascanius, 1772

FEATURES: Silvery with variably spaced spots and wavy lines on sides. Fins reddish. First 10–12 dorsal-fin rays elongate. Pelvic fins long and trailing. Body very long, laterally compressed. Tail blunt with few or no caudal-fin rays. HABITAT: Circumglobal in tropical to warm temperate seas. Found from shore to well offshore from surface to about 5,300 ft. BIOLOGY: Use undulating fins and body to swim vertically in the water column. Feed on a variety of invertebrates and fishes.

Opahs

Spotted Opah
to 6 ft.

Crestfishes

North Pacific Crestfish
to 4 ft.

Ribbonfishes

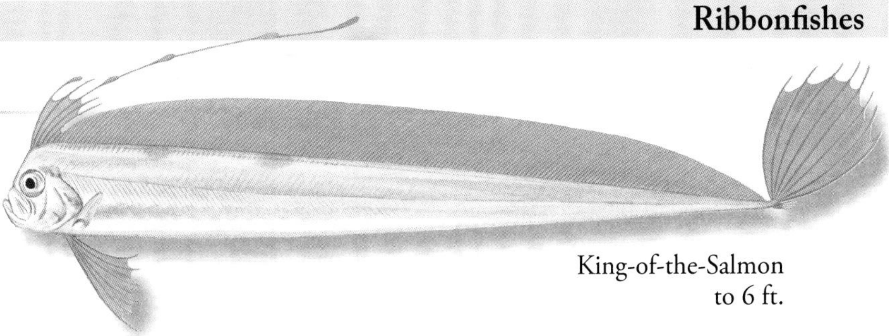

King-of-the-Salmon
to 6 ft.

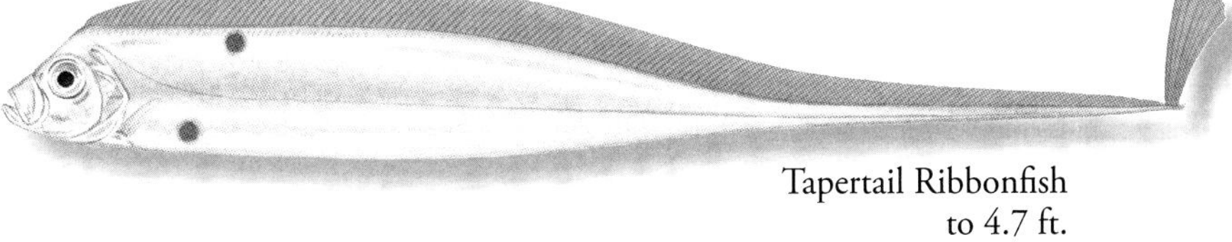

Tapertail Ribbonfish
to 4.7 ft.

Oarfishes

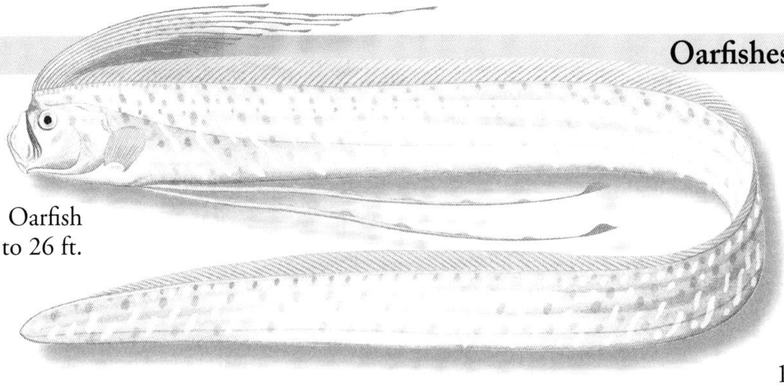

Oarfish
to 26 ft.

Macrouridae - Grenadiers

Giant Grenadier - *Coryphaenoides pectoralis* (Gilbert, 1892)

FEATURES: Head, body, and fins brownish. Snout short, blunt. Chin with small barbel. Body robust anteriorly, tapering to a pointed tail. First dorsal fin tall, with a serrated spine. Second dorsal fin low. Anal-fin origin posterior to second dorsal-fin origin. HABITAT: Bering Sea to Baja California. Also to Sea of Okhotsk and off northern Japan. Demersal over sandy mud bottoms from about 460 to 7,180 ft.

Merlucciidae - Merlucciid Hakes

Pacific Hake - *Merluccius productus* (Ayres, 1855)

FEATURES: Color varies. Metallic brownish to grayish dorsally. Silvery on sides and below. Lower jaw slightly protruding. Mouth large. Chin barbel absent. Head with V-shaped ridge on top. First dorsal fin tall. Second dorsal and anal fins deeply notched. HABITAT: Aleutian Islands (rare), Gulf of Alaska to southern Baja California and Gulf of California. Oceanic and coastal from near surface to about 4,590 ft. BIOLOGY: Form large schools and make complex spawning migrations. Important commercially.

Gadidae - Cods

Arctic Cod - *Boreogadus saida* (Lepechin, 1774)

FEATURES: Metallic brownish dorsally, fading to silvery below. Small black flecks on back and upper sides. Vertical fins with pale outer margins. Chin barbel minute. Lateral line interrupted with wavy undulations at midbody. Third dorsal-fin base longer than second dorsal-fin base. Caudal fin shallowly forked. HABITAT: Circumpolar. From NE AK to Bristol Bay and Pribilof Islands in eastern Bering Sea. Occur from brackish lagoons and river mouths to well offshore past continental shelf and under ice packs.

Saffron Cod - *Eleginus gracilis* (Tilesius, 1810)

FEATURES: Brownish to gray green with diffuse dark blotches dorsally. Ventral area silvery to white. Chin barbel not longer than pupil diameter. Lateral line complete to about second-dorsal fin origin; interrupted to caudal fin. Moderately long interspace between second and third dorsal and anal fins. Caudal-fin margin straight. HABITAT: Beaufort Sea to Bering Sea and SE Gulf of Alaska. Also to Dease Strait, Canada, and to Siberian and Yellow seas. Demersal in shallow coastal waters to about 1,100 ft.

Walleye Pollock - *Gadus chalcogrammus* Pallas, 1814

FEATURES: Oivaceous to brownish with greenish to brassy spots and blotches dorsally. Blotches on sides form irregular rows. Ventral area pale. Chin barbel minute or absent. First anal-fin origin under second-dorsal fin origin. Caudal-fin margin slightly forked to almost straight. Lateral line continuous to second dorsal fin. HABITAT: Bering Sea, Aleutian Islands, and Gulf of Alaska to Carmel, CA. Also to Commander Islands and Seas of Okhotsk and Japan. Usually demersal, but also found near surface and pelagically to about 3,920 ft. Sometimes occur in surf zone. BIOLOGY: Schooling.

Grenadiers

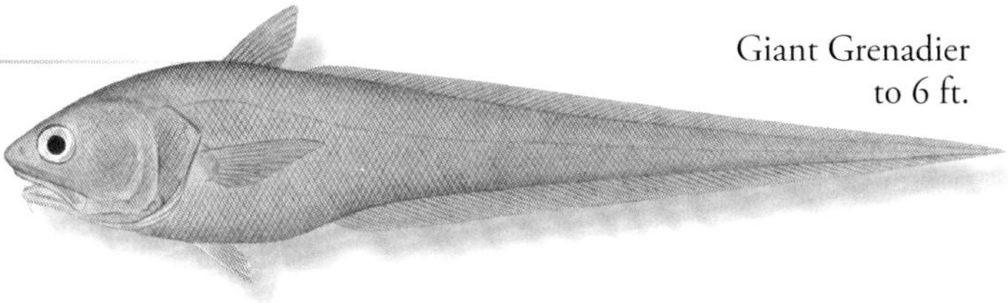

Giant Grenadier
to 6 ft.

Merlucciid Hakes

Pacific Hake
to 3 ft.

Cods

Arctic Cod
to 16 in.

Saffron Cod
to 24 in.

Walleye Pollock
to 3 ft.

Pacific Cod - *Gadus macrocephalus* Tilesius, 1810

FEATURES: Grayish to brownish with greenish to brownish spots and vermiculations dorsally. Lateral line and ventral area pale. Dorsal, caudal, pelvic, and anal fins dark with pale margins. Chin barbel about as long as eye diameter. Lateral line complete nearly to caudal peduncle. Caudal-fin margin straight. HABITAT: Bering Sea, Aleutian Islands, and Gulf of Alaska to Santa Monica, CA. Also to Yellow Sea, China. Found near surface, demersally, and pelagically in deep water to about 2,800 ft. Sometimes occur in surf zone. BIOLOGY: Important commercially.

Pacific Tomcod - *Microgadus proximus* (Girard, 1854)

FEATURES: Tannish with irregular brownish blotches dorsally. May be very dark. Others described as uniformly colored dorsally. Pale below. Upper jaw protrudes slightly. Chin barbel about as long as pupil diameter. Lateral line complete to about third dorsal-fin origin. Interspace between anal fins very short. Caudal-fin margin nearly straight. HABITAT: SE Bering Sea and Aleutian Islands to central CA. Demersal over soft bottoms to about 900 ft. Sometimes enter brackish water.

Ophidiidae - Cusk-eels

Spotted Cusk-eel - *Chilara taylori* (Girard, 1858)

FEATURES: Tannish dorsally and on sides with numerous irregular brown spots and blotches that may form rows. A few unspotted specimens reported. Abdomen pale. Margins of dorsal, caudal, and anal fins dark. Pelvic fins long, filamentous, and located under the chin. HABITAT: Willapa Bay, WA, to southern Baja California and Ecuador. Demersal over soft bottoms from intertidal zone to about 2,400 ft. May enter estuaries. BIOLOGY: Burrow in soft sediment at night; forage during the day.

Basketweave Cusk-eel - *Ophidion scrippsae* (Hubbs, 1916)

FEATURES: Tannish dorsally and on sides. Abdomen pale. Margins of dorsal, caudal, and anal fins dark. Pelvic fins long, filamentous, and located under the chin. Scales form a cross-hatched or "basket weave" pattern. HABITAT: Central CA to southern Baja California. Demersal over soft bottoms from shore to about 850 ft. Sometimes enter estuaries. BIOLOGY: Maintain burrows in sediment at night; forage during the day.

Bythitidae - Viviparous Brotulas

Red Brotula - *Brosmophycis marginata* (Ayres, 1854)

FEATURES: Head and body purplish red to orange red. May have some mottling. Margins of fins dark orange. Pelvic fins long, filamentous, and located near throat. Caudal fin free from dorsal and anal fins. HABITAT: Petersburg, AK, to Ensenada, Baja California. Demersal around rocky bottoms and in caves and crevices from about 10 to 840 ft. BIOLOGY: Reclusive and likely nocturnal. Reportedly form mated pairs. Feed on fishes, crustaceans, and gastropods.

Cods

Pacific Cod
to 4 ft.

Pacific Tomcod
to 15 in.

Cusk-eels

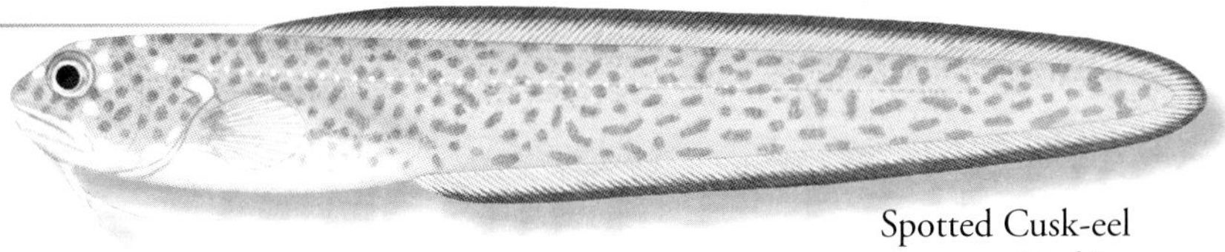

Spotted Cusk-eel
to 15.6 in.

Basketweave Cusk-eel
to 11 in.

Viviparous Brotulas

Red Brotula
to 18 in.

Batrachoididae - Toadfishes

Specklefin Midshipman - *Porichthys myriaster* Hubbs & Schultz, 1939

FEATURES: Grayish brown with irregular dark patches dorsally. Yellowish to white below. Dorsal and pectoral fins banded. Jaws wide and gaping. Head broad and flattened. Rows of photophores on scaleless head and body. HABITAT: Point Arguello, CA, to southern Baja California. Occur near bottom from intertidal zone in bays and estuaries to about 415 ft. BIOLOGY: Nocturnal. Prey on invertebrates and fishes.

Plainfin Midshipman - *Porichthys notatus* Girard, 1854

FEATURES: Grayish brown to blackish with a metallic sheen dorsally. Pale to white below. Fins unmarked. Jaws wide and gaping. Head broad and flattened. Rows of photophores on scaleless head and body. HABITAT: Southern British Columbia to southern Baja California. Reported absent between Puget Sound and Humboldt Bay. Found near bottom from intertidal zone to about 1,250 ft. BIOLOGY: Males guard eggs.

Lophiidae - Goosefishes

Threadfin Goosefish - *Lophiodes spilurus* (Garman, 1899)

FEATURES: Shades of brown with dark blotches and irregular pale, wavy spots and lines dorsally. Jaws wide, gaping, and toothy. Head very broad and flattened, tapering to narrow tail. First three dorsal-fin spines elongate. First dorsal-fin spine with a fleshy lure. Numerous fleshy tabs on body and fins. HABITAT: Central CA to Peru, including Gulf of California. Demersal over soft bottoms from about 260 to 1,560 ft.

ALSO IN THE AREA: Spottedtail Goosefish, *Lophiodes caulinaris*, page 349.

Antennariidae - Frogfishes

Roughjaw Frogfish - *Fowlerichthys avalonis* (Jordan & Starks, 1907)

FEATURES: Highly variable in mottled, spotted, and blotched shades of brown, black, yellow, white to red. All with an ocellated spot at second dorsal-fin base. First dorsal-fin spine with small fleshy lure. Head and body with numerous fleshy tabs. HABITAT: Southern CA to Peru, including Gulf of California. Demersal over sandy, muddy, and hard bottoms from intertidal zone to about 985 ft.

Ogcocephalidae - Batfishes

Roundel Batfish - *Zalieutes elater* (Jordan & Gilbert, 1882)

FEATURES: Tannish, with numerous spots and blotches dorsally. Pale ventrally. Two yellow, ocellated spots at shoulder. Dorsal surface covered with short spines and tubercles. Body flattened, disk-shaped. Pelvic fins limb-like. HABITAT: Southern OR to Peru, including Gulf of California. Occur over soft bottoms from about 30 to 420 ft. BIOLOGY: Pelvic and pectoral fins are used to "walk" over bottom.

Toadfishes

Specklefin Midshipman
to 20 in.

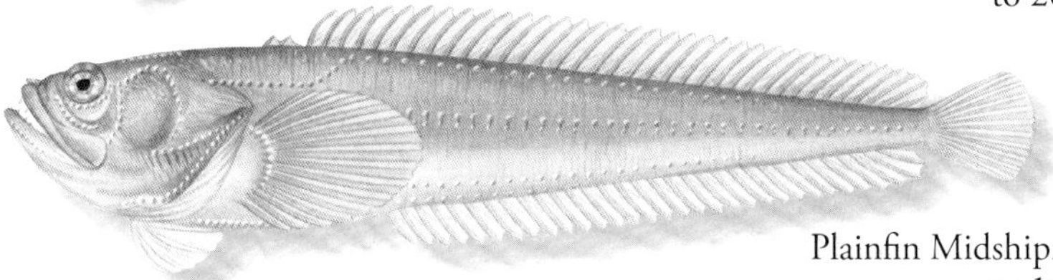

Plainfin Midshipman
to 16 in.

Goosefishes

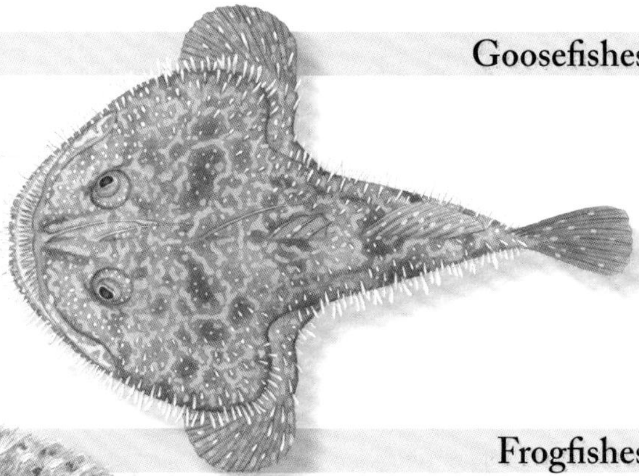

Threadfin Goosefish
to 14 in.

Frogfishes

Roughjaw Frogfish
to 14 in.

Batfishes

Roundel Batfish
to 6.5 in.

Mugilidae - Mullets

Striped Mullet - *Mugil cephalus* Linnaeus, 1758

FEATURES: Grayish olive to grayish brown dorsally. Silvery on sides and below. Dark spots on scales form stripes along sides. Dorsal and caudal fins dusky. Pectoral fins with dark spot at base. Second dorsal and anal fins with small scales on lower anterior portions only. Gray to blackish spots form rows along scales. HABITAT: Circumglobal in tropical to warm temperate coastal, brackish, and fresh waters. In eastern Pacific from San Francisco Bay to Chile, including Gulf of California. Occur from intertidal zone to about 400 ft. BIOLOGY: Form large schools. Catadromous. Feed on plankton and detritus. Caught and raised commercially. Commonly used as bait.

Atherinopsidae - New World Silversides

Topsmelt - *Atherinops affinis* (Ayres, 1860)

FEATURES: Translucent greenish to bluish dorsally. Pale below. Silvery stripe on sides. Sides often with yellow blotches. Jaws with single row of forked teeth. Interspace between first and second dorsal fins with five to eight scales. Anal fin with 19–25 rays; originates under first dorsal fin. HABITAT: Vancouver Island, British Columbia, to Gulf of California. Absent from central and southern Gulf of California. Found near shore from intertidal zone to about 85 ft. Also in estuaries and sometimes in fresh water. BIOLOGY: Form large schools. Feed on wide variety of invertebrates and small fishes.

Jacksmelt - *Atherinopsis californiensis* Girard, 1854

FEATURES: Bluish green dorsally. Silvery below. Silvery stripe on sides. Golden blotch on opercle. Jaws with bands of unforked teeth. Interspace between first and second dorsal fins with 10–12 scales. Anal fin with 21–26 rays; originates behind first dorsal fin. HABITAT: Yaquina Bay, OR, to southern Baja California, and W and NW Gulf of California. Found near shore from surf zone to about 95 ft. Also in estuaries and hypersaline pools. BIOLOGY: Schooling. Important prey item.

California Grunion - *Leuresthes tenuis* (Ayres, 1860)

FEATURES: Greenish dorsally. Silvery below. Silvery stripe on sides. Jaws with minute teeth or no teeth. Interspace between first and second dorsal fins with seven to nine scales. Anal fin with 21–24 rays; originates under first dorsal fin. HABITAT: San Francisco Bay to southern Baja California. Occur from surf zone and tidal pools to about 60 ft. BIOLOGY: Form large nighttime spawning groups on sandy beaches during high tide after new and full moon during spring and summer.

Inland Silverside - *Menidia beryllina* (Cope, 1867)

FEATURES: Translucent yellowish to olivaceous dorsally. Pale below. Silvery stripe on sides. May have reddish tinge during breeding. Jaws with two rows of small teeth interspace between first and second dorsal fin with about five scales. Anal fin with 15–19 rays; originates behind first dorsal fin. HABITAT: Introduced into the area from east coast of US. Occur in San Francisco Bay and freshwater systems of CA.

Mullets

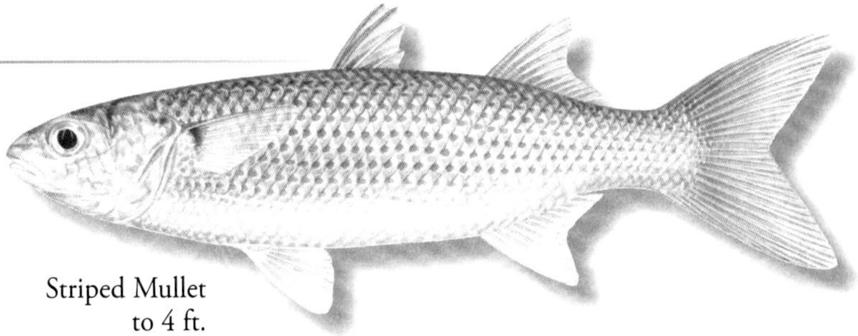

Striped Mullet
to 4 ft.

New World Silversides

Topsmelt
to 14.8 in.

Jacksmelt
to 19 in.

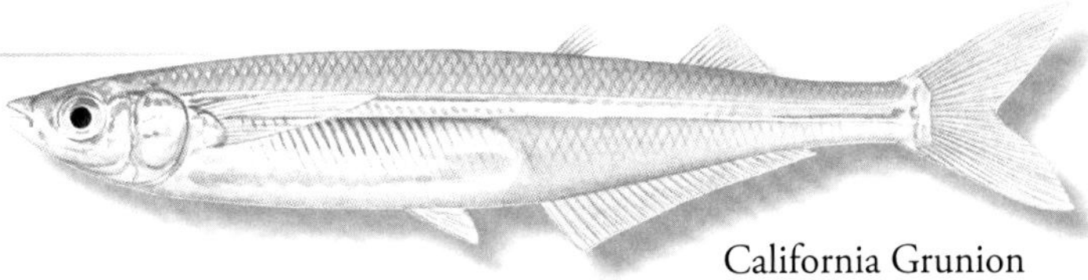

California Grunion
to 7.5 in.

Inland Silverside
to 4 in.

Exocoetidae - Flyingfishes

Blotchwing Flyingfish - *Cheilopogon heterurus* (Rafinesque, 1810)

FEATURES: Dark iridescent blackish blue dorsally. Silvery white below. Pectoral fins coppery to grayish with pale inner band and pale margins. When pressed against body, pectoral fins extend to end of dorsal fin. Dorsal fin pale, low, and with 12–14 rays. HABITAT: Circumglobal in tropical seas. In eastern Pacific from southern CA to central Baja California and Chile. Found at surface from near shore to well offshore.

Smallhead Flyingfish - *Cheilopogon pinnatibarbatus* (Bennett, 1831)

FEATURES: Dark iridescent blackish blue dorsally. Silvery white below. Pectoral fins grayish with pale margins. When pressed against the body, pectoral fins extend past end of dorsal fin. Dorsal fin pale, low, and with 9–13 rays. HABITAT: May be circumglobal. In eastern Pacific from Astoria, OR, to southern Baja California, including Gulf of California. Also to Kuril Is. Found at or near surface from near shore to well offshore.

Sharpchin Flyingfish - *Fodiator acutus* (Valenciennes, 1847)

FEATURES: Iridescent greenish blue dorsally. Silvery white below. Dorsal fin dark gray to blackish. Snout long; lower jaw protrudes. When pressed against the body, pectoral fins extend to about middle of dorsal fin. Dorsal fin dark, tall, and with 9–10 rays. HABITAT: Circumglobal in tropical seas. In eastern Pacific from Goleta, CA, to Chile, including Gulf of California. Found at surface from coastal waters and surf zone to well offshore.

Hemiramphidae - Halfbeaks

Ribbon Halfbeak - *Euleptorhamphus viridis* (van Hasselt, 1823)

FEATURES: Bluish to greenish dorsally. Silvery stripe on sides. Silvery white below. Lower jaw and pectoral fins very long. Body slender, ribbon-like. HABITAT: Widely distributed in tropical to subtropical waters of the Pacific and Indian oceans. Southern CA to Easter Island, Chile. Also lower Gulf of California. Found at or near surface.

Pacific Silvertripe Halfbeak - *Hyporhamphus naos* Banford & Collette, 2001

FEATURES: Translucent bluish green to greenish dorsally. Scales on back with dark margins. Silvery stripe on sides. Silvery white below. Tip of elongated lower jaw red. Pelvic-fin origin located at about midbody line. Bases of dorsal and anal fins scaled in adults. HABITAT: San Diego, CA, to Paita, Peru, including Gulf of California. Found near surface along coastal and estuarine waters. BIOLOGY: Schooling.

California Halfbeak - *Hyporhamphus rosae* (Jordan & Gilbert, 1880)

FEATURES: Translucent greenish dorsally. Scales on back with dark margins. Silvery stripe on sides. Silvery white below. Tip of elongated lower jaw red. Pelvic-fin origin located posterior to midbody line. Bases of dorsal and anal fins lack scales. HABITAT: Santa Ana River, CA, to Gulf of California and Bahia Nonura, Peru. Occur near surface of nearshore and intertidal waters. Enter fresh water. BIOLOGY: Form schools.

Flyingfishes

Blotchwing Flyingfish
to 13.7 in.

Smallhead Flyingfish
to 19 in.

Sharpchin Flyingfish
to 9.8 in.

Halfbeaks

Ribbon Halfbeak
to 20 in.

Pacific Silverstripe Halfbeak
to 12 in.

California Halfbeak
to 8 in.

Belonidae - Needlefishes

California Needlefish - *Strongylura exilis* (Girard, 1854)

FEATURES: Bluish green dorsally, silvery on sides and below. Tips of dorsal and caudal fins may be orangish. Both jaws elongate, toothy. Anal-fin base longer than dorsal-fin base. HABITAT: Northern CA to Peru, including Gulf of California and Galápagos Islands. Occur at or near surface from intertidal zone and estuaries to about 330 ft. BIOLOGY: Schooling. Females attach eggs to floating objects. Feed on fishes.

Scomberesocidae - Sauries

Pacific Saury - *Cololabis saira* (Brevoort, 1856)

FEATURES: Dark blue, bluish black, to greenish dorsally, with an iridescent band below. Silvery on sides and below. Mouth small; lower jaw slightly protrudes. Dorsal and anal fins followed by series of small finlets. HABITAT: SE Bering Sea, south of Aleutian Islands and offshore Gulf of Alaska to Islas Revillagigedo, Mexico. Also to Yellow Sea and Sea of Japan. Found from surface to about 750 ft in oceanic and nearshore waters.

Fundulidae - Topminnows

California Killifish - *Fundulus parvipinnis* Girard, 1854

FEATURES: Pale to dark olivaceous dorsally, with iridescent highlights. Pale below. Sides usually with irregular blackish spots and bars. Dorsal fin unmarked. Breeding males develop elongate anal-fin rays and become blackish brown dorsally and golden below, with dark stippling on body. HABITAT: Morro Bay, CA, to southern Baja California. Occur inshore in bays, estuaries, and coastal lagoons, from fresh water to hypersaline waters. BIOLOGY: Short lived and resilient. Form schools.

Rainwater Killifish - *Lucania parva* (Baird & Girard, 1855)

FEATURES: Tannish, greenish, to silvery above. Abdomen silvery. Scales cross-hatched. Females: less colorful than males, lack dark pigment on fins. Males: more colorful than females; dorsal, anal, and pelvic fins may have orange and black pigment. Black anterior spot on dorsal fin may be present. HABITAT: Introduced into the area. Populations in Yaquin Bay, OR, and San Francisco Bay. In brackish bays and estuaries.

Poeciliidae - Livebearers

Sailfin Molly - *Poecilia latipinna* (Lesueur, 1821)

FEATURES: Olivaceous to brownish dorsally. Pale below. Scales form diamond-shaped pattern. Spots on scales form rows. Females: dorsal fin low; peduncle moderately deep. Males: dorsal fin tall, sail-like, and banded with yellow margin; peduncle deep. HABITAT: Introduced into the area from east coast of US. Found in intertidal wetlands, estuaries, sloughs, and fresh waters of Ventura, Los Angeles, and San Diego, CA. BIOLOGY: Feed on algae, detritus, insects, and crustaceans.

Needlefishes

California Needlfish
to 3 ft.

Sauries

Pacific Saury
to 17.3 in.

Topminnows

California Killifish
to 4.5 in.

Rainwater Killifish
to 2.4 in.

Livebearers

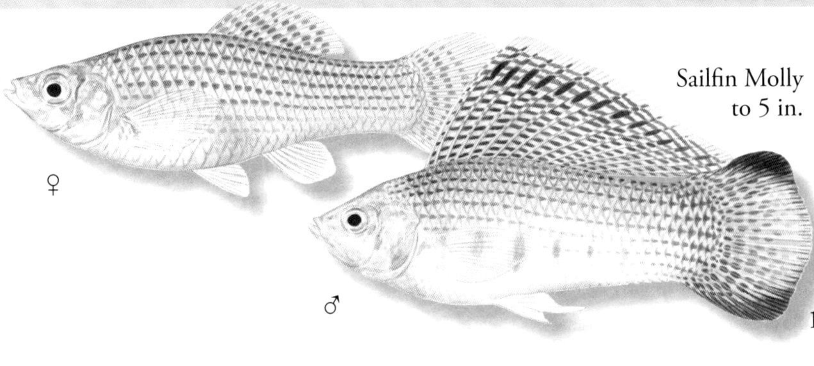

Sailfin Molly
to 5 in.

Anoplogastridae - Fangtooths

Longhorn Fangtooth - *Anoplogaster cornuta* (Valenciennes, 1833)

FEATURES: Adults brownish to blackish. Juveniles grayish. Mouth large with long, pointed teeth. Head with ridges, grooves, and spines. Body deep and compressed. Scales not overlapping, with each on a short pedicel. HABITAT: Nearly circumglobal in tropical to temperate seas. In eastern Pacific from southern British Columbia to Chile. Pelagic from about 250 to 16,400 ft. BIOLOGY: Juveniles found in shallow water; move to deeper water with age. Feed on crustaceans and fishes.

Zeidae - Dories

Mirror Dory - *Zenopsis nebulosa* (Temminck & Schlegel, 1845)

FEATURES: Silvery with a grayish blotch midlaterally. Juveniles with numerous dark blotches on sides. Mouth nearly vertical, protrusible. Two to four pairs of bucklers precede pelvic-fin origin. Rows of bucklers along abdomen and dorsal- and anal-fin bases. Dorsal fin tall with trailing filaments that shorten with age. HABITAT: Central CA to Chile. Also in North and South Pacific. Pelagic from about 100 to 2,620 ft.

Aulorhynchidae - Tubesnouts

Tubesnout - *Aulorhynchus flavidus* Gill, 1861

FEATURES: Greenish to brownish, with a dark stripe on head. Silvery to golden below. Breeding males with bright blue on snout, blue patches on body, and reddish pelvic fins. Snout long, tubular. Second dorsal fin preceded by 23–27 short spines. HABITAT: Pavlof Bay and Kodiak Island, AK, to central Baja California. Found near surface to about 100 ft from shore to well offshore in open water and weedy habitats.

Gasterosteidae - Sticklebacks

Threespine Stickleback - *Gasterosteus aculeatus* Linnaeus, 1758

FEATURES: Dark bluish or blackish dorsally while in salt water, becoming greenish, grayish to brownish after spawning in fresh water. Silvery to reddish below. Dorsal fin with three stout spines. Body with conspicuous plates. HABITAT: Found in nearly all coastal waters of the Northern Hemisphere. In E Pacific from Beaufort Sea and Gulf of Alaska to central CA. Found from open ocean to shallow estuaries and freshwater habitats.

Ninespine Stickleback - *Pungitius pungitius* (Linnaeus, 1758)

FEATURES: Olive brown, grayish to blackish dorsally. Ventral area silvery to blackish. May be barred or blotched. Dorsal fin with 9–13 stout spines. Caudal peduncle very slim. HABITAT: Circumboreal. In E Pacific from Beaufort Sea and Gulf of Alaska to NE British Columbia. In shallow, vegetated estuaries and marshes. Also in fresh water. BIOLOGY: Males build nests and court multiple females.

Fangtooths

Longhorn Fangooth
to 6.3 in.

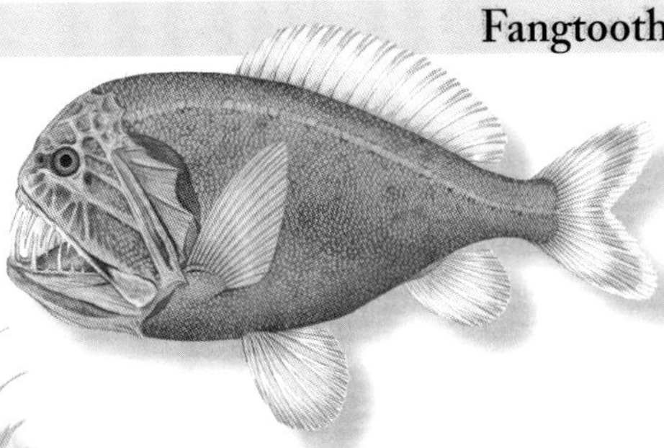

Dories

Mirror Dory
to 21.7 in.

Tubesnouts

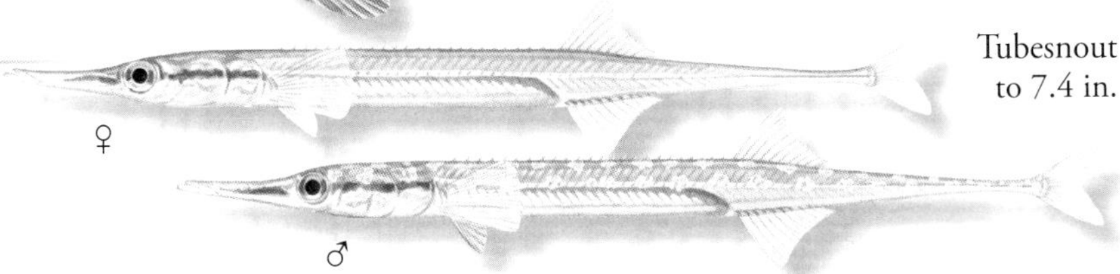

Tubesnout
to 7.4 in.

Sticklebacks

Threespine Stickleback
to 4 in.

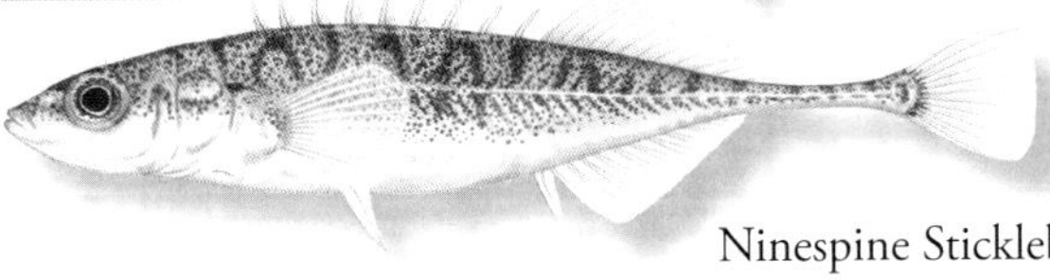

Ninespine Stickleback
to 3 in.

Pacific Seahorse - *Hippocampus ingens* Girard, 1858

FEATURES: Color highly variable. Shades of olive, yellow, tan, gray, brown, black, or red. Often with small dark spots, pale wavy lines, or whitish cross-bands. May have numerous fleshy tabs on head and body. HABITAT: Southern CA to Chile, including Gulf of California and Galápagos Islands. Found over reefs and grass and algae beds from surface to about 350 ft. BIOLOGY: Males perform courtship displays; brood eggs.

Snubnose Pipefish - *Cosmocampus arctus* (Jenkins & Evermann, 1889)

FEATURES: Color variable. Tannish, reddish, or grayish. May have paler and darker bars. Often with dark spots along abdomen. Snout short. Fleshy tabs usually on head and body. Trunk with 14–16 rings; tail with 36–41 rings. HABITAT: Northern CA to Peru, including Gulf of California. Over algal and rocky bottoms to about 60 ft.

Barred Pipefish - *Syngnathus auliscus* (Swain, 1882)

FEATURES: Greenish to brownish, often with a dark midlateral stripe. All also have dark bars on sides, or widely spaced narrow white bars. Snout moderately long. Trunk with 14–16 rings; tail with 34–39 rings. HABITAT: Southern CA to Peru, including Gulf of California. Found over vegetated intertidal and shallow embayments to about 65 ft. Occasionally found with *Sargassum* seaweed.

Kelp Pipefish - *Syngnathus californiensis* Storer, 1845

FEATURES: Olivaceous, yellowish, or brownish. May have pale blotches or dark speckles. Snout long, compressed. Trunk with 19–22 rings; tail with 46–52 rings. HABITAT: Northern CA to southern Baja California. From intertidal zone to about 50 ft. Usually associated with kelp beds; rarely in bays.

Chocolate Pipefish - *Syngnathus euchrous* Fritzche, 1980

FEATURES: Reddish to greenish brown with bluish speckles on head. Silvery streak on cheek behind eye. Brood pouch on males brownish. Snout long, round in cross-section. Trunk with 18–20 rings; tail with 42–49 rings. HABITAT: Southern CA to southern Baja California. Intertidal zone to about 60 ft. NOTE: May not be a valid species.

Barcheek Pipefish - *Syngnathus exilis* (Osburn & Nichols, 1916)

FEATURES: Tannish, greenish, or brownish. Pale speckles on head. May have variable pale spots and blotches on body. Always with dark streak behind lower eye. Snout long, compressed. Trunk with 17–21 rings; tail with 43–49 rings. HABITAT: Central CA to southern Baja California. Found in drift algae along open coastal sandy beaches.

Bay Pipefish - *Syngnathus leptorhynchus* Girard, 1854

FEATURES: Shades of green to brown. Often with pale or dark speckles or mottling. Snout long, round in cross-section. Trunk with 16–21 rings; tail with 36–46 rings. HABITAT: Prince William Sound to southern Baja California. Occur from surface and intertidal zone to about 10 ft. Commonly in seagrass beds in embayments.

♀ Pacific Seahorse
to 12 in.

♀ Snubnose Pipefish
to 5 in.

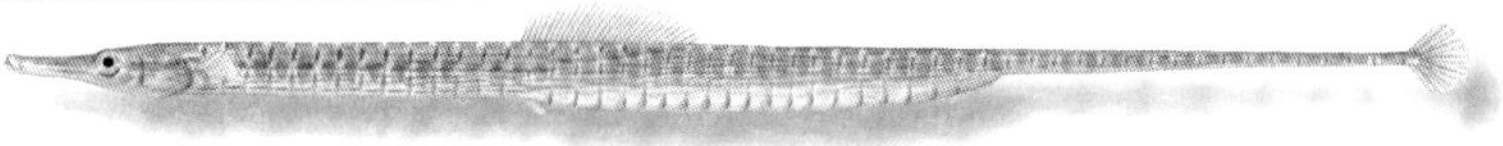

♂ Barred Pipefish
to 7.5 in.

♀ Kelp Pipefish
to 19.5 in.

♂ Chocolate Pipefish
to 11.7 in.

♀ Barcheek Pipefish
to 9.8 in.

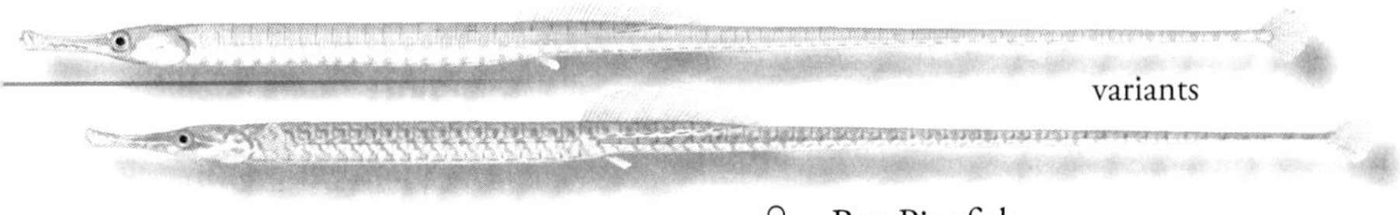

♀ Bay Pipefish
to 15.2 in.

Fistulariidae - Cornetfishes

Deepwater Cornetfish - *Fistularia corneta* Gilbert & Starks, 1904

FEATURES: Orange to reddish or brown dorsally, sometimes with brownish or blackish blotches. Pale below. Snout, head, and body elongate. First dorsal fin absent. Caudal fin forked with long, trailing filament. HABITAT: Huntington Beach, CA, to Peru, including Gulf of California and offshore islands. Found over rocky bottoms from surface to about 165 ft. BIOLOGY: Deepwater Cornetfish feed on small fishes.

Macroramphosidae - Snipefishes

Longspine Snipefish - *Macroramphosus scolopax* (Linnaeus, 1758)

FEATURES: Reddish to orange dorsally. Sides and abdomen silvery pink. Snout long, tube-like. Body deep and compressed. First dorsal fin with elongate, serrated spine that is about as long as snout. Scales finely toothed. HABITAT: Circumglobal in warm seas. Southern CA to Chile. Occur from surface to near bottom to about 1,970 ft. Adults near bottom; juveniles pelagic. BIOLOGY: Feed on plankton and small fishes. Schooling.

Scorpaenidae - Scorpionfishes

California Scorpionfish - *Scorpaena guttata* Girard, 1854

FEATURES: Color highly variable. Mottled, blotched, and saddled in shades of red, brown, or orange to lavender. All usually covered with dark speckles and spots on head and body. Teeth on sides of roof of mouth. Head spiny. Fleshy tabs around eyes, mouth, and on body. First dorsal fin with 12 spines. HABITAT: Santa Cruz, CA, to Gulf of California. Demersal over a variety of bottoms from tidepools and surf zone to about 600 ft. BIOLOGY: Form spawning aggregations. Feed on fishes, crustaceans, squids, and octopi. Dorsal, anal, and pelvic spines venomous. Other name: Sculpin.

Stone Scorpionfish - *Scorpaena mystes* Jordan & Starks, 1895

FEATURES: Color highly variable. Mottled and blotched in shades of brown, green, gray, or red to lavender. Underside of pectoral fins dark with white spots. Caudal fin distinctly banded. Teeth on sides of roof of mouth. Head spiny. Numerous fleshy tabs on head and body. First dorsal fin with 12 spines. HABITAT: Redondo Beach, CA, to Chile, including Gulf of California. Demersal over a variety of bottoms from intertidal zone to about 280 ft. BIOLOGY: Dorsal, anal, and pelvic spines venomous.

Rainbow Scorpionfish - *Scorpaenodes xyris* (Jordan & Gilbert, 1882)

FEATURES: Mottled and blotched in shades of red to brown. Often with irregular dark bars on sides. Lower opercle with a distinct blackish blotch. Teeth absent from sides of roof of mouth. Head moderately spiny. Fleshy tabs on head. First dorsal fin with 13 spines. HABITAT: Anacapa Island and Santa Barbara, CA, to Peru, including Gulf of California and offshore islands. Demersal over reefs and rocky bottoms from intertidal zone to about 165 ft. BIOLOGY: Dorsal, anal, and pelvic spines venomous.

Cornetfishes

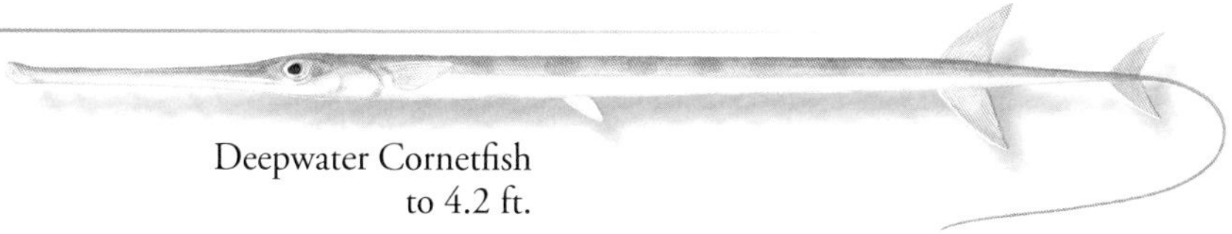

Deepwater Cornetfish
to 4.2 ft.

Snipefishes

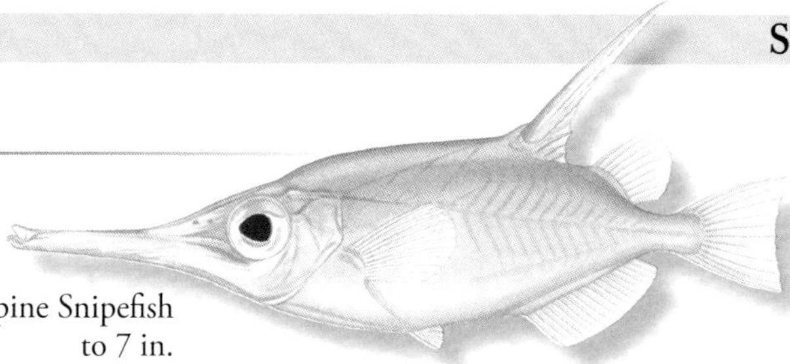

Longspine Snipefish
to 7 in.

Scorpionfishes

California Scorpionfish
to 18.5 in.

Stone Scorpionfish
to 17.7 in.

Rainbow Scorpionfish
to 6 in.

Rougheye Rockfish - *Sebastes aleutianus* (Jordan & Evermann, 1898)

FEATURES: Reddish to pinkish or reddish orange. Often with brown to bronze flecks saddles, and blotches on flanks. Usually with a dark patch on opercle. Margins of fins usuall blackish. Dark spots on spiny dorsal fin usually absent. Head with eight pairs of spines an a series of 2–10 spines under eyes. Gill rakers short and slender on first arch. Second anal fin spine usually shorter than third. HABITAT: Eastern Aleutian Islands to southern OR Found near bottom from about 80 to 2,950 ft. BIOLOGY: Feed on shrimps, crabs, an fishes. May live to 205 years. Often confused with *Sebastes melanostictus.*

Pacific Ocean Perch - *Sebastes alutus* (Gilbert, 1890)

FEATURES: Bright red to pale pinkish dorsally. Blackish, brownish, to olive blotches o back and caudal peduncle. Dark blotch usually present on opercle. Deep-water specimen darker than shallow-water specimens. Lower jaw protrudes with a well-defined knob Second anal-fin spine slightly shorter than third in adults. HABITAT: Bering Sea a Navarin Canyon and Aleutian Islands to southern California. Rarely to central Baj California. Also to Japan and Sea of Okhotsk. Found from near surface to about 2,70 ft. BIOLOGY: Feed on shrimps, amphipods, and small fishes. May live to 100 years.

Kelp Rockfish - *Sebastes atrovirens* (Jordan & Gilbert, 1880)

FEATURES: Mottled and blotched in shades of olive, tan, brown, reddish, or whitis to blackish. Usually darker above. Faint to distinct bands radiate from below eye Conspicuous dark opercular blotch absent. Eyes large. Head spines relatively large. Gi rakers long and slender on first arch. Pectoral fins wispy. Caudal-fin margin rounded HABITAT: Albion, CA, to central Baja California, around kelp beds and hard structure from surface to about 190 ft. BIOLOGY: Occur in large schools and singly in kelp bed Feed on a variety of invertebrates and fishes. May live to 25 years. Other name: Sugar Bas

Brown Rockfish - *Sebastes auriculatus* Girard, 1854

FEATURES: Brownish, pinkish, to dusky with dark brownish mottling dorsally and o sides. A blackish blotch is present on opercle; may be faint in large specimens. Pectora caudal, and anal-fin membranes pinkish. Gill rakers on first arch long. Top of hea flattened to slightly convex. Spines on head strong. HABITAT: Northern Gulf of Alask to central Baja California. Occur over hard and soft bottoms and around structures fro surf zone to about 480 ft. BIOLOGY: Frequently found with other rockfishes. Toleran of turbid water. Feed on a variety of invertebrates and small fishes. May live to 34 year

Aurora Rockfish - *Sebastes aurora* (Gilbert, 1890)

FEATURES: Red with irregular pink to white saddles and blotches dorsally and o sides. Ventral area pink to red. Lateral line may lie in a pale area. Small knobs at ti of upper jaw. Spines above maxilla simple and blunt. Eyes comparatively small. Gi rakers on upper arch number 24–28. Anal fin usually with six rays (rarely five); secon spine longer than third. HABITAT: Langara Island, British Columbia, to central Baj California. Commonly from OR to southern CA. Demersal over deep-water soft an hard slopes from about 260 to 2,930 ft; most abundant between 990 and 1,650 ft.

Rougheye Rockfish
to 3.2 ft.

Pacific Ocean Perch
to 21.6 in.

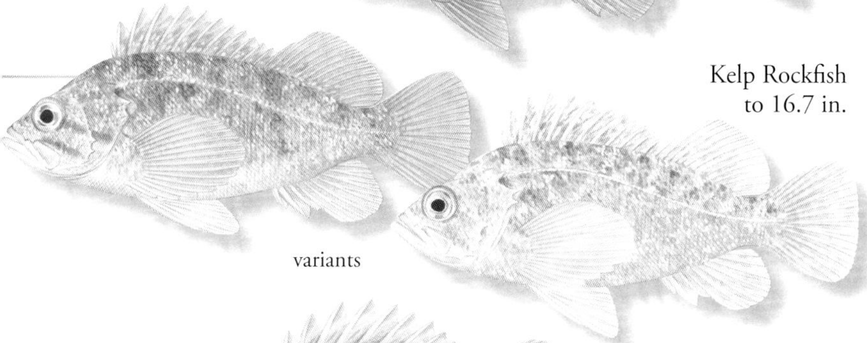

Kelp Rockfish
to 16.7 in.

variants

Brown Rockfish
to 22 in.

Aurora Rockfish
to 16 in.

Redbanded Rockfish - *Sebastes babcocki* (Thompson, 1915)

FEATURES: Whitish to pale pink with four reddish bars on body that extend onto fins. First bar covers only upper portion of opercle. Second bar usually convex to head. Bars less distinct in large specimens. A reddish band extends from below eyes. Some may have dusky to black coloring on dorsal and pectoral fins. Maxilla scaled. Head comparatively rounded and spiny. Body comparatively deep. HABITAT: Bering Sea from Zhemchug Canyon and Amchitka Island, Aleutian Islands, to San Diego, CA. Found over hard bottoms from about 160 to 3,750 ft. Also found over mixed mud, cobble, and pebble bottoms. BIOLOGY: Often solitary, but may form small groups. May live to 106 years.

Shortraker Rockfish - *Sebastes borealis* Barsukov, 1970

FEATURES: Orange, pink, or whitish to reddish. May be almost uniformly colored or with dark or pale blotches or saddles. Fins may have some black, and dorsal-fin margin often whitish. Mouth and gill cavities red with dark blotches. Large pores present on lower jaw. Ridge below eyes absent or low with no more than one spine. Gill rakers short, tipped with small knobs. Body large, robust. HABITAT: Western Bering Sea at Navarin Canyon and Aleutian Islands to southern CA. Also to northern Hokkaido, Japan. Found near or on bottom from about 80 to 3,940 ft. BIOLOGY: May live to 115 years.

Silvergray Rockfish - *Sebastes brevispinis* (Bean, 1884)

FEATURES: Greenish to grayish with bronze highlights dorsally. May be pale to very dark. Sides may be flecked with dark-rimmed scales. Pinkish to pale orange or whitish ventrally. Lower pectoral, pelvic, and anal fins pinkish. Lower jaw strongly protrudes and bears a distinct knob. Second anal-fin spine shorter or about the same length as third. HABITAT: SE Bering Sea to central Baja California. Found around reefs and over soft bottoms from surface to about 1,440 ft. BIOLOGY: Sought commercially.

Gopher Rockfish - *Sebastes carnatus* (Jordan & Gilbert, 1880)

FEATURES: Color variable. Shades of bronze to brownish or blackish dorsally, with whitish to pinkish blotches and saddles that extend into fins. Ventral area whitish to pinkish with dark speckles and mottling. Dark bands extend from below eyes. Pale area around posterior portion of lateral line broken, not distinctly defined. Anal fin with five to seven rays; usually six. Body stout. HABITAT: Cape Blanco, OR, to southern Baja California. Found over reefs and rocky bottoms from intertidal zone to about 280 ft. BIOLOGY: Shelter in holes and crevices. Territorial.

Copper Rockfish - *Sebastes caurinus* Richardson, 1844

FEATURES: Brownish, reddish, or blackish to greenish, with few small pale blotches and speckles dorsally. Ventral area pinkish to yellowish or whitish with dark speckles and blotches. Golden to coppery bands extend from below eyes. Unbroken pale area around posterior two-thirds of lateral line typically present. Some specimens from OR to CA largely reddish orange with few pale blotches dorsally. Dorsal-fin spines comparatively tall. Anal fin with five to seven rays. Body stout. HABITAT: Western Gulf of Alaska to central Baja California. Occur near bottom from intertidal zone to about 600 ft.

Redbanded Rockfish
to 2 ft.

Shortraker Rockfish
to 4 ft.

Silvergray Rockfish
to 2.4 ft.

Gopher Rockfish
to 15.6 in.

Copper Rockfish
to 2 ft.

Greenspotted Rockfish - *Sebastes chlorostictus* (Jordan & Gilbert, 1880)

FEATURES: Pinkish to reddish dorsally, with three to five large white or pink blotches and numerous irregular greenish to bronze spots that extend onto bases of dorsal fins. Yellowish with pinkish wash on sides; pink to white below. Spiny dorsal fin tall, deeply incised, and with whitish margins. Lower jaw protrudes slightly, with a small reddish knob under tip; scales absent on underside. HABITAT: Southern Vancouver Island to southern Baja California. Demersal over soft to hard bottoms from about 100 to 1,240 ft. Also found in caves and crevices. BIOLOGY: Usually solitary, but also form groups.

Black-and-Yellow Rockfish - *Sebastes chrysomelas* (Jordan & Gilbert, 1881)

FEATURES: Black to dark brown dorsally with three or more irregular pale to bright or dusky yellow blotches. Ventral area shades of yellow with black to brown speckles. Dark bands radiate from below eyes. Pale area around posterior portion of lateral line broken, not distinctly defined. Fins generally streaked. Anal fin with five to seven rays; usually six. Body stout. HABITAT: Cape Blanco, OR, to central Baja California. Found in holes and crevices and over rocky bottoms from intertidal zone to about 120 ft. BIOLOGY: Territorial. Feed on a variety of invertebrates, octopi, and fishes.

Dark Rockfish - *Sebastes ciliatus* (Tilesius, 1813)

FEATURES: Dark bluish gray to blackish and often with darker mottling that forms an oblique band dorsally; somewhat paler ventrally. Deepwater specimens paler. Faint dark bands radiate from eyes. All fins dark; may have dark flecks or streaks. Knob at tip of lower jaw small to absent. Spines on head weak. Second anal-fin spine shorter to about equal to third. HABITAT: Western Aleutian Islands and eastern Bering Sea to Johnston Strait, BC. Found near bottom around kelp beds and over hard bottoms from about 20 to 530 ft. BIOLOGY: Feed on a variety of invertebrates. May live to 67 years.

Starry Rockfish - *Sebastes constellatus* (Jordan & Gilbert, 1880)

FEATURES: Uniformly bright orange to brownish, with three to five irregular whitish blotches along the back. Numerous yellow, whitish, or pale blue speckles scattered over head and body. Fins may be sparsely to densely speckled. Juveniles bright yellow. Jaws usually lack scales. Head spiny, with moderately to deeply concave interorbital space. HABITAT: Cordell Bank, CA, to southern Baja California. Found near bottom over deep reefs and rocky areas from about 50 to 900 ft. BIOLOGY: Feed on shrimps, crabs, and small fishes. May live to 32 years. Sought commercially and recreationally.

Darkblotched Rockfish - *Sebastes crameri* (Jordan, 1897)

FEATURES: Shades of white to pink, with four to five irregular brownish to reddish brown saddles that extend into fins. Saddles may be faint in large specimens. Dark blotch on upper opercle. Lower jaw with knob at tip. Second anal-fin spine shorter than the third. Body comparatively deep. HABITAT: Eastern Bering Sea and Aleutian Islands to near Santa Catalina Island and Laguna Beach, CA. Demersal over soft and mixed bottoms near cobbles or rocks from about 95 to 2,985 ft. BIOLOGY: Feed on a variety of invertebrates, octopi, and fishes. May live to 105 years.

Greenspotted Rockfish
to 19 in.

Black-and-Yellow Rockfish
to 15.3 in.

Dark Rockfish
to 21 in.

juvenile

Starry Rockfish
to 18 in.

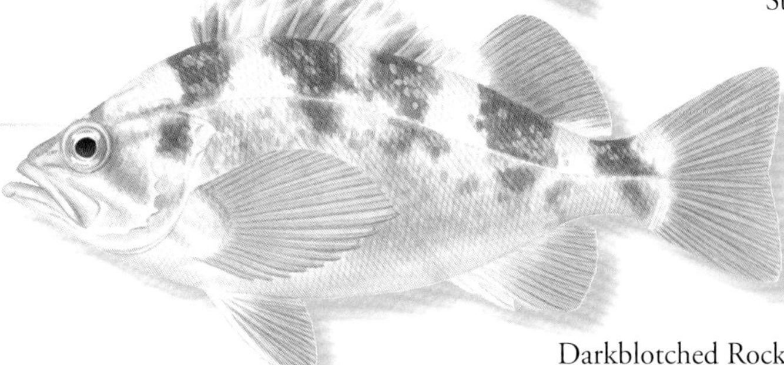

Darkblotched Rockfish
to 23 in.

Sunset Rockfish - *Sebastes crocotulus* Hyde, 2008

FEATURES: Shades of orange with highly variable pinkish gray blotches and mottling dorsally and on sides. Bands radiate from below eyes. Lateral line in pale area that is broken behind head. Color of fins darker than body color. Interorbital space comparatively wide. Eyes comparatively small. Pelvic fins long, extend to anus. Anal fin bluntly pointed with second spine about as long as third. HABITAT: Montague Island, AK, to central Baja California. Juveniles found in shallow water; adults occur over high-relief hard bottoms with caves and crevices to about 1,440 ft. NOTE: Previously grouped with *Sebastes miniatus*. Formal description pending.

Calico Rockfish - *Sebastes dallii* (Eigenmann & Beeson, 1894)

FEATURES: Dusky to pale orange on head and body with oblique brownish to reddish brown bands that extend into fins. Often densely mottled with dark spots and speckles. Caudal fin with dark spots and streaks. Jaws reach to rear of pupil. Head spines strong. Body comparatively slender. HABITAT: San Francisco, CA, to central Baja California. Demersal over a wide range of mixed habitats from intertidal zone to about 1,000 ft. BIOLOGY: Solitary or in large groups. Feed on a variety of invertebrates and fishes.

Splitnose Rockfish - *Sebastes diploproa* (Gilbert, 1890)

FEATURES: Shades of white to pink with variable pinkish to red saddles and blotches dorsally. Saddles may form an upside down U-shaped pattern at midbody. Some almost uniformly red. Many with a dark mark on upper spiny dorsal-fin membranes. Tip of upper jaw notched, with a distinct knob on either side. Spines above maxilla simple, with first forward-facing. Eyes very large. HABITAT: Sanak Islands, AK, to central Baja California. Juveniles near surface under drifting kelp. Adults usually demersal over low-relief muddy bottoms, often near rocks, cobbles, or shells from about 260 to 2,930 ft.

Greenstriped Rockfish - *Sebastes elongatus* Ayres, 1859

FEATURES: Whitish to pinkish with olive green blotches loosely arranged in rows that form three or four stripes on sides. Some with blackish blotches. Ventral area white to pinkish. Knob on tip of lower jaw small or absent. Spines on head strong. Body slender, comparatively elongate. Second anal-fin spine longer than third. HABITAT: Western Gulf of Alaska to central Baja California. Demersal over sandy, silty, mixed, and rocky bottoms from about 40 to 3,250 ft. BIOLOGY: Feed on a variety of invertebrates and fishes in the water column and on bottom. May live to 54 years. Taken commercially.

Puget Sound Rockfish - *Sebastes emphaeus* (Starks, 1911)

FEATURES: Reddish orange to brownish orange, with dark patches dorsally. Patches may be very faint. May also have a dark, broken stripe on lower sides. Knob on tip of lower jaw small or absent. Spines on head moderate to strong. Body slender, comparatively elongate. Second anal-fin spine longer than third. HABITAT: Northern Gulf of Alaska to Punta Gorda, CA. Occur over irregular rocky bottoms, often in high currents, from about 10 to 1,200 ft. Found demersally and in the water column. BIOLOGY: Schooling during the day; largely inactive at night.

Sunset Rockfish
to 2.5 ft.

Calico Rockfish
to 10 in.

Splitnose Rockfish
to 18 in.

Greenstriped Rockfish
to 18 in.

Puget Sound Rockfish
to 7 in.

Swordspine Rockfish - *Sebastes ensifer* Chen, 1971

FEATURES: Shades of red to pink or orange dorsally with three to five whitish blotches on back. White ventrally. Top of head may be blotched. Lower jaw slightly projecting with knob at tip and scales on underside. Eyes large. Second anal-fin spine strong, white, and as long or longer than soft rays. HABITAT: San Francisco, CA, to central Baja California. Demersal over rocky, cobble, and muddy bottoms and in caves and crevices from about 165 to 1,420 ft. BIOLOGY: Females may produce up to 38,000 eggs annually.

Widow Rockfish - *Sebastes entomelas* (Jordan & Gilbert, 1880)

FEATURES: Golden brown to dusky brown dorsally, pale below. May have a pale blotch on sides and indistinct darker flecks, mottling, or saddles. Dark band radiates from below eyes. Fin membranes dark. Knob at tip of lower jaw absent. Head spines reduced. Anal fin with eight soft rays; second spine shorter than third. HABITAT: Albatross Bank, western Gulf of Alaska, to northern Baja California. Occur from near surface and midwater to near bottom over reefs, seamounts, and rocky and cobble bottoms to about 2,625 ft. BIOLOGY: Solitary and in large schools; often with other Rockfishes.

Pink Rockfish - *Sebastes eos* (Eigenmann & Eigenmann, 1890)

FEATURES: Orange to almost white, with dark vermiculations and five irregular pale blotches dorsally. Blotches fade with age. Margins of dorsal, caudal, and anal fins white. Dark bands radiate from eyes. Lower jaw does not protrude; small rounded knob at tip and scales on underside present. Head broad, craggy, and spiny. Pectoral fin usually with 18 rays. HABITAT: Central OR to southern Baja California. Demersal over rocky bottoms and reefs from about 150 to 1,200 ft. More common in deeper waters. BIOLOGY: Usually solitary. May seek shelter in crevices.

Yellowtail Rockfish - *Sebastes flavidus* (Ayres, 1862)

FEATURES: Brownish to greenish brown, with irregular pale blotches dorsally. Sides whitish to yellowish, with brownish flecks on scales. Yellowish bands usually radiate from eyes. Fins yellowish. Knob at tip of lower jaw small to moderate. Spines on head reduced. Anal fin usually with eight soft rays; second spine shorter than third. HABITAT: Eastern Aleutian Islands to northern Baja California. Occur from intertidal zone and near surface over high-relief hard bottoms to about 1,800 ft. BIOLOGY: May form large schools or occasionally rest in caves and crevices. Feed on krill and fishes.

Bronzespotted Rockfish - *Sebastes gilli* (Eigenmann, 1891)

FEATURES: Shades of pink to orange, with irregular darker mottling and brown to bronze spots that extend into fins. May also display large pale blotches. Back usually with several clear or unmarked patches. Area surrounding lateral line largely unmarked. Mouth large, oblique. Spines on head strong. Anal fin usually with seven soft rays; second spine about as long as third. Body deep, robust. HABITAT: Monterey Bay, CA, to northern Baja California. Found near bottom over high-relief rocky bottoms from about 250 to 1,350 ft. BIOLOGY: Apparently reclusive; rarely leave shelter. Taken by gill net and hook-and-line. Other name: Warthog.

Swordspine Rockfish
to 12 in.

Widow Rockfish
to 23 in.

Pink Rockfish
to 22 in.

Yellowtail Rockfish
to 2 ft.

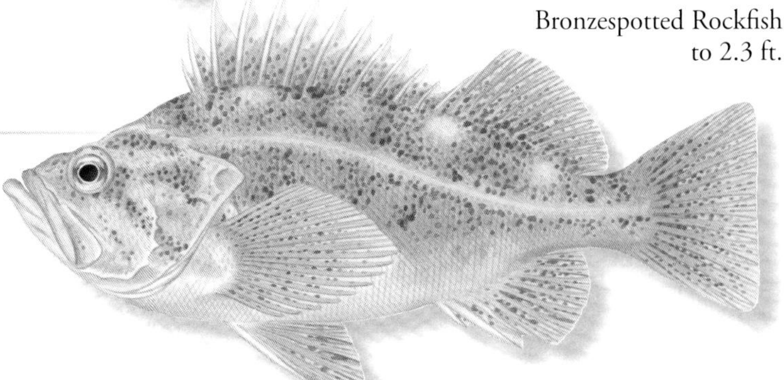

Bronzespotted Rockfish
to 2.3 ft.

Chilipepper - *Sebastes goodei* (Eigenmann & Eigenmann, 1890)

FEATURES: Shades of red to reddish brown dorsally, whitish below. Quickly become variably blotched while resting. Distinct spots on back absent. Area surrounding lateral line reddish. Lower jaw projects. Maxilla extends to about middle of eye. Head spines reduced. Anal fin small, usually with eight rays. HABITAT: Eastern Gulf of Alaska to southern Baja California. Occur over boulders and high-relief bottoms from near surface to about 1,610 ft. BIOLOGY: Form very large to small schools. Feed on midwater krill, squids, and small fishes. Females may produce up to 538,000 eggs per year.

Rosethorn Rockfish - *Sebastes helvomaculatus* Ayres, 1859

FEATURES: Pinkish to yellow orange or red, with irregular yellowish to olive bands on back and sides. Four to five white blotches present on back. Bands radiate from below eyes. Opercle with faint, dusky blotch. Some may be almost entirely greenish to yellowish. Lower jaw with knob at tip. Eyes large. Pectoral fins usually with 16 rays. Second anal-fin spine strong and longer than third. Body comparatively slender. HABITAT: Western Gulf of Alaska to central Baja California. Found over a variety of soft, mixed, and hard bottoms from about 195 to 3,750 ft.

Squarespot Rockfish - *Sebastes hopkinsi* (Cramer, 1895)

FEATURES: Shades of brown to tan dorsally, with several irregular dark blotches on back and sides. Blotches may be very faint to absent. Ventral area yellowish to whitish. Dark bands radiate from below eyes. Lower jaw protrudes slightly with small knob at tip. Anal fin with six to seven soft rays; second spine longer than third. Body comparatively slender. HABITAT: Southern OR to northern Baja California. Occur over hard bottoms, around wrecks, and in crevices from about 60 to 1,000 ft. BIOLOGY: Often in large schools, but may also be solitary. Feed in midwater on a variety of invertebrates.

Shortbelly Rockfish - *Sebastes jordani* (Gilbert, 1896)

FEATURES: Silvery to pinkish or bronze, with variable dark and pale blotches, saddles, and mottling. Blotches become darker while resting. Lower jaw protrudes. Eyes large. Anus located about midway between pelvic and anal fins. Caudal fin forked. Body comparatively long and slender. HABITAT: Southern British Columbia to southern Baja California around kelp beds and over soft and hard bottoms from about 90 to 1,610 ft. Occur near bottom and in midwater. BIOLOGY: Usually in very large, dense schools; occasionally solitary. Feed primarily on Euphausiids but also other invertebrates.

Freckled Rockfish - *Sebastes lentiginosus* Chen, 1971

FEATURES: Shades of brownish pink to orange brown, with dark greenish brown speckles and vermiculations dorsally; speckles become paler on sides. Back with five to six whitish blotches. Tip of upper jaw is notched, with a prominent, blunt knob on either side. Eyes large. Second anal-fin spine strong, longer than third. HABITAT: Central CA to central Baja California; rare north of Santa Monica Bay. Found over reefs, over boulder fields, in crevices, and around oil rigs from about 70 to 950 ft. BIOLOGY: Solitary or in small groups with other Rockfishes. May live to 22 years.

Chilipepper
to 23 in.

Rosethorn Rockfish
to 16 in.

Squarespot Rockfish
to 11.5 in.

Shortbelly Rockfish
to 13.8 in.

Freckled Rockfish
to 9 in.

Cowcod - *Sebastes levis* (Eigenmann & Eigenmann, 1889)

FEATURES: Color highly variable; capable of rapid change. Whitish, pinkish, orange, or yellowish, with four or five narrow, faint to dark bars on sides. Bars distinct in juveniles, fade with age. Lower jaw with knob under tip. Eyes comparatively small. First dorsal fin deeply incised. Second anal-fin spine about as long as third. Body deep, robust. HABITAT: Newport, OR, to central Baja California. Found demersally over soft, mixed, hard, and complex bottoms with caves, crevices, and other shelter from about 130 to 1,610 ft. BIOLOGY: Usually solitary, but may congregate. Feed on a variety of invertebrates and fishes. May live to 55 years. Populations low due to overfishing.

Mexican Rockfish - *Sebastes macdonaldi* (Eigenmann & Beeson, 1893)

FEATURES: Mottled reddish black dorsally, becoming reddish below. Larger specimens sometimes with irregular pale blotches on back and sides. Bars may radiate from below eyes. Lateral line pink to pinkish red. Lower jaw protrudes; knob at tip present. A flat, rear-facing spine is present below anterior portion of eyes. Maxilla extend to near rear of eyes. Anal fin with seven to eight rays; second spine shorter than third. HABITAT: Point Sur, CA, to southern Baja California, and central Gulf of California. Occur over rocky outcrops and oil platforms from about 250 to 1,150 ft.

Quillback Rockfish - *Sebastes maliger* (Jordan & Gilbert, 1880)

FEATURES: Brownish with a large area of yellow on head, irregular yellowish blotches on back that extend into spiny dorsal fin, and a few yellowish speckles on posterior body. Lower jaw protrudes slightly. Maxilla extends to rear of eyes. Spiny dorsal fin tall with deeply incised membranes. Anal fin with six to seven rays; second spine longer than third. HABITAT: Kodiak Island, AK, to southern CA at Anacapa Passage. Found demersally over low- and high-relief rocky bottoms from intertidal zone to about 900 ft. BIOLOGY: Occupy home ranges. May be territorial. Often solitary.

Black Rockfish - *Sebastes melanops* Girard, 1856

FEATURES: Gray, with blackish speckles and mottling dorsally. Pale ventrally. Dark speckles extend into dorsal fin. Some with dense speckles on lower sides that form a loose stripe. Faint dark band may radiate from below eyes. Lower jaw protrudes slightly; knob at tip absent. Maxilla extend to rear of eyes. Head spines reduced to almost absent. Anal fin rounded at lower edge. HABITAT: Amchitka Island, AK, to Huntington Beach, CA. Found demersally to near surface over high-relief rocky bottoms to about 1,200 ft. BIOLOGY: Form small to large schools. May leap from water in pursuit of prey.

Semaphore Rockfish - *Sebastes melanosema* Lea & Fitch 1979

FEATURES: Reddish, with irregular pale blotches dorsally. May be very pale with reddish blotches. All with a blackish mark on each upper spiny dorsal-fin membrane. Second dorsal, anal, and pectoral fins may have black marks. Lateral line pinkish to reddish. Spines above maxilla are single-pronged and forward-facing. Eyes large. Head spiny. Anal fin with six rays. HABITAT: Central OR to central Baja California. Recorded over rocky bottoms and adjacent areas from about 320 to 1,600 ft.

Cowcod
to 3.3 ft.

juvenile

Mexican Rockfish
to 2.2 ft.

Quillback Rockfish
to 2 ft.

Black Rockfish
to 2.3 ft.

Semaphore Rockfish
to 15.4 in.

Blackspotted Rockfish - *Sebastes melanostictus* (Matsubara, 1934)

FEATURES: Reddish, pinkish, to dusky orange, with variable dark speckles and mottling on sides. Others pink to red with greenish wash and heavy mottling on body. Dark blotch present on opercle. Few to many dark flecks always present on spiny portion of dorsal fin and often on soft portion. Three pale areas often under dorsal-fin base. Fins often dark along margins. Head with eight pairs of spines and two or more spines under eyes. Gill rakers long and slender on first arch. Second anal-fin spine slightly shorter or equal to third. HABITAT: Aleutian Islands to southern CA at Coronado Bank. Also to eastern coast of Japan. Found near bottom or in midwater from about 290 to 2,100 ft. NOTE: Previously known as and may be confused with *Sebastes aleutianus.*

Blackgill Rockfish - *Sebastes melanostomus* (Eigenmann & Eigenmann, 1890)

FEATURES: Body and fins red to pink. White blotches or dark saddles often present on back. May have scattered blackish flecks on body and black streaks on fin membranes. First dorsal-fin membranes with white tips. Tip of lower jaw, inside of mouth, folds above jaws, and gill membranes mostly black. Lower jaw projects; moderate knob under tip. Second anal-fin spine slightly shorter than third. Eyes large, body robust. HABITAT: Northern British Columbia to southern Baja California. Occur demersally over rocky, muddy, and high-relief hard bottoms from about 290 to 2,520 ft; often found in holes and crevices. BIOLOGY: Feed primarily on fishes. May live to 90 years. Historically important in commercial fisheries.

Vermilion Rockfish - *Sebastes miniatus* (Jordan & Gilbert, 1880)

FEATURES: Shades of red to reddish orange with highly variable grayish blotches and mottling dorsally and on sides. Bands radiate from below eyes. Lateral line in pale area that is broken behind head. Color of fins similar to color of body. Margins of second dorsal, caudal, and anal fins may be blackish. Interorbital space comparatively narrow. Eyes comparatively large. Pelvic fins long, extend to anus. Anal fin bluntly pointed with second spine about as long as third. HABITAT: Montague Island, AK, to central Baja California. Juveniles found in shallow water; adults occur over high-relief hard bottoms with caves and crevices to about 1,440 ft. BIOLOGY: Form schools near bottom and in water column, often with other Rockfishes. Some are territorial. Other name: Red Snapper. NOTE: Previously confused with *Sebastes crocotulus.*

Whitespotted Rockfish - *Sebastes moseri* Eitner, 1999

FEATURES: Pale pinkish to reddish, with numerous small pale speckles on head and body. May be almost uniformly colored or display dark and pale bars, particularly while resting. Lateral line area often in a dark red area. Lower jaw projects, with small knob at tip. Maxilla extends to anterior portion of pupil. Anal fin with nine soft rays; second spine strong, longer than third. Body elongate, slender. HABITAT: Point Arguella, CA, to northern Baja California. Found demersally over rocky reefs and around oil platforms from about 165 to 725 ft. BIOLOGY: May form schools with other Rockfishes. A poorly known species.

Blackspotted Rockfish
to 2.4 ft.

Blackgill Rockfish
to 24 in.

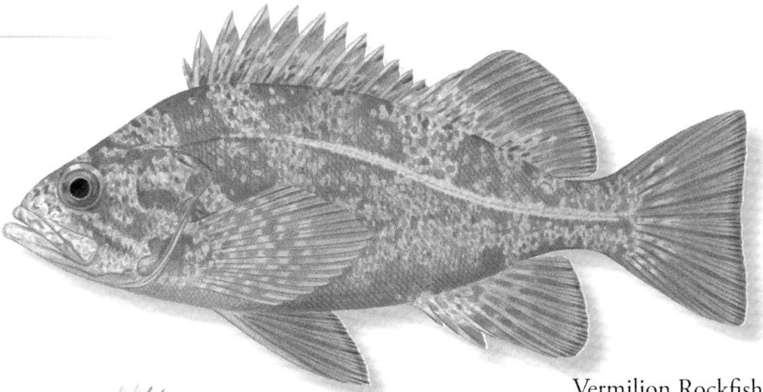

Vermilion Rockfish
to 2.5 ft.

Whitespotted Rockfish
to 8.3 in.

Blue Rockfish - *Sebastes mystinus* (Jordan & Gilbert, 1881)

FEATURES: Two forms. Blue-sided form grayish to bluish gray, with darker flecks and speckles on body that may be very dense, a dark band across nape, and bands radiating from below eyes; lower jaw protrudes; maxilla extend to about rear of pupil; body comparatively slender. Blue-blotched form grayish to bluish gray, with darker irregular mottling on body, a dark band across nape, and bands radiating from below eyes; lower jaw protrudes slightly; maxilla extends to about mid-eye; body comparatively deep. Both with small knob at tip of lower jaw. Juveniles with orange, red, or brown vermiculations. Anal fin angular, with 8–10 rays; second spine slightly shorter than third. HABITAT: Chatham Strait, AK, to northern Baja California. Blue-sided form apparently occurs in northern portion of range; blue-blotched form commonly found in southern portion of range. Both occur around kelp beds and over rocky and reef bottoms from near shore and surface to about 1,800 ft. BIOLOGY: Form large schools with other Rockfishes. Seek shelter at night. Other name: Blue Bass. NOTE: Recent evidence suggests the blue-sided form is a distinct species, *Sebastes diaconus*, with the common name Deacon Rockfish.

China Rockfish - *Sebastes nebulosus* Ayres, 1854

FEATURES: Blackish to bluish black, with numerous irregular yellowish speckles and spots on body and fins. Blotches and bands extend from eyes. A variable oblique yellow band runs from third and forth dorsal-fin spines to a yellow stripe along lateral line. Stripe may be broad or narrow. Head spiny. Anal fin with 7 rays; second spine longer than third. HABITAT: Kodiak Island, AK, to Redondo Beach and San Nicholas Island, CA. More common from Prince William Sound, AK, to northern CA. Demersal over high-relief rocky bottoms and around kelp beds from intertidal zone to about 420 ft. BIOLOGY: Solitary and territorial. May defend home range by erecting spines.

Tiger Rockfish - *Sebastes nigrocinctus* Ayres, 1859

FEATURES: Pinkish to whitish with five red, burgundy, brown, or black bars that extend onto fins. Dark bands radiate from eyes. Lower jaw protrudes slightly. Maxilla extend to rear of eyes. Head spiny. Body deep. Anal fin with six or seven rays; second spine strong, slightly longer than third. HABITAT: Unalaska Island, AK, to Tanner and Cortes Banks, CA. More common from SE AK to northern CA. Occur over rocky outcrops with sheltering caves and crevices from about 55 to 980 ft. BIOLOGY: Usually solitary and nocturnal. May migrate to deeper water in winter. May live to about 116 years.

Speckled Rockfish - *Sebastes ovalis* (Ayres, 1862)

FEATURES: Tannish dorsally and on sides, with small brown to blackish speckles that merge onto dorsal fins. May display irregular pale bars or blotches while resting. Ventral area whitish. Lower jaw protrudes, with a distinct knob at tip. Maxilla extends to middle of eyes. Anal fin with seven to eight rays; second spine not longer than third. Body comparatively deep. HABITAT: Northern WA to central Baja California. More common in southern part of range. Found over high-relief hard bottoms, boulder fields, and cobbles from about 100 to 1,200 ft. BIOLOGY: Form schools with other Rockfishes. Sought commercially and recreationally.

Blue Rockfish
to 21 in.

blue-sided
form

blue-blotched
form

juvenile

China Rockfish
to 17.7 in.

Tiger Rockfish
to 24 in.

Speckled Rockfish
to 22 in.

Bocaccio - *Sebastes paucispinis* Ayres, 1854

FEATURES: Color varies. Pinkish brown to gray or reddish. May have dark speckles or pale blotches. Juveniles with scattered dark spots on upper sides. Mouth large. Lower jaw strongly projecting, with a prominent knob at tip. Maxilla extends to rear of eyes. Head spines reduced. Anal fin usually with nine rays; second spine shorter than third. HABITAT: Western Gulf of Alaska to central Baja California. Juveniles inshore and near surface; adults demersal and in the water column over hard and soft bottoms from about 70 to 1,580 ft. BIOLOGY: Form schools with other Rockfishes. Critically Endangered.

Chameleon Rockfish - *Sebastes phillipsi* (Fitch, 1964)

FEATURES: Shades of pink to red. May be uniformly colored or with whitish blotches that contain red speckles and mottling on head and dorsal area. Capable of rapid color change. Tip of upper jaw with a distinct knob on either side. Spines above maxilla multi-pronged and forward-facing. Eyes large. Head spiny. Anal fin with six rays; second spine slightly longer than third. HABITAT: Point St. George to Nine Mile Bank, CA. Occur demersally over high-relief bottoms from about 570 to 900 ft.

Canary Rockfish - *Sebastes pinniger* (Gill, 1864)

FEATURES: Orange to yellow or reddish with dense grayish mottling. Dark bands radiate from eyes. Lateral line in a broad pale area that is continuous behind head. Juveniles with a black blotch on dorsal fin. Lower jaw protrudes slightly, with a smooth underside and a small knob at tip. Anal fin angular, pointed; second spine about as long as third. Caudal fin with concave margin. HABITAT: Western Gulf of Alaska to northern Baja California. Pelagic northern juveniles in shallow waters. Adults occur near bottom over high-relief and mixed muddy and rocky bottoms from about 60 to 2,750 ft. BIOLOGY: Adults and subadults aggregate and often associate with other Rockfishes. May live to 84 years.

Northern Rockfish - *Sebastes polyspinis* (Taranetz & Moiseev, 1933)

FEATURES: Reddish to pinkish, with brown to gray mottling. Dark bands radiate from eyes. Lower portion of pectoral fins whitish. Lower jaw protrudes, with a distinct knob at tip. Maxilla extends to rear portion of eyes. First dorsal fin with 14 spines. Anal fin with seven to nine rays; second spine not longer than third. HABITAT: Aleutian Islands to Graham Island, British Columbia. Also to Kuril Islands. Demersal over rocky bottoms from about 30 to 2,430 ft. Juveniles found more inshore than adults. BIOLOGY: Schooling. Migrate from shallow water in summer to deep water in fall.

Redstripe Rockfish - *Sebastes proriger* (Jordan & Gilbert, 1880)

FEATURES: Tannish to reddish or pinkish, with a pale red to grayish pink lateral line bordered by darker mottling. May display irregular mottling and streaks on body. Vague olive bands radiate from eyes. Lips and knob at tip of lower jaw blackish. Lower portion of pectoral fins pale. Anal fin with six to seven rays; second spine longer than third. HABITAT: SE Bering Sea and Aleutian Islands to central Baja California. Usually demersal over boulders, ridges, and cobbles from about 40 to 1,450 ft. BIOLOGY: Solitary or in small to large schools. May live to 55 years. Females larger than males.

Bocaccio
to 3 ft.

juvenile

Chameleon Rockfish
to 20 in.

Canary Rockfish
to 2.5 ft.

juvenile

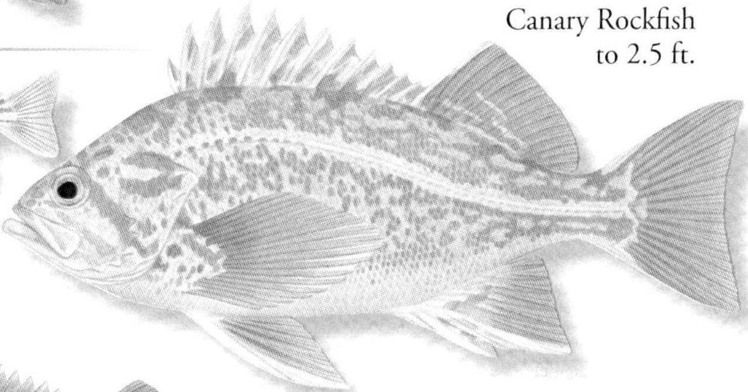

Northern Rockfish
to 19 in.

Redstripe Rockfish
to 2 ft.

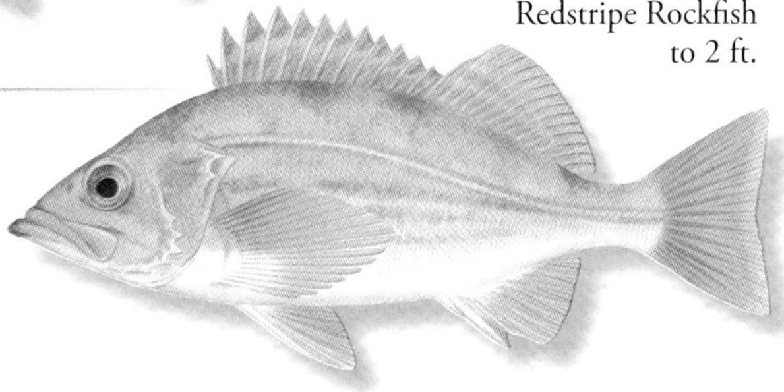

Grass Rockfish - *Sebastes rastrelliger* (Jordan & Gilbert, 1880)

FEATURES: Color varies. Shades of dark green to greenish brown or gray, with variable blackish spots and mottling on body and fins. Some specimens may be mottled orange. Dark blotch on gill cover sometimes present. Capable of rapid color change. Lower jaw protrudes slightly. Maxilla extends to or past rear margin of eyes. Gill rakers on first arch short, stubby. Top of head flattened. Head spines strong. HABITAT: Westport, WA, to central Baja California. Demersal over mixed hard bottoms and around kelp beds and other vegetation from intertidal zone to about 150 ft.

Rosy Rockfish - *Sebastes rosaceus* Girard, 1854

FEATURES: Shades of orange to red, with pale to dark purplish streaks on head and a saddle on nape. Body with variable purplish blotches and mottling. Usually with four to six purple-rimmed white blotches on back. Fins streaked. Lower jaw protrudes slightly, with moderate knob at tip. Maxilla extends to or beyond rear of eyes. Pectoral fins usually with 17 rays. Anal fin with five to seven soft rays; second spine longer than third. HABITAT: Strait of Juan de Fuca, WA, to southern Baja California. Demersal over reefs, high-relief and mixed bottoms, and mussel mounds from about 25 to 865 ft.

Greenblotched Rockfish - *Sebastes rosenblatti* Chen, 1971

FEATURES: Pinkish, orange, or reddish to whitish, with olivaceous to brownish vermiculations dorsally. Back with four or five white blotches. Spiny dorsal fin tall, deeply incised, streaked, and tipped with white. Lower jaw does not protrude; scales present on underside. Pectoral fins usually with 17 rays. Head spiny. Body robust. HABITAT: Point Delgada, CA, to central Baja California. Demersal over high-relief rocks, over mixed rock and mud, near caves and crevices, and around oil platforms from about 180 to 1,610 ft. BIOLOGY: Usually solitary. May ascend a few yards above bottom.

Yelloweye Rockfish - *Sebastes ruberrimus* (Cramer, 1895)

FEATURES: Reddish orange to orange brown dorsally and on sides. May display blotches on back. Fins may have whitish tips or black on outer membranes. Lateral line may lie in a pale pink to yellowish area. Irises reflective yellow. Juveniles reddish with two distinct white stripes on sides; tips of fins white or blackish. Lower jaw protrudes slightly with rounded knob under tip. Maxilla extends to rear of eyes. Ridges over eyes rasp-like in adults. HABITAT: Umnak Island, AK, to northern Baja California. Demersal over high-relief hard bottoms with caves and crevices from about 50 to 1,800 ft.

Flag Rockfish - *Sebastes rubrivinctus* (Jordan & Gilbert, 1880)

FEATURES: Whitish to pinkish white with four reddish orange to brownish bars on body that extend onto fins. Juveniles with blackish bars. First bar covers most of opercle. Second bar usually concave to head. Head profile comparatively steep. Spines on head reduced. HABITAT: Haceta Bank, OR, to central Baja California. Demersal over high-relief hard bottoms, around oil platforms, and over cobbles from about 100 to 1,370 ft. Often in caves and crevices. BIOLOGY: Usually solitary but may form small to large aggregations. May live to 38 years. Feed on a variety of fishes and invertebrates.

Grass Rockfish
to 22 in.

Rosy Rockfish
to 14 in.

Greenblotched Rockfish
to 19 in.

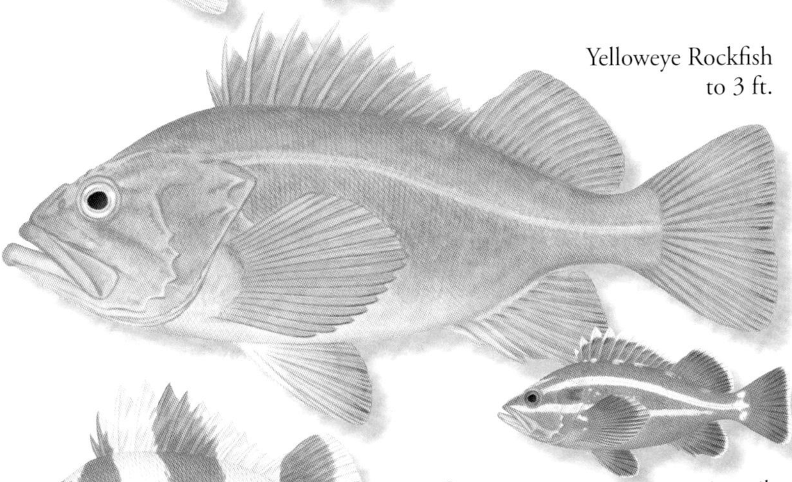

Yelloweye Rockfish
to 3 ft.

juvenile

Flag Rockfish
to 17.2 in.

variants

Dwarf-red Rockfish - *Sebastes rufinanus* Lea & Fitch, 1972

FEATURES: Grayish pink dorsally, becoming paler below, with whitish to pink or reddish on lateral line. May display faint saddles. Lower pelvic-fin rays usually white. Body otherwise unmarked. Lower jaw projects slightly, with a small knob under tip. Maxilla extends to about middle of eyes. Spines above eyes absent. Body slender. HABITAT: Channel Islands, CA, to northern Baja California. Occur demersally over a variety of bottoms. Commonly found over high-relief cobbles and rocks from about 360 to 660 ft. BIOLOGY: Form small aggregations with Pygmy and Whitespotted Rockfishes.

Bank Rockfish - *Sebastes rufus* (Eigenmann & Eigenmann, 1890)

FEATURES: Color varies. Pinkish brown, brownish, or blackish to whitish, with irregular blotches that form faint to distinct saddles. Upper body with small to large blackish flecks and spots that often extend onto dorsal fin. Dark bands radiate from below eyes. Lateral line pinkish to reddish. Fin membranes dark. Snout short, mouth small. Lower jaw protrudes with distinct knob at tip. Anal fin with eight to nine rays. HABITAT: Queen Charlotte Sound, British Columbia, to central Baja California. Occur over high-relief boulder fields and steep cliffs with caves and crevices from about 100 to 1,500 ft. BIOLOGY: Often reclusive but may form aggregations.

Stripetail Rockfish - *Sebastes saxicola* (Gilbert, 1890)

FEATURES: Shades of pink, red, or pinkish brown. May display irregular whitish or brownish blotches and saddles that can be distinct or faint. Dorsal fin with dark blotches; caudal fin always with dark blotches and stripes. Lower jaw protrudes, with a knob at tip. Eyes large. Second anal-fin spine strong. HABITAT: Yakutat Bay, AK, to southern Baja California. Usually demersal over soft and mixed low-relief bottoms and shell mounds from about 80 to 1,800 ft. BIOLOGY: Usually solitary. May live to 38 years.

Halfbanded Rockfish - *Sebastes semicinctus* (Gilbert, 1897)

FEATURES: Pinkish brown with darker flecks dorsally, becoming mottled reddish to orange on sides. More pale and uniformly patterned while in water column, becoming blotched while resting. All with a large, diamond-shaped blotch near midbody and an irregular dark blotch under soft dorsal fin. Dorsal fins banded; caudal fin blotched and striped. Lower opercular margin with two spines. Second anal-fin spine strong, very long. HABITAT: Northern WA to central Baja California. Demersal and in water column over a variety of soft, hard, and mixed bottoms from about 50 to 1,320 ft.

Olive Rockfish - *Sebastes serranoides* (Eigenmann & Eigenmann, 1890)

FEATURES: Dark brown to greenish brown dorsally, with irregular pale blotches. Sides buff to pale grayish green; may have some faint mottling. Yellowish bands behind eyes absent. Fins greenish yellow to grayish yellow. Knob at tip of lower jaw small to moderate. Spines on head reduced. Anal fin usually with nine rays; second spine shorter than third. HABITAT: Southern OR to central Baja California. Found demersally and in water column over high-relief hard bottoms and around kelp beds from near surface to about 565 ft. BIOLOGY: School with Blue Rockfish. Feed on fishes and octopi.

Dwarf-red Rockfish
to 6.8 in.

Bank Rockfish
to 21.5 in.

Stripetail Rockfish
to 16 in.

Halfbanded Rockfish
to 10 in.

Olive Rockfish
to 24 in.

Treefish - *Sebastes serriceps* (Jordan & Gilbert, 1880)

FEATURES: Greenish yellow to grayish yellow with five to six broad blackish to brown or dark green bars on body. Bars may appear double. Dark bands radiate from eyes. Adults with pink to red lips. Juveniles typically bright yellow with black markings. Lower jaw protrudes slightly. Maxilla extends to rear of eyes. Head spiny. Body deep. Anal fin with five to seven rays; second spine strong, slightly longer than third. HABITAT: San Francisco Bay to central Baja California. Found over complex hard bottoms from intertidal zone to about 320 ft. Often in caves and crevices. BIOLOGY: May be solitary and territorial, but also reported to form small schools. Active at night, dusk, and dawn.

Pinkrose Rockfish - *Sebastes simulator* Chen, 1971

FEATURES: Bright reddish orange dorsally, with five to six white blotches on back. Often with pale mottling anteriorly. Dorsal, caudal, and anal fins usually with whitish tips. Lower jaw does not protrude; small knob at tip present. Pectoral fins usually with 17 rays. Second anal-fin spine strong. HABITAT: Carmel Submarine Canyon, CA, to northern Baja California and Isla Guadalupe. Demersal around oil platforms, cobbles, and rocky mud bottoms with sheltering caves and crevices to about 1,470 ft.

Honeycomb Rockfish - *Sebastes umbrosus* (Jordan & Gilbert, 1882)

FEATURES: Shades of brown to pinkish, with four to six white or pink blotches on back. Area above lateral line may be blotched or with dark vermiculations. Scales below lateral line with dark greenish to brownish rims, forming a "honeycomb" pattern. Some specimens pale with dark blotches. Fins may have white tips. Tip of upper jaw slightly notched, with rounded knobs on either side. Second anal-fin spine strong, longer than third. HABITAT: Point Pinos, CA, to southern Baja California. Found near or on bottom over boulder fields and low-relief hard bottoms from about 60 to 890 ft.

Light Dusky Rockfish - *Sebastes variabilis* (Pallas, 1814)

FEATURES: Golden brown to greenish brown dorsally and on sides. May have irregular paler and darker mottling and speckles. Abdomen white, pink, or reddish; distinctly contrasts with dorsal color. Faint bands radiate from eyes. Lateral line pale. Lower jaw protrudes, with well-developed knob at tip. Spines on head weak. Second anal-fin spine shorter than third. HABITAT: Aleutian Islands to Johnston Strait, British Columbia; one record from OR. More common north of Sitka, AK. Also to Japan and Kamchatka. Found demersally over high- and low-relief bottoms from about 20 to 2,230 ft.

Harlequin Rockfish - *Sebastes variegatus* Quast, 1971

FEATURES: Reddish or pinkish to tannish pink, with faint to dark saddles or blotches on back and sides. Dark blotch on opercle. Posterior two-thirds of lateral line pale. Soft dorsal, caudal, and anal fins with blackish membranes and red outer margins. Caudal fin with clear trailing edge. Lower jaw protrudes; knob at tip weak to absent. Second anal-fin spine strong, longer than third. HABITAT: SE Bering Sea and Aleutian Islands to 59 miles southwest of Newport, OR. Demersal over high-relief bottoms from about 20 to 1,830 ft. BIOLOGY: Solitary or in small aggregations with other Rockfishes.

Treefish
to 16 in.

juvenile

Pinkrose Rockfish
to 16.4 in.

Honeycomb Rockfish
to 11 in.

Light Dusky Rockfish
to 23 in.

Harlequin Rockfish
to 15 in.

Pygmy Rockfish - *Sebastes wilsoni* (Gilbert, 1915)

FEATURES: Two forms. Red-stripe form tannish to pinkish, with red blotches dorsally and on sides; lower blotches form a dark broken stripe; some (presumably females) with bright yellow on abdomen. White-stripe form pinkish, with dark saddles and a distinct white stripe on lower sides. Lower jaw projects; knob at tip weak. Anal fin with six rays; second spine strong. Body slender. HABITAT: Kodiak Island, AK, to 60 Mile Bank, southern CA. Found demersally over boulders, cobbles, and occasionally over soft bottoms with nearby shelter from about 95 to 1,250 ft. BIOLOGY: Adults and juveniles form large schools during the day; seek shelter at night. Sometimes found singly. NOTE: While two color forms are currently recognized, some authors speculate that these may constitute separate species.

Sharpchin Rockfish - *Sebastes zacentrus* (Gilbert, 1890)

FEATURES: Pink to buff or reddish, with dark red, orange, or brownish saddles and blotches dorsally. Dark bands radiate from eyes; bands may be dark to faint. Dark blotch on opercle usually present. Lower jaw protrudes, with a distinct knob at tip. Lower edge of opercle with one or two small spines. Anal fin with six to seven soft rays (usually seven); second spine longer than third. Body moderately deep. HABITAT: Saint George Island and Attu Island, AK, to San Diego, CA. More common from Kodiak Island to central CA. Demersal over mixed bottoms and low- to high-relief hard bottoms from about 80 to 2,100 ft.

Shortspine Thornyhead - *Sebastolobus alascanus* Bean, 1890

FEATURES: Pinkish to reddish with variable irregular darker blotches. Some also with white blotches dorsally. Tips of spiny dorsal fin pale. Tip of upper jaw notched. Lower jaw does not project. Head spines strong. Dorsal fin usually with 16 spines; fourth to fifth spines tallest. Pectoral fins bilobed. Head broad, body slender. HABITAT: Bering Sea and Aleutian Islands to southern Baja California. Also to Sea of Okhotsk and off Kamchatka. Demersal over muddy and low hard-relief bottoms from about 55 to 5,000 ft. Rarely found over boulders. BIOLOGY: Usually solitary. Glandular tissue on posterior portion of fin spines believed to be venomous. Listed as Endangered.

Longspine Thornyhead - *Sebastolobus altivelis* Gilbert, 1896

FEATURES: Head and body pinkish to red or brownish. Fins may have black at margins; spiny dorsal fin may have white tips. Juveniles often with white blotches and streaks. Tip of upper jaw notched. Lower jaw does not project. Head spines strong. Dorsal fin deeply incised and usually with 15 spines; third spine tallest. Pectoral fins bilobed. Head broad, body slender. HABITAT: Shumagin Islands, AK, to southern Baja California. Most common from Vancouver Island to southern CA; rare in Bering Sea. Demersal over muddy bottoms near small rocks and other hard material from about 660 to 5,760 ft.

ALSO IN THE AREA: Brickred Rockfish, *Sebastes baramenuke*; Gray Rockfish, *Sebastes glaucus*; Yellowmouth Rockfish, *Sebastes reedi*; Broadfin Thornyhead, *Sebastolobus macrochir*, page 349.

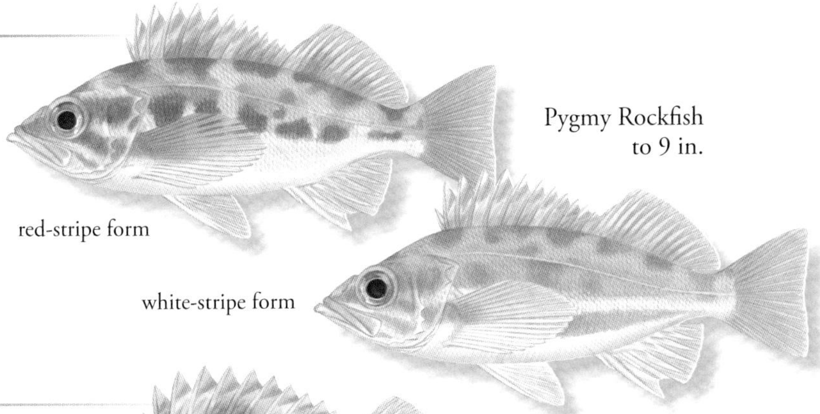

Pygmy Rockfish
to 9 in.

red-stripe form

white-stripe form

Sharpchin Rockfish
to 18 in.

Shortspine Thornyhead
to 2.7 ft.

Longspine Thornyhead
to 15 in.

Triglidae - Searobins

Lumptail Searobin - *Prionotus stephanophrys* Lockington, 1881

FEATURES: Shades of gray to grayish brown, with brown spots and blotches dorsally. Ventral area white. Spiny dorsal fin with a dark blotch between the fourth and fifth spines. Snout long, mouth large. Head with numerous spines and ridges. Pectoral fins broad and wing-like with three free rays. HABITAT: Columbia River, WA, to Chile, including Gulf of California. More common in southern portion of range. Found demersally over sandy and sandy mud bottoms from shore to about 835 ft. BIOLOGY: Use free pectoral rays to "walk" along the bottom. Feed on a variety of invertebrates.

Anoplopomatidae - Sablefishes

Sablefish - *Anoplopoma fimbria* (Pallas, 1814)

FEATURES: Shades of greenish to gray or blackish dorsally, often with darker blotches that form a lattice-like pattern. Pale below. Snout long. Spiny dorsal fin with 17–30 spines. Spiny and soft dorsal fins widely separated, with soft dorsal-fin base about the same length or shorter than spiny dorsal-fin base. Scales absent from fins. Body elongate. HABITAT: Aleutian Islands and Bering Sea south of St. Lawrence Island to central Baja California. Also to Japan. Adults found near bottom over a variety of substrates to about 9,000 ft; usually found above 3,000 ft. Juveniles often found in shallow waters near shore. BIOLOGY: May live to 113 years. Sought commercially. Other name: Black Cod.

Skilfish - *Erilepis zonifer* (Lockington, 1880)

FEATURES: Gray to black or grayish green, with variable irregular blotches and bars on head and body that fade with age. Snout blunt, rounded. Spiny dorsal fin with 12–14 spines. Dorsal fins set close together, with soft dorsal-fin base longer than spiny dorsal-fin base. Small scales present on soft-rayed fins. Body robust. HABITAT: South of Aleutian Islands and Gulf of Alaska to Monterey Bay, CA. Adults occur near bottom usually offshore from about 655 to 3,380 ft. Juveniles and young adults found near surface and sometimes close to shore. BIOLOGY: Adults eat squids and octopi.

Hexagrammidae - Greenlings

Kelp Greenling - *Hexagrammos decagrammus* (Pallas, 1810)

FEATURES: Females grayish to tannish, with yellow to brown spots that may form clusters on head and body; fins yellowish; body may be darkly blotched. Males uniformly to mottled brownish to reddish purple with bluish blotches bordered by small black spots; scattered black spots may be present or absent posteriorly. All highly variable. All with golden yellow irises, a single cirrus above each eye, and a pair of small cirri on back of head. Five lateral lines with the fourth extending well beyond anal-fin origin. HABITAT: Aleutian Islands to La Jolla, CA. Found over rocky reefs and kelp beds from intertidal zone to about 425 ft. Some reported to about 795 ft. Females occur in shallow waters, males in deeper waters. BIOLOGY: Usually solitary and territorial. Males guard nests, with up to 11 separate egg masses. Feed on a wide variety of invertebraates and fishes.

Searobins

Lumptail Searobin
to 17 in.

Sablefishes

Sablefish
to 3.7 ft.

Skilfish
to 6 ft.

Greenlings

♀

Kelp Greenling
to 21 in.

♂

Rock Greenling - *Hexagrammos lagocephalus* (Pallas, 1810)

FEATURES: Females mottled dark greenish brown to blackish, with variable bluish to pinkish blotches on body; blotches may be large to very small. Males reddish brown to blackish, with dark reddish to orange blotches that may be distinct or obscure. All highly variable. All usually with bluish mouths and a large, irregular dark blotch above pectoral-fin base. Single fleshy tab present above each eye. Cirri absent from top of head. Five lateral lines, with the fourth extending well beyond anal-fin origin. HABITAT: Eastern Bering Sea and Aleutian Islands to Point Conception, CA. Also to western Bering Sea and Commander Islands to Okhotsk, Japan, and Yellow seas. Found over rocky bottoms with vegetation and along exposed coasts from intertidal zone to about 1,955 ft. BIOLOGY: Feed primarily at night on a wide variety of invertebrates and algae.

Masked Greenling - *Hexagrammos octogrammus* (Pallas, 1814)

FEATURES: Color and pattern variable. Reddish brown to greenish brown with irregular darker spots and mottling; a small white blotch usually present on rear soft dorsal fin. Preguarding and guarding males primarily dark brown with tiny white spots. All with red irises. Single fleshy cirrus present above each eye. Cirri absent from top of head. Juveniles dark green dorsally, silvery white on sides. Five lateral lines with the fourth not extending past pelvic fins. Caudal peduncle comparatively deep. HABITAT: Bering Sea and Aleutian Islands to Banks Island, British Columbia. Also to Commander Islands and Seas of Okhotsk and Japan. Found over nearshore rocky areas, among vegetation, and in lagoons from about 20 to 100 ft. BIOLOGY: Males guard and aerate up to 20 egg masses until hatching. Feed on a wide variety of invertebrates.

Whitespotted Greenling - *Hexagrammos stelleri* Tilesius, 1810

FEATURES: Color variable. Brownish to greenish, with darker saddles, mottling, and distinct white spots about equal to size of pupil on body; body often tinged with red. Spawning and guarding males golden yellow to golden brown. A large blackish blotch usually present on anterior portion of spiny dorsal fin. Single cirrus present above each eye. Cirri absent from top of head. Five lateral lines, with the fourth extending just past pelvic-fin base. Caudal peduncle comparatively slender. HABITAT: Bering Sea and Aleutian Islands to Puget Sound, WA. Also to Sea of Japan and Commander Islands. Occur over reefs and kelp beds from intertidal zone to about 575 ft. BIOLOGY: Usually solitary.

Lingcod - *Ophiodon elongatus* Girard, 1854

FEATURES: Color and pattern highly variable. Greenish, bluish, or grayish to brownish with dense, irregular spots and mottling on head and body. Spots often cluster to form rosettes and oblique bands. Some almost uniformly dark. Often with an overall bluish green hue. Mouth large, toothy. A small fleshy cirrus present over each eye; cirri absent from top of head. One lateral line present. Body comparatively elongate. HABITAT: SW Gulf of Alaska to northern Baja California. Found near bottom over high-relief reefs, hard bottoms, and adjacent soft bottoms from intertidal zone to about 1,560 ft. BIOLOGY: Found singly or in closely stacked groups. Males defend nests with up to seven egg masses until hatching, sometimes returning to same nest site in sequential years.

♀

♂

Rock Greenling
to 24 in.

Masked Greenling
to 11 in.

♀

spawning ♂

Whitespotted Greenling
to 19 in.

Lingcod
to 5 ft.

Painted Greenling - *Oxylebius pictus* Gill, 1862

FEATURES: Color highly variable. Body gray, tan, or ochre to whitish with orange, reddish, or dark brown alternating bars that extend onto fins. Often with dense or diffuse mottling. Some with small white spots on body and fins. Males become very dark to almost black during spawning. All with irregular dark bands radiating from eyes. Snout pointed, elongate. Fringed cirri present above eyes and on top of head. Anal fin notched. HABITAT: Prince William Sound, AK, to central Baja California. Occur over boulders, hard bottoms, and in caves and crevices from intertidal zone to about 740 ft; usually above 160 ft. Also found perched on sea anemones or clinging to steep rock faces. BIOLOGY: Territorial. Males guard nests, with up to 10 egg masses that may be deposited by multiple females. Feed primarily on crustaceans but may also eat eggs from their own clutches. May live to eight years.

Atka Mackerel - *Pleurogrammus monopterygius* (Pallas, 1810)

FEATURES: Color and pattern somewhat variable. Females and non-breeding males grayish to greenish gray with dark brownish gray to blackish bars and mottling. Other males yellow to orange with dark bars and mottling; lower portion of pectoral fins, pelvic fins, and anal fin may become blackish. Snout bluntly pointed; mouth moderately large. Dorsal fin continuous. Caudal fin deeply forked. Five lateral lines present. HABITAT: Bering Sea and Aleutian Islands to Redondo Beach, CA. Also to Sea of Japan, Sea of Okhotsk, and Commander Islands. Found demersally and near surface over reefs and hard bottoms from intertidal zone to about 2,360 ft. BIOLOGY: Travel in schools. Females and non-guarding males make vertical migrations to feed. Important prey item for other fishes, sea birds, and marine mammals.

Shortspine Combfish - *Zaniolepis frenata* Eigenmann & Eigenmann, 1889

FEATURES: Pale gray to tannish, with irregular and variable brownish spots, mottling, and saddles. Specimens in complex habitats more densely and darkly patterned than those on soft bottoms. Spots on dorsal fins form bands. Dark band through eyes. Fleshy cirrus present above each eye. Anterior dorsal-fin spines tall and incised, with second spine slightly longer than third. Anal fin notched. Skin covered with tiny prickles. Body slender, elongate. HABITAT: Southern OR to southern Baja California. Found demersally and in the water column over mixed and soft bottoms from about 180 to 800 ft. BIOLOGY: Feed on a variety of invertebrates and fish eggs.

Longspine Combfish - *Zaniolepis latipinnis* Girard,1858

FEATURES: Pale gray to tannish with irregular spots and blotches generally arranged in rows. Spots on dorsal fins form bands. Dark band through eyes. Cirri over eyes absent or very small. Anterior dorsal-fin spines deeply incised, with first and second through third very tall; second spine tallest. Anal fin notched. Skin covered with tiny prickles. Body slender, elongate. HABITAT: Vancouver Island, British Columbia, to southern Baja California. Found demersally over muddy to hard, low-relief bottoms and in small depressions from about 120 to 660 ft. BIOLOGY: May migrate into water column to feed. Prey includes a wide variety of invertebrates.

Painted Greenling
to 10 in.

Atka Mackerel
to 20 in.

Shortspine Combfish
to 10 in.

Longspine Combfish
to 12 in.

Rhamphocottidae - Grunt Sculpin

Grunt Sculpin - *Rhamphocottus richardsonii* Günther, 1874

FEATURES: Buff to tan or reddish, with irregular oblique brownish bars and blotches that are bordered by blackish lines and spots. Caudal peduncle with a bright red to orange blotch on posterior portion. Fin rays yellowish orange to red. Snout long, pointed. Lower pectoral-fin rays free. Head large with deep ridges. Most of body covered with prickly scales. HABITAT: Gulf of Alaska to Santa Monica Bay, CA. Also to Japan. Occur over rocky and mixed sandy bottoms from intertidal zone and tidepools to about 845 ft. BIOLOGY: May jump and crawl across the bottom. Spawning believed to occur between January and April. Females guard nests inside crevices, abandoned shells, and other hollows. Produce grunting sounds when disturbed.

Cottidae - Sculpins

Scaled Sculpin - *Archistes biseriatus* (Gilbert & Burke, 1912)

FEATURES: Body tinged with salmon or pink, with irregular brownish to reddish or bluish saddles and pale spots and marbling. Abdomen whitish. Spiny dorsal fin marbled with a dark blotch at posterior margin and fleshy tabs at tips. Soft dorsal fin banded. Pelvic fins white in females, dark in males. Snout short and steep. Long fringed cirrus above each eye. A patch of scales present behind head. Two rows of scales along back and a single row along lateral line. Spines above eyes and on head absent. Lower pectoral-fin rays deeply incised. HABITAT: Known from southern Bering Sea north of Petrel Bank and Seguam Island. Also central and northern Kuril Islands. May be more widespread. Found demersally from about 245 to 475 ft.

Hamecon - *Artediellus scaber* Knipowitsch, 1907

FEATURES: Pale grayish to tannish, with darker spots and mottling. Spots dense on head, large and diffuse on lower body. Fins with orangish bands. Spiny dorsal fin with large black blotch in males. Snout short. Numerous fleshy cirri on head and along anterior portion of lateral line. Preopercle with two spines: upper is hooked upward, lower is directed downward. Spiny dorsal fin low in females, tall in males. Scales on body absent. HABITAT: Eastern Bering Sea from Beaufort and Chukchi seas to south of St. Lawrence Island. Also in western Bering Sea to Cape Navarin and in Arctic Ocean to Somerset Island, Canada. Found demersally from about 30 to 330 ft.

Coralline Sculpin - *Artedius corallinus* (Hubbs, 1926)

FEATURES: Color highly variable. Pink, gray, tan, to whitish or dark red with variable alternating saddles and irregular pale and dark spots and mottling dorsally. Large pale spots on lower sides. Snout comparatively long. Cirri present in front of eyes, on maxilla margin, on head, and along lateral line. A moderately wide band of 39–49 small oblique scale-rows present along upper back to rear of soft dorsal fin. A few scales between top of pectoral-fin base and lateral line present. Scales on head absent. HABITAT: Orcas Island, WA, to northern Baja California. Occur over rocky, reef, and mixed hard bottoms, under boulders, and in crevices, from lower intertidal zone to about 230 ft. BIOLOGY: Feed on small fishes and shrimps.

Grunt Sculpin

Grunt Sculpin
to 3.7 in.

Sculpins

♂ Scaled Sculpin
to 6.3 in.

♂ Hamecon
to 3.5 in.

Coralline Sculpin
to 5.5 in.

variants

Padded Sculpin - *Artedius fenestralis* Jordan & Gilbert, 1883

FEATURES: Brownish or grayish to dusky orange or greenish with dark saddles dorsally and pale spots ventrally. May be heavily mottled and spotted. Spiny dorsal fin often with a dark spot that may appear ocellated on anterior and posterior margins. Snout blunt. Maxilla extend to below pupils. Small cirri present at maxilla margin, on head, and on lateral line but absent from in front of eyes. Scales present below front of eyes and on top of head. Broad band of 18–29 comparatively large oblique scale rows present along back to caudal peduncle, where they form a dense patch. HABITAT: Unalaska Island, AK, to Diablo Cove, CA. Occur demersally over a variety of hard bottoms and sandy areas with vegetation from intertidal zone and tidepools to about 180 ft. Tolerant of brackish water.

Scalyhead Sculpin - *Artedius harringtoni* (Starks, 1896)

FEATURES: Color highly variable. Heavily mottled or marbled in shades of brownish to olive or pinkish, with five to seven irregular dark saddles dorsally and pale spots ventrally. Underside of head orange. Females with banded anal fin; males with small speckles or lattice-work on anal fin. Snout blunt. Maxilla extend to below pupils. Cirrus in front of eyes small in females, large and tree-like in males. Numerous cirri present on head and along lateral line. Scales present on head, sometimes on cheeks, absent from snout. Broad band of 38–51 small oblique scale rows present along back to top of caudal peduncle. HABITAT: Unalaska Island, AK, and from Kodiak Island to San Miguel Island, CA. Found demersally over rocky bottoms and around pilings from intertidal zone to about 70 ft.

Smoothhead Sculpin - *Artedius lateralis* (Girard, 1854)

FEATURES: Color highly variable. Mottled and marbled in shades of red, purplish, greenish, brown, or gray dorsally, with about six dark saddles. Ventral area yellowish, creamy, or pale green to red and with or without spots. Snout moderately long. Maxilla extend to rear of eyes. Cirri present on head and along lateral line but absent from in front of eyes. Scales absent from head. Narrow band of 18–29 small oblique scale rows present along back to near rear of soft dorsal fin. Scales above pectoral-fin base absent. HABITAT: Sanak Island, AK, to northern Baja California. Found demersally over hard bottoms and sandy areas with vegetation from intertidal zone and in tidepools to about 50 ft.

Bonyhead Sculpin - *Artedius notospilotus* Girard, 1856

FEATURES: Olivaceous to brownish, with oblique dark saddles and pale mottling dorsally. Ventral area dusky with pale spots. Spiny dorsal fin with a black ocellated spot between first and second spines, often followed by a dull black spot. A large black spot present between sixth and eighth spines. Snout blunt. Maxilla extend beyond pupils. Small spiny knobs behind eyes and tubercles present on head bearing short, simple cirri. Minute cirri on rear maxilla and preopercular margin. Scales present on top of head and upper opercle, absent below front of eyes. Narrow band of relatively large oblique scale rows present along back to rear of soft dorsal fin. HABITAT: Puget Sound, WA, to central Baja California. Demersal over soft bottoms from intertidal zone to about 170 ft.

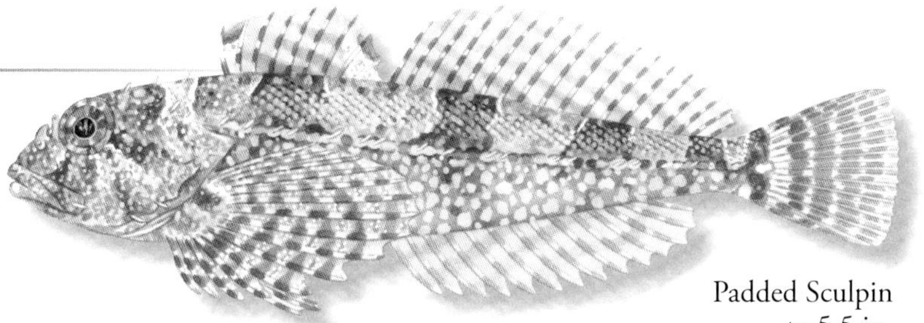

Padded Sculpin
to 5.5 in.

♀

Scalyhead Sculpin
to 4 in.

variants

♂

Smoothhead Sculpin
to 5.5 in.

Bonyhead Sculpin
to 10 in.

Rosylip Sculpin - *Ascelichthys rhodorus* Jordan & Gilbert, 1880

FEATURES: Grayish to brownish or olivaceous, with variable dark saddles and a broad pale to bright white saddle below spiny dorsal fin and below rear soft dorsal-fin base. Lips pale to bright red. Spiny dorsal-fin margin red. Maxilla extend to rear of eyes. A tufted cirrus present behind each eye. Spiny and soft dorsal fins continuous with a shallow to deep notch. Head and body scaleless. Pelvic fins absent. HABITAT: Prince William Sound, AK, to Pillar Point, CA. More commonly to Bodega Bay, CA. Found demersally in tidepools, along rocky, gravel, and exposed shores, and in grass and kelp beds from subtidal zone to about 50 ft. BIOLOGY: Capable of breathing air.

Roughback Sculpin - *Chitonotus pugetensis* (Steindachner, 1876)

FEATURES: Mottled and blotched in shades of pale gray to tannish with irregular dark brown to blackish saddles dorsally. Ventral area creamy with diffuse areas of pink to red. Maxilla extend to rear of pupil. Simple cirri present at maxilla margin, in front of and on top of eyes. Flap-like cirrus behind eyes. Upper spine on preopercle multi-pronged and antler-like. Spiny dorsal fin notched between third and fourth spines; first spine tall to very tall. Scales present on top of head. A broad band of oblique scale rows cover back and top of caudal peduncle. HABITAT: Mears Pass and Coco Harbor, AK, to southern Baja California. Demersal. Usually found over sandy, silt, and mud bottoms but also over gravel, low-relief rocks, and grass beds from intertidal zone to about 475 ft.

Sharpnose Sculpin - *Clinocottus acuticeps* (Gilbert, 1896)

FEATURES: Color highly variable. Shades of blue gray, olive, brownish, or emerald green with variable pale vermiculations and saddles dorsally. An oblique pale bar usually present at nape. May be very dark overall. Ventral area variably spotted. Anterior portion of spiny dorsal fin with an irregular dark blotch or mottling. Maxilla with one or two cirri at margin. Simple, slender cirri present above eyes, on head, and along lateral line. Snout moderately long and acutely pointed when viewed from above. Preopercle with a single spine at upper margin. Inner pelvic-fin ray attached to body by a broad membrane. Head and body scaleless. HABITAT: Aleutian Islands to Big Sur River, CA. Occur demersally over rocky, sandy, and vegetated bottoms from intertidal zone and tidepools to about 65 ft. Also found in brackish water; occasionally in fresh water.

Woolly Sculpin - *Clinocottus analis* (Girard, 1858)

FEATURES: Color highly variable. Marbled in shades of green, brown, or red, with five to six variable darker saddles bordered with whitish spots and speckles dorsally. Ventral area with variable pale and dark spots. Overall color darker in deep water, paler in shallow water. May be very dark overall or prominently marked. Snout steep, short. Small, slender cirri present on nasal spine, above eyes, at rear of maxilla, and along preopercular margin. Numerous cirri on head, on back, and along lateral line, where they may form tufts. Prickles present on head and back. Preopercle with a two- or three-pronged spine at upper margin. HABITAT: Cape Mendocino, CA, to central Baja California. Occur demersally over a variety of hard, soft, and mixed bottoms from intertidal zone and tidepools to about 60 ft.

Rosylip Sculpin
to 5.9 in.

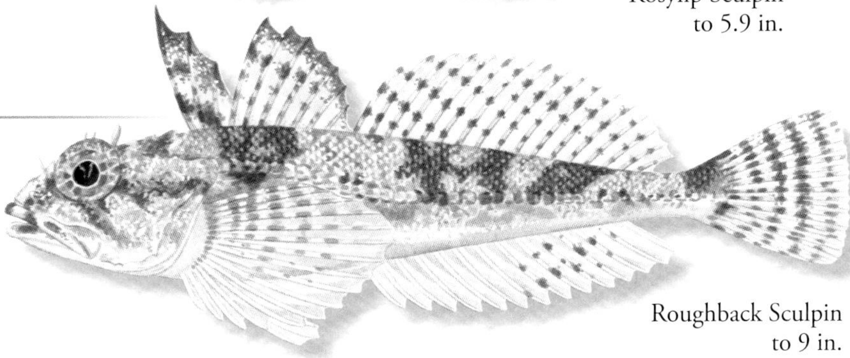

Roughback Sculpin
to 9 in.

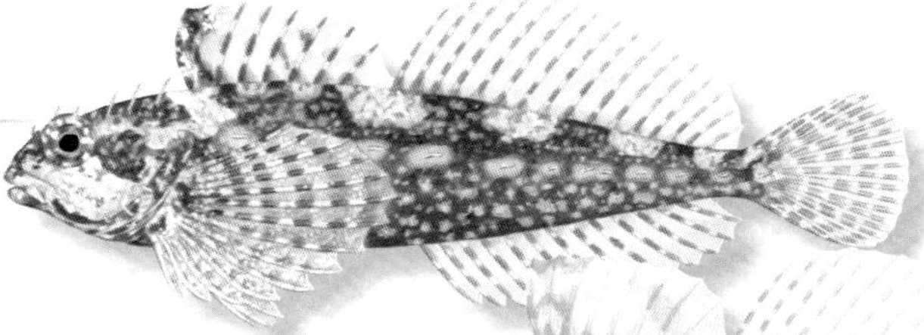

Sharpnose Sculpin
to 2.5 in.

variants

Woolly Sculpin
to 7 in.

Calico Sculpin - *Clinocottus embryum* (Jordan & Starks, 1895)

FEATURES: Color varies. Dark gray, greenish, brown, or burgundy to pink, with distinct, darker, wedge-shaped saddles bordered by paler spots and marbling dorsally. Fins variably banded. Small fleshy tubercle present above tip of upper lip. Snout moderately long. Tufts of cirri present behind eyes, on head, at upper opercular margin, and along anterior lateral line. Simple cirri present at nasal spine and tips of dorsal-fin spines. Cirri absent from maxilla margin and on back. Preopercle with single, simple spine. Head and body scaleless. HABITAT: Aleutian Islands to northern Baja California. Found demersally over rocky, shelly, and sandy bottoms in intertidal zone.

Mosshead Sculpin - *Clinocottus globiceps* (Girard, 1858)

FEATURES: Mottled in shades of olive to reddish brown or almost black with variable dark bars. May have broad pale bar at midbody and below rear of soft dorsal fin. Snout very short, round in profile. Mouth small, nearly horizontal. Dense tufts of cirri present behind eyes, on top and sides of head, above pectoral-fin base, and along anterior lateral line. Adults with tufts of cirri between eyes. Cirri absent from maxilla. Preopercle with a single, simple spine. Head and body scaleless. HABITAT: Western Gulf of Alaska to Gaviota, CA. Occur demersally over soft and hard bottoms in intertidal zone.

Bald Sculpin - *Clinocottus recalvus* (Greeley, 1899)

FEATURES: Shades of brown with dense, lighter vermiculations and four or five dark, wedge-shaped blotches along back. Some almost uniformly blackish brown or pale green, others prominently marked. Snout short, round in profile. Mouth small, nearly horizontal. Several rows of cirri that are joined at base present behind eyes and on anterior portion of head. Cirri present on preopercular margin, above pectoral fin, and along anterior portion of lateral line. Cirri absent from maxilla. Preopercular spine reduced or absent in adults. Head and body scaleless. HABITAT: Mill Beach, OR, to central Baja California. Occur from intertidal zone and shallow subtidal areas.

Coastrange Sculpin - *Cottus aleuticus* Gilbert, 1896

FEATURES: Heavily mottled in shades of brown, gray, or greenish, with dark saddles that are usually more distinct posteriorly. Spawning males with orange on anterior spiny dorsal fin. Dark blotches on spiny dorsal fin absent. Cirri absent. Both pairs of nostrils tubular. Scales absent; small prickles sometimes present behind pectoral-fin base. HABITAT: Bristol Bay and Aleutian Islands to Oso Flaco Creek, CA. Also in Kobuk River, AK. Found demersally in freshwater rivers and streams and in brackish estuaries.

Prickly Sculpin - *Cottus asper* Richardson, 1837

FEATURES: Variable shades of brown, gray, or greenish with irregular dark saddles, spots, and mottling. Fins banded. Spawning males darker than females. Spiny dorsal fin with vague dark blotches and sometimes a narrow orange margin. Cirri absent. Anterior nostrils tubular. Scales absent, but body partially to nearly covered in small prickles. HABITAT: Steward, AK, to Ventura River, CA. Demersal in quiet coastal and inland rivers, streams, and lakes; occasionally in estuaries and nearshore marine waters near river mouths.

Calico Sculpin
to 2.75 in.

variants

Mosshead Sculpin
to 7.5 in.

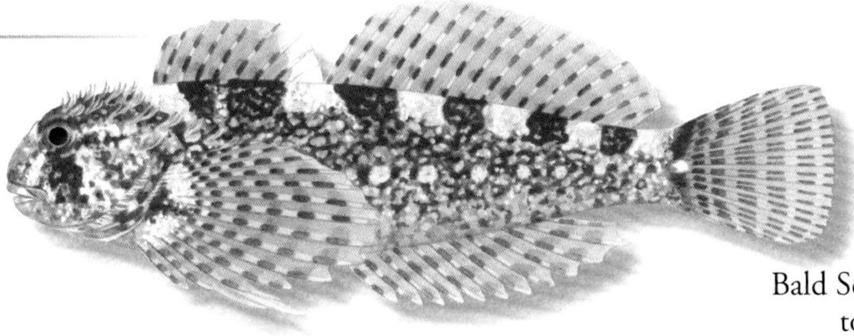

Bald Sculpin
to 5 in.

Coastrange Sculpin
to 6.8 in.

Prickly Sculpin
to 7.7 in.

Buffalo Sculpin - *Enophrys bison* (Girard, 1854)

FEATURES: Color, pattern highly variable. Heavily mottled and blotched in bright to dark shades of brown, green, black, pink, yellow, and white. May be partially or wholly covered in dark flecks and tiny spots. Three or four variable dark saddles usually present. Fins variably mottled, with irregular, diffuse banding. Eyes comparatively small. Head very broad and rounded, with low ridges that are covered with minute tubercles. Maxilla margin with one or more small cirri. Preopercle with four strong spines—upper spine very long and serrated at anterior base. Spiny dorsal fin with cirri at margin. Anal fin with 8–10 rays. Lateral line with 30–35 enlarged, keelless plates. HABITAT: Uyak Bay and Kodiak Island, AK, to Monterey Bay, CA. Demersal over inshore rocky and sandy bottoms from intertidal zone to about 85 ft.

Antlered Sculpin - *Enophrys diceraus* (Pallas, 1787)

FEATURES: Color and pattern highly variable. Heavily mottled, spotted, and speckled in shades of brown to olive or pink. Often with three or four dark saddles dorsally. Anal fin spotted, other fins variably banded. Eyes large. Head very large and broad, with raised knobs and ridges that are covered with minute tubercles. A prominent forked knob present over maxilla. Preopercle with four strong spines—upper spine very long with two to eight prongs. Cirri present on chin, maxilla margin, preopercular margin, and below lateral line but absent on spiny dorsal-fin margin. Anal fin with 11–13 rays. Lateral line with 34–38 enlarged, keeled plates. HABITAT: Point Barrow to Fort Tongass, AK, including Aleutian Islands. Also to Seas of Okhotsk and Japan. Found demersally over rocky bottoms from about 35 to 395 ft.

Leister Sculpin - *Enophrys lucasi* (Jordan & Gilbert, 1898)

FEATURES: Heavily mottled and spotted in shades of brown to olive and lavender. Often with three to four vague dark saddles dorsally. Anal fin spotted, other fins banded. Eyes large. Head very large and broad with raised knobs and ridges that are covered with minute tubercles. A blunt, rounded knob present over maxilla. Preopercle with four strong spines—upper spine very long, with one to four prongs. Spiny dorsal fin with numerous cirri at margin. Anal fin with 9–11 rays. Lateral line with 34–37 enlarged, keeled plates. HABITAT: Bering Strait and Aleutian Islands to British Columbia near Port McNeill. Found demersally over rocky bottoms from about 55 to 650 ft.

Bull Sculpin - *Enophrys taurina* Gilbert, 1914

FEATURES: Heavily mottled, marbled, and spotted in variable shades of brown, gray, pink, and white. Often with distinct, alternating dark and pale bars posteriorly. Fins variably spotted and banded. Eyes very large. Head broad, with moderate knobs and ridges that are covered with minute tubercles. An irregular ridge present above maxilla. A single fleshy cirrus present at rear of maxilla. Preopercle with four strong spines—upper spine very long and unpronged. Cirri absent on spiny dorsal-fin margin. Anal fin with 6–7 rays. Lateral line with 24–29 low plates—anterior plates with low keels. HABITAT: San Francisco to Santa Catalina and San Nicholas islands, CA. Found demersally over soft and mixed bottoms from about 35 to about 840 ft.

Buffalo Sculpin
to 14.6 in.

Antlered Sculpin
to 11.2 in.

Leister Sculpin
to 9.8 in.

Bull Sculpin
to 6.8 in.

Armorhead Sculpin - *Gymnocanthus galeatus* Bean, 1881

FEATURES: Brownish to olivaceous or grayish, with darker saddles and mottling dorsally. Pale ventrally. Dorsal, pectoral, and caudal fins irregularly banded. Preopercle with four spines; upper spine long with one to three prongs on upper margin. Rough, comparatively large plate-like scales on top of head that extend to rim of eyes. Head with one or two pairs of low tubercles. Modified T-shaped scales behind pectoral fins; body otherwise scaleless. Pelvic-fin rays elongate in males. Lateral line with 41–65 embedded tubules. HABITAT: Bering Sea and Aleutian Islands to central British Columbia. Also to northern Sea of Japan and Commander Islands. Found demersally over soft bottoms from near shore to about 2,000 ft; more commonly to about 540 ft.

Threaded Sculpin - *Gymnocanthus pistilliger* (Pallas, 1814)

FEATURES: Brownish with darker reticulations and faint dark saddles dorsally. Yellow to white below. Females with banded spiny dorsal fin. Males with dark brownish to blackish spiny dorsal fin bearing white spots, thread-like pelvic-fin rays, and long white pistil-shaped cirri behind pectoral fins. Single cirrus usually absent behind eye. Preopercle with four spines; upper spine long with one to four prongs on upper margin. Weak plate-like scales above eyes and on nape; body otherwise scaleless. Head with three or four pairs of well-developed, knob-like tubercles. Lateral line with 34–37 embedded tubules. HABITAT: Bering Sea and Aleutian Islands to southeastern AK. Also to South Korea. Found demersally over soft bottoms from near shore to about 1,070 ft; more commonly to about 490 ft.

Arctic Staghorn Sculpin - *Gymnocanthus tricuspis* (Reinhardt, 1830)

FEATURES: Brownish, with small, dense, darker spots and reticulations dorsally. Faint to very dark saddles under spiny and soft dorsal fins. Several dark blotches and wavy lines below lateral line. Ventral area pale to yellowish. Females with banded spiny dorsal fin. Males with dark brown to blackish spiny dorsal fin bearing white blotches and very long pelvic-fin rays. Rough, comparatively small plate-like scales on head that when present between eyes do not extend to eye rims. Protuberances or tubercles on head weak to absent. A few scales present behind pectoral-fin base; body otherwise scaleless. Cirri absent from head and body. Lateral line with 39–45 embedded tubules. HABITAT: Circumpolar. Beaufort and Chukchi seas to northern Bering Sea. Found demersally from near shore to about 1,480 ft.

Red Irish Lord - *Hemilepidotus hemilepidotus* (Tilesius, 1811)

FEATURES: Color highly variable. Heavily mottled and speckled in shades of red, orange, brown, pink, and white, with four irregular saddles dorsally. Fins variably speckled and banded. Cirrus on rear maxilla short and broad. Fleshy cirri present on snout, behind eyes, on chin, and on preopercle. Spiny dorsal fin notched anteriorly. Three bands of scales present: dorsal scale band with four or five rows; ventral band with six or seven rows. Single rows of scales above anal fin absent. HABITAT: Aleutian Islands to Monterey Bay, CA. Also to Commander Islands. Occur demersally over rocky bottoms from intertidal zone and tidepools to about 290 ft. Prefer bottoms with coralline algae.

Armorhead Sculpin
to 14 in.

Threaded Sculpin
to 11.5 in.

Arctic Staghorn Sculpin
to 11.8 in.

Red Irish Lord
to 20 in.

Yellow Irish Lord - *Hemilepidotus jordani* Bean, 1881

FEATURES: Mottled and speckled in shades of dark greenish gray to brown, with darker saddles dorsally. Often with a golden sheen. Whitish to yellowish ventrally. Gill membranes yellow. Juveniles brightly colored. Males with numerous dots on abdomen and dark pelvic fins. Females with pale pelvic fins. Cirrus on rear maxilla narrow. Pelvic fins in males usually longer than in females. Spiny dorsal fin notched anteriorly. Three bands of scales present: dorsal scale band with four or five rows; ventral band with about eight rows. Single row of scales above anal fin usually absent. HABITAT: SE Chukchi Sea, Bering Sea, and Aleutian Islands to Port Conclusion, AK. Also Gulf of Anadyr to Commander Islands and Sea of Okhotsk off Hokkaido, Japan. Found demersally over soft bottoms from near shore to about 3,000 ft; usually above about 820 ft.

Butterfly Sculpin - *Hemilepidotus papilio* (Bean, 1880)

FEATURES: Color and pattern vary slightly. Females brownish to reddish. Males yellowish to grayish with reddish pectoral fins. All with four irregular dark saddles that extend into fins and broad, banded pectoral fins. Cirri on maxilla very short or absent. First three dorsal-fin spines increasing in height and very tall and incised in males. Pelvic fins with long, free, trailing rays; very long and banded in males. Four bands of scales present: dorsal scale band with two to four rows; ventral band with about four rows; row above anal fin single. HABITAT: SE Chukchi Sea and Bering Sea to Aleutian Islands. Also to Commander Islands and Sea of Okhotsk to off Hokkaido, Japan. Occur demersally from intertidal zone to about 1,050 ft; usually above about 490 ft.

Brown Irish Lord - *Hemilepidotus spinosus* Ayres, 1854

FEATURES: Color and pattern vary slightly. Marbled and speckled in shades of brown with four to five irregular saddles dorsally. Top of head often with patches of pink to brick red. Cirrus on rear maxilla broad, paddle-shaped. Spiny dorsal fin notched anteriorly. Three bands of scales present: dorsal scale band with six to eight rows; ventral band with four or five rows. Single row of scales above anal fin absent. Pectoral fins comparatively short. HABITAT: Southern Bering Sea and northern Gulf of Alaska to Santa Barbara Islands, CA. Found demersally over rocky bottoms from intertidal zone to about 335 ft. Often over weed-covered rocks of outer tidepools.

Northern Sculpin - *Icelinus borealis* Gilbert, 1896

FEATURES: Mottled and blotched in shades of tan to brown with irregular grayish spots dorsally. Often with three dark saddles. Ventral area with pale spots. Fins with spots that form bands. Males with two dark blotches on spiny dorsal fin. Cirri present at nostrils, behind eyes, on maxilla, on head, and along lateral line. Preopercle with elongate spine bearing several prongs on upper margin. Spiny dorsal fin with cirri at tips of spines. Pelvic fins short. Two rows of scales on upper back that begin under first or second dorsal-fin spine and extend to peduncle. Lateral line with plate-like scales. HABITAT: Aleutian Islands and Bristol Bay to southern Puget Sound, WA. Found demersally over soft, gravel, pebble, and shelly bottoms from near shore to about 815 ft.

♂ Yellow Irish Lord
to 20.5 in.

Butterfly Sculpin
to 14.6 in.

Brown Irish Lord
to 11.3 in.

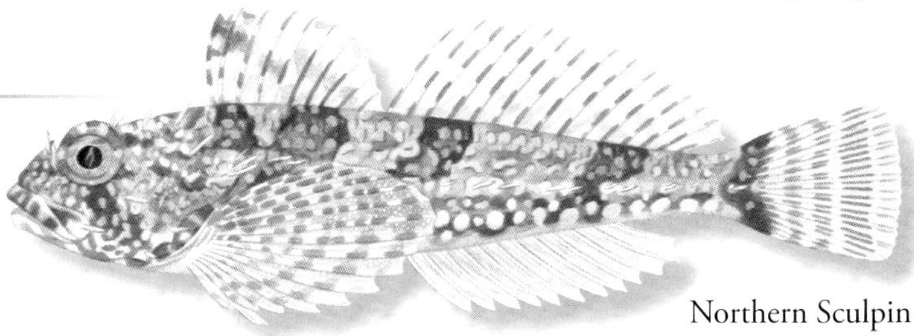

Northern Sculpin
to 4 in.

Dusky Sculpin - *Icelinus burchami* Evermann & Goldsborough, 1907

FEATURES: Tannish to pale olive with darker saddles bordered by whitish spots and wavy lines dorsally. Spiny dorsal fin with dark blotches. Other fins banded or spotted. Dark area at pectoral-fin base. Nasal spine reduced. Pores on head enlarged. Dark, branched cirri present behind eyes. Small slender cirri on maxilla margin, nape, and at tips of spiny fin rays. Preopercle with an elongate upper spine bearing three prongs. Two dorsal rows of scales extend from fifth dorsal-fin spine to near end of soft dorsal fin where lower rows end, and upper row continues for three scales. Pelvic fins minute. HABITAT: Behm Canal off Loring, AK, to La Jolla, CA. Occur demersally over a variety of bottoms from about 200 to 1,860 ft.

Pit-head Sculpin - *Icelinus cavifrons* Gilbert, 1890

FEATURES: Color and pattern highly variable. Marbled in shades of gray, brown, red, yellow, or olive with four variable dark saddles dorsally. May be almost uniformly colored. Dark band below eyes. Ventral area with irregular pale spots. Gill membranes usually dusky. Spiny dorsal fin with a diffuse to distinct black blotch. Dark area at pectoral-fin base usually present. Flap-like cirri present behind eyes. Cirri on maxilla margin, head, and lateral line reduced. Head comparatively large with a distinct "pit" on top bordered by ridges and spines. Preopercle with elongate upper spine bearing four or five prongs. Two rows of dorsal scales extend from about fifth dorsal spine to third-to-last soft ray. First two dorsal-fin spines variably tall in males. HABITAT: Monterey Bay, CA, to southern Baja California. Found demersally over soft and mixed bottoms from about 35 to 360 ft.

Threadfin Sculpin - *Icelinus filamentosus* Gilbert, 1890

FEATURES: Tannish to pale reddish brown or olive with three diffuse dark areas dorsally and brownish to reddish spots on back and along lateral line. An irregular dark bar radiates below eyes. Spiny dorsal fin with wavy bands and a narrow blackish margin. Body robust. Numerous small cirri on head and along lateral line. Dark, flap-like cirrus present behind eyes. Thread-like cirri present at nasal-spine base, at maxilla margin, on head, and along lateral line. First two dorsal-fin spines elongate and about equal in length. Preopercle with elongate upper spine bearing about four prongs. Two rows of dorsal scales extend from third or fourth dorsal-fin spine to last soft ray. HABITAT: Near Chirikof Island and Petersburg, AK, and southern British Columbia to Cortez Bank and Point Loma, CA. Occur demersally over sandy and muddy bottoms from about 60 to 3,750 ft.

Fringed Sculpin - *Icelinus fimbriatus* Gilbert, 1890

FEATURES: Densely marbled and speckled in shades of brown to reddish brown with diffuse dark saddles dorsally. Ventral area with pale spots. Fins banded. Fringed cirri present at base of nasal spine and behind eyes. Clusters of thread-like cirri present on maxilla margin, on head, and along lateral line. Tips of spiny dorsal fin with short tufts. Preopercle with elongate upper spine bearing three prongs. Two rows of dorsal scales extend from third or fourth dorsal-fin spine to near end of soft dorsal fin, where they become a single row on caudal peduncle. HABITAT: Pendrell Sound, BC, to San Diego, CA. Found demersally over soft and hard bottoms from about 100 to 870 ft.

Dusky Sculpin
to 5 in.

Pit-head Sculpin
to 3.5 in.

Threadfin Sculpin
to 10.6 in.

Fringed Sculpin
to 6.4 in.

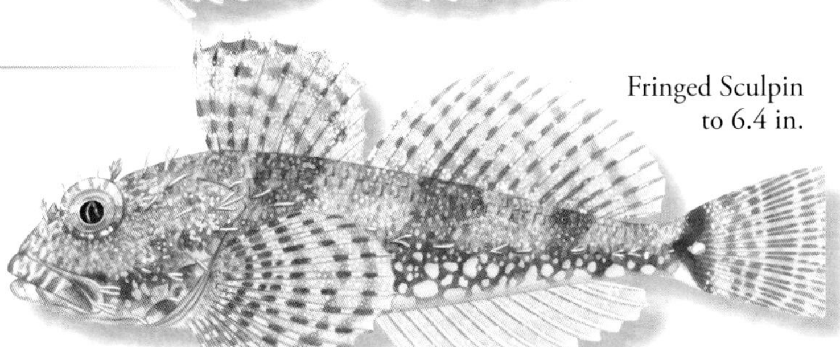

Yellowchin Sculpin - *Icelinus quadriseriatus* (Lockington, 1880)

FEATURES: Pale tannish, with irregular wavy spots and lines dorsally. Irregular dark blotches on back and along lower sides. Some with orange on top of head. Ventral area pale. Underside of lower jaw washed with yellow. Gill membranes with some black in females and non-breeding males. Breeding males with deep black gill membranes and blackish pelvic and anal fins. Spiny dorsal fin with a dark spot anteriorly and a black posterior margin. Cirri minute when present. Preopercle with elongate upper spine bearing about five prongs. Two rows of dorsal scales interrupted under rear portion of soft dorsal fin. HABITAT: Sonoma County, CA, to southern Baja California. Found demersally from intertidal zone to about 660 ft.

Spotfin Sculpin - *Icelinus tenuis* Gilbert, 1890

FEATURES: Marbled in shades of brown to reddish brown, with pale wavy spots and lines and three irregular dark saddles dorsally. Lower sides pale, with variable dark spots and marbling. Spiny dorsal fin with a diffuse to very dark brownish to blackish spot. Flap-like cirri present behind eyes. Other cirri reduced. First dorsal-fin spine tall and filamentous in all. Second dorsal-fin spine usually shorter than first, but may be prolonged. Both spines may have irregular flaps along upper margins. Cirri present at nasal spine and on head. Flap-like cirri present behind eyes. Preopercle with elongate upper spine bearing three to five prongs. Two rows of dorsal scales extend from under fifth dorsal-fin spine to about first third of soft dorsal fin. HABITAT: Near Ketchikan and west of Noyes Island, AK, to central Baja California. Occur demersally over sandy and silty bottoms from about 20 to 1,225 ft.

Blacknose Sculpin - *Icelus canaliculatus* Gilbert, 1896

FEATURES: Head blackish brown with dark blue luster. Body dark blackish brown with vague, darker saddles. Fins similar to body color; pectoral fins with bluish margins and pale blue spots. Cirri present behind eyes, on nape, on opercle, and at maxilla margin. Eyes large. Preopercle with four spines—uppermost may be forked. Dorsal row of scales and lateral-line scales with spinules. HABITAT: Along continental slope of Bering Sea from Navarin Canyon to Bristol Bay, AK, and Aleutian Islands. Also to Sea of Okhotsk off Hokkaido, Japan. Found demersally from about 65 to 3,300 ft.

Spatulate Sculpin - *Icelus spatula* Gilbert & Burke, 1912

FEATURES: Shades of brown dorsally with four or five vague to dark saddles. May be pale to dark and with or without freckles and spots. Spiny dorsal fin with dark blotches. Dark blotch at pectoral-fin base. Cirri behind eyes present. Cirri on head and operculum present or absent. Cirri on maxilla absent. Top of head with two pairs of prominent, variably shaped spines. Back appears humped in profile. Preopercle with four spines—uppermost simple or forked. Dorsal row of scales bear spinules. Lateral line scales extend almost to caudal fin. HABITAT: Beaufort, Chukchi, and Bering seas to Glacier Bay, AK. Also to Greenland, western Chukchi Sea, off Kamchatka, and Sea of Okhotsk. Demersal from about 40 to 1,200 ft.

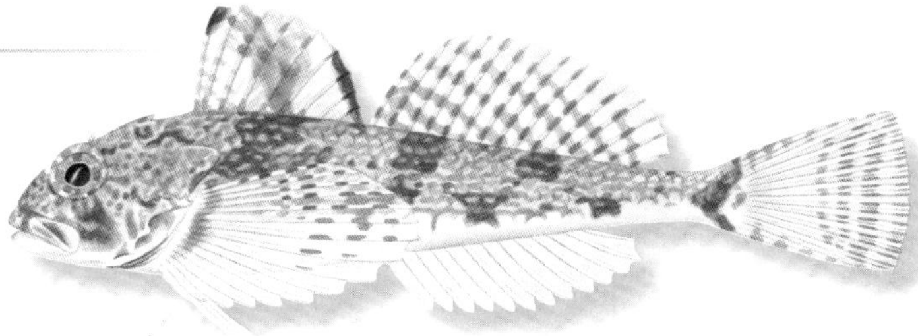

♀ Yellowchin Sculpin
to 4 in.

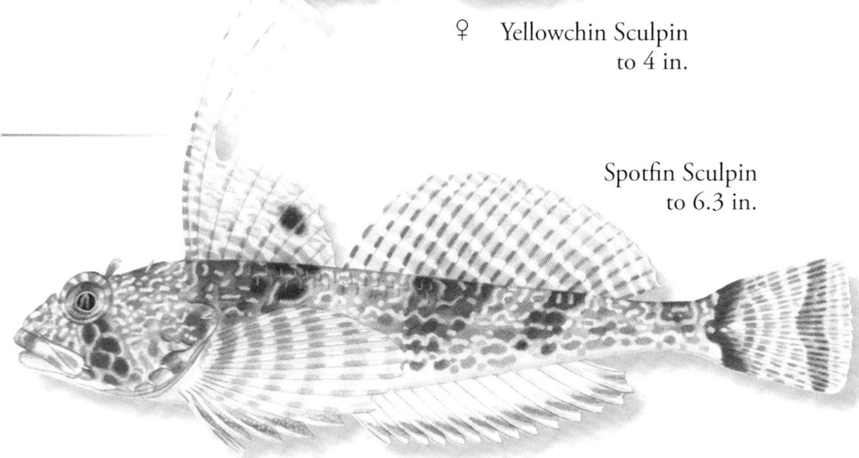

Spotfin Sculpin
to 6.3 in.

Blacknose Sculpin
to 9 in.

Spatulate Sculpin
to 7 in.

Thorny Sculpin - *Icelus spiniger* Gilbert, 1896

FEATURES: Shades of brown with irregular darker spots and four dark saddles dorsally. Spiny dorsal fin often with dark blotches or bands. Dark area at pectoral-fin base. Cirri present behind eyes and on cheeks. Other cirri absent. Top of head with one pair of prominent spines. Preopercle with four spines—uppermost forked. Dorsal row of scales with a single, strong spine. Lateral-line scales extend onto caudal fin. HABITAT: Bering Sea from Cape Navarin and Aleutian Islands to La Perouse Bank, British Columbia. Also Sea of Okhotsk. Occur demersally from about 100 to 2,525 ft.

Longfin Sculpin - *Jordania zonope* Starks, 1895

FEATURES: Color highly variable. Shades of pale blue, gray, or whitish, with tan, yellow, orange, red, or brown bands on head and wavy lines and spots on body. Often with irregular dark blotches dorsally and along lateral line. May be very bright to darkly colored. Females darken and males become dark brown during spawning. Display dark saddles at night. Anal fin shades of yellow to orange. All with a dark bar below eyes and five pairs of broad cirri on head. Cirri behind eyes may be stalked. Spiny and soft dorsal fins long-based and similarly shaped. Upper body covered with large, oblique scales. HABITAT: Danger Island, AK, to central CA at Diablo Canyon. Found demersally over rocky bottoms and around kelp from intertidal zone to about 125 ft. BIOLOGY: Observed cleaning parasites from Lingcod. May cling upside down. Territorial and active.

Lavender Sculpin - *Leiocottus hirundo* Girard, 1856

FEATURES: Color and pattern somewhat variable. Head and anterior body with numerous small whitish spots and wavy lines overlying lavender, tannish, brownish, or grayish saddles. Posterior body reddish with whitish spots and alternating oblique wavy bands. Caudal fin reddish, other fins banded. Mouth nearly horizonal. Cirri present on nasal-spine base, at maxilla margin, at opercular margin, and along anterior lateral line. Anterior portion of spiny dorsal fin tall and pointed, with first two spines elongate. Scales absent on head and body. HABITAT: Gaviota Pier, southern CA, to northern Baja California. Occur demersally over sandy, shelly, rocky, and mixed bottoms from near shore to about 120 ft.

Pacific Staghorn Sculpin - *Leptocottus armatus* Girard, 1854

FEATURES: Color and pattern vary. Shades of gray to tan or olive, with darker marbling dorsally. Often with a golden cast. May display several irregular dark saddles or may appear almost uniformly colored. Ventral area yellowish to whitish. Spiny dorsal fin with a black blotch posteriorly. Pectoral, soft dorsal, and caudal fins variably banded. Head broad and flattened. Nasal spine, cirri, and scales absent. Preopercle with elongate spine bearing two to six prongs. Lateral line distinct. HABITAT: SE Bering Sea at Port Moller and Gulf of Alaska to north-central Baja California. Found demersally over a wide variety of soft, mixed, and hard bottoms from intertidal zone to about 300 ft. Often found in lower reaches of rivers and streams. BIOLOGY: Sometimes buries in soft sediments.

Thorny Sculpin
to 11 in.

Longfin Sculpin
to 6 in.

variants

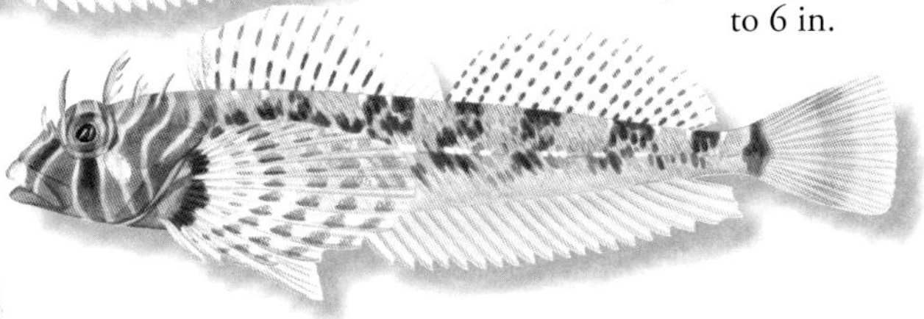

Lavender Sculpin
to 10 in.

Pacific Staghorn Sculpin
to 19 in.

Belligerent Sculpin - *Megalocottus platycephalus* (Pallas, 1814)

FEATURES: Dark olive brown with dense blackish marbling on head and body. Ventral area with large pale bands anteriorly and pale spots posteriorly. Fins spotted and banded. Mouth oblique with protruding lower jaw. Head broad, depressed, and with ridges, low spines, and small prickles. Upper preopercular spines strong. Irregular plate-like scales present above and below lateral line. HABITAT: Point Barrow to Herendeen Bay, AK. Also to western Chukchi Sea, Sea of Okhotsk, Sea of Japan, and Peter the Great Bay. Found demersally from inshore brackish waters to about 100 ft. Enter lower reaches of rivers.

Brightbelly Sculpin - *Microcottus sellaris* (Gilbert, 1896)

FEATURES: Dark blackish brown to blackish, with two oblique whitish saddles on back and a third pale band at caudal-fin base. Pectoral-fin base with two pale blotches. Ventral area in females pale grayish with whitish mottling; anal fin with pale brownish spots. Ventral area in males yellow with large whitish spots; anal fin with reddish orange spots. Snout short and steep. Head large, with ridges and spines. Upper preopercular spine curved upward. Pelvic fins attached to abdomen by broad membrane. Simple cirri present behind eyes and at tips of dorsal-fin spines. Weak prickles present below lateral line; otherwise, scales absent. HABITAT: Chukchi Sea to southern Bering Sea and Aleutian Islands. Also to Gulf of Anadyr, Commander Islands, Kuril Islands, and southern Sea of Okhotsk and northern Sea of Japan. Found demersally from inshore brackish waters to about 165 ft.

Plain Sculpin - *Myoxocephalus jaok* (Cuvier, 1829)

FEATURES: Brownish to grayish with irregular spots and mottling dorsally. Spots and mottling cluster but do not usually form distinct saddles or bands. Lower sides with pale spots. Ventral area pale. Fins banded. Head comparatively narrow and flattened with low ridges and numerous small wart-like knobs. Jaws large. Cluster of short ridges present behind eyes. Peropercle with three spines, uppermost long. Cirri absent. Numerous spiny plates present above and a few prickles present below lateral line. HABITAT: Chukchi Sea to eastern Aleutian Islands and Cook Inlet, AK. Also to Sea of Okhotsk, Sea of Japan, and northern North Korea. Occur demersally over sandy and muddy bottoms from intertidal zone to about 2,230 ft. Rarely below 500 ft.

Warthead Sculpin - *Myoxocephalus niger* (Bean, 1881)

FEATURES: Specimens over black lava bottoms blackish with brown mottling dorsally; may have pale spots ventrally and vague pale blotches on fins. Membranes on caudal fin pale. Specimens over granitic bottoms sand-colored. All with fleshy cirri behind eyes, on top of head, and at tips of spiny dorsal fin. Numerous small protuberances present on top and sides of head. Lateral line with three rows of pores. Skin thick, partially conceals nasal spines, opercular spines, and fin bases. HABITAT: Aleutian Island chain to Sanak and Shumagin islands, western Gulf of Alaska, and Pribilof Islands. Also to Commander Islands, Sea of Okhotsk, and northern Sea of Japan. Occur over hard bottoms in intertidal zone. Typically found in rocky pools and crevices.

Belligerent Sculpin
to 16.5 in.

♀

♂

Brightbelly Sculpin
to 5 in.

Plain Sculpin
to 2.3 ft.

Warthead Sculpin
to 10.6 in.

Great Sculpin - *Myoxocephalus polyacanthocephalus* (Pallas, 1814)

FEATURES: Color and pattern variable. Shades of gray to brown with several irregular alternating dark and pale saddles, with numerous paler spots and dark vermiculations dorsally; often with areas of lavender, orange, or white. Lower sides with distinct, variable white spots. Fins variably banded. Head comparatively broad, moderately flattened, and with low ridges and small wart-like knobs. Cluster of short ridges present directly behind eyes. Preopercle with three spines; uppermost long. Cirri absent. Prickly scales sometimes present above lateral line in large males. HABITAT: Bering Strait to Puget Sound, WA, including Aleutian Islands. Also to Commander Islands, Sea of Okhotsk, and Sea of Japan. Found demersally over soft bottoms and round rocks from intertidal zone to about 2,700 ft.

Fourhorn Sculpin - *Myoxocephalus quadricornis* (Linnaeus, 1758)

FEATURES: Color variable. Shades of gray, olive, brown, to almost black and with or without darker blotches or faint saddles dorsally. Pectoral, pelvic, and anal fins may be pinkish to reddish and with or without dark bands. Head broad and flattened with two pairs of large, rough, wart-like protuberances. Preopercle with four spines, uppermost straight to slightly curved. Cirri absent. Irregular row of spiny plates present above and below lateral line in males; poorly developed or absent in females and juveniles. HABITAT: Circumpolar. Beaufort Sea to St. Lawrence Island and Norton Sound, AK. Found demersally from coastal marine and estuarine waters. Ascend rivers as far as 62 miles inland. Reported to inhabit landlocked lakes.

Arctic Sculpin - *Myoxocephalus scorpioides* (Fabricius, 1780)

FEATURES: Blackish, purplish blue, or olivaceous, with paler mottling and dark saddles dorsally. Ventral area in females with pale spots and blotches. Ventral area in males orange red with a wide white stripe along midline and silvery to white spots along ventral area. Fins variably banded. Top of head with numerous, small warty protuberances. Spines behind eyes and on nape weak, each with a fleshy cirrus. Dorsal area with scattered, raised spiny plates. A few spiny plates present below lateral line. HABITAT: Beaufort Sea to Norton Sound, Bering Sea. Also to Gulf of Anadyr, east Siberian Sea, and Canadian coast to Greenland and Gulf of St. Lawrence. Demersal. Occur along shallow coastal waters from shore to about 130 ft.

Shorthorn Sculpin - *Myoxocephalus scorpius* (Linnaeus, 1758)

FEATURES: Mottled and blotched in shades of brown, olive, gray, tan, to black with variable darker saddles dorsally. Ventral area in females with irregular pale spots. Ventral area in males bright orange to reddish with large white spots. Fins variably blotched and banded. Top of head usually with two pairs of raised spines or protuberances (one or both may be absent) with or without cirri. May also have numerous small warty knobs on head. Preopercle with three simple spines. Spiny, plate-like scales with depressed centers present above lateral line. HABITAT: Beaufort Sea and Aleutian Islands to northern British Columbia. Also to Laptev Sea, Kamchatka Bay, Greenland, and NW Atlantic. Occur demersally over a variety of bottoms from near shore to about 1,800 ft.

Great Sculpin
to 2.5 ft.

Fourhorn Sculpin
to 14.4 in.

♀ Arctic Sculpin
to 9.4 in.

♀

♂

Shorthorn Sculpin
to 23.6 in.

Frog Sculpin - *Myoxocephalus stelleri* Tilesius, 1811

FEATURES: Shades of dark olive to brown or almost black, with dense vermiculations and speckles and several distinct to vague pale blotches dorsally. Lower sides and ventral area with pale spots. Mouth, chin, and gill membranes with dark speckles, becoming pale in adults. Cluster of short ridges behind eyes. Top of head with small warty protuberances. Single cirri present behind eyes, on nape, and on maxilla margin. HABITAT: Unalaska Island to Port Conclusion, AK. Also from Commander Islands to South Korea. Demersal from near shore to about 180 ft. Also in lower streams and river mouths.

Tidepool Sculpin - *Oligocottus maculosus* Girard, 1856

FEATURES: Color highly variable in both sexes. Dorsal area with alternating pale and dark saddles in shades of green, greenish and reddish, to gray and blackish. All are sparsely to densely covered with pale spots, speckles, or marbling. Ventral area with pale spots. Spiny dorsal fin with prominent to vague black to bluish black spot between first and second spines. Cirri present behind eyes, on top of head, at rear of maxilla, on preopercular and upper opercular margins, at tips of spiny dorsal fin, and along lateral line, where they may form small clusters anteriorly. Cirri absent from nasal spine and from back. Preopercle with a short, forked upper spine. Anterior anal-fin rays elongate in males. HABITAT: Pribilof and Shumargin Islands, AK, to Palos Verdes Peninsula, CA. Demersal over shallow, intertidal rocky bottoms and in tidepools.

Saddleback Sculpin - *Oligocottus rimensis* (Greeley, 1899)

FEATURES: Dorsal area with alternating pinkish and dark reddish brown saddles that are bordered with whitish to pale bluish spots and wavy lines. Lateral area dark and broken, with numerous pale spots. Fins finely spotted and banded. Simple cirri present behind eyes, on top of head, at maxilla margin, at peropercular margin, and along lateral line. Preopercle with a simple, upturned spine. Body covered with minute prickles. First two anal-fin rays elongate in males. Body comparatively slender in profile. HABITAT: Southeastern AK at Kakul Narrows to northern Baja California. Found demersally over sandy bottoms, in shallow rocky areas, and in tidepools to about 25 ft.

Fluffy Sculpin - *Oligocottus snyderi* Greeley, 1898

FEATURES: At least three variants: pale green with faint bands below eyes; dark greenish to reddish brown with pale blotches dorsally, irregular pale spots along lateral line, and pale bands below eyes; or marbled and spotted in shades of pink, gray, brown, to green, with darker saddles that merge along the back and a similarly colored dark stripe along lateral line that is interrupted by large, irregular pale spots. All usually with white spots on underside of head. Cirri present at nasal spines, preopercle, and tips of dorsal-fin spines; absent from maxilla margin. Tufts of cirri present behind eyes, on head, along dorsal profile, and along anterior lateral line. Cirri also present on back above lateral line but may be obscure. Preopercle with a short, forked, upper spine. First two anal-fin spines separate and elongate in males. HABITAT: Chernabura Is. and Samsing Cove, AK, to central Baja California. Found demersally over shallow rocky areas and in tidepools to about 20 ft. Typically in very shallow water. Often found in algal beds.

Frog Sculpin
to 22.8 in.

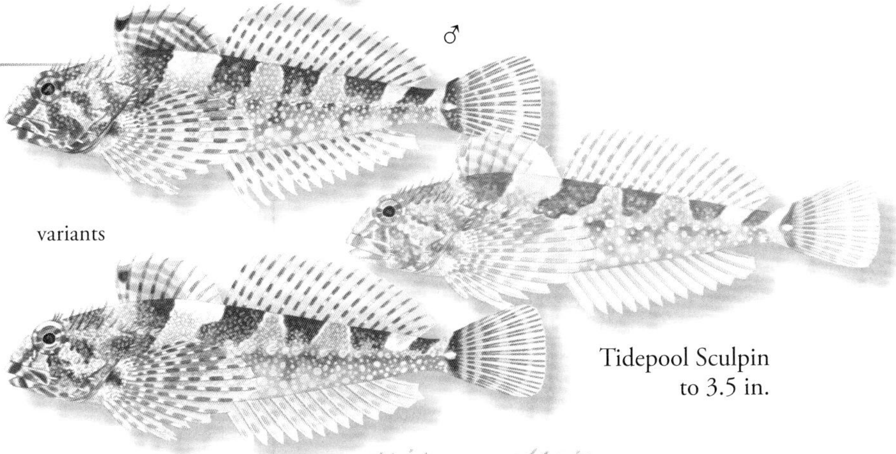

♂

variants

Tidepool Sculpin
to 3.5 in.

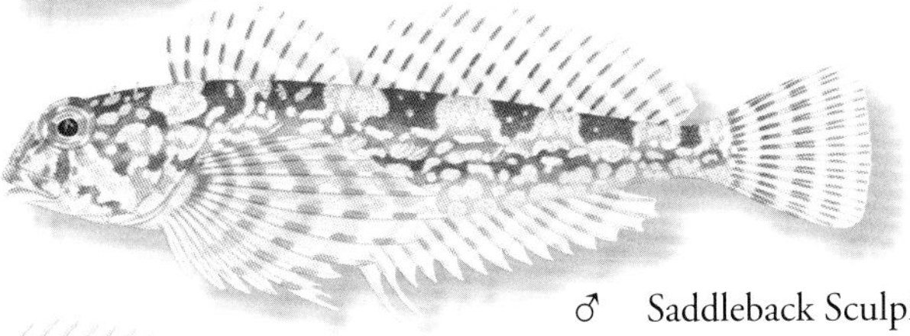

♂ Saddleback Sculpin
to 2.7 in.

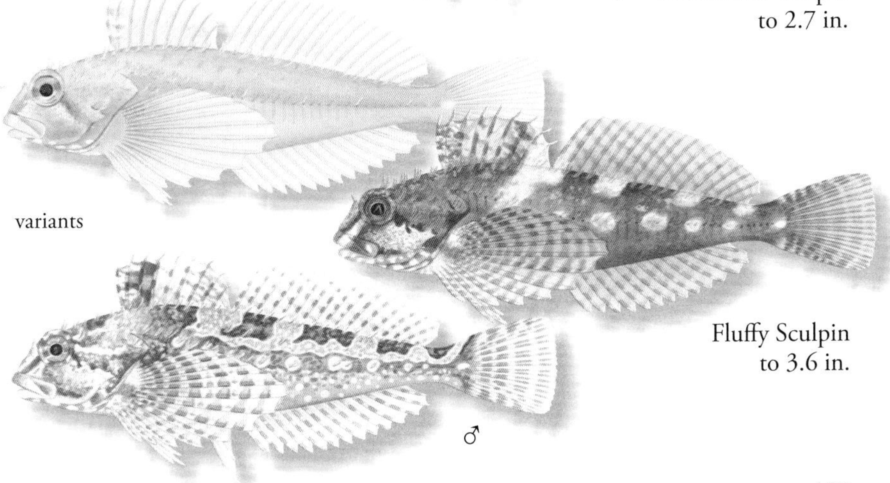

variants

Fluffy Sculpin
to 3.6 in.

♂

Snubnose Sculpin - *Orthonopias triacis* Starks & Mann, 1911

FEATURES: Color and pattern highly variable. Dorsal area with alternating saddles in shades of red, pink, yellow, gray, green, orange, or white, with scattered whitish spots and mottling. Ventral area with numerous, irregular pale spots. Fins variably spotted and banded. Anterior margin of spiny dorsal fin may be dark. Snout short and very steep. Mouth small. Scales present on top of head. Simple cirri present on nasal spine, maxilla margin, preopercle margin, and top of head and along lateral line. Preopercle with a short, forked upper spine. Back almost completely covered with oblique rows of relatively large scales. HABITAT: Farallon Islands, CA, to central Baja California. Found demersally over hard, soft, and mixed bottoms from intertidal zone to about 100 ft.

Thornback Sculpin - *Paricelinus hoplititicus* Eigenmann & Eigenmann, 1889

FEATURES: Tan to grayish or olivaceous with four vague to distinct dark saddles and reddish brown and whitish spots dorsally. Ventral area with reddish brown blotches. Single, fleshy cirrus present behind eyes. Cirri present behind maxilla, on lower cheeks, and at tips of dorsal-fin spines. Preopercle with three strong spines. Lower pectoral-fin rays free and elongate. Anal fin long-based. A single row of enlarged, sharp, and recurved scales present along dorsal profile. Body covered with tiny prickles. HABITAT: Hecate Strait, British Columbia, to Cortez Bank, CA. Demersal over rocky and mixed bottoms from nearshore to about 600 ft; typically below 65 ft.

Slim Sculpin - *Radulinus asprellus* Gilbert, 1890

FEATURES: Tannish to grayish, with three dark brown saddles and pale to dark brownish vermiculations that extend to lower sides. Ventral area whitish. Dorsal, caudal, and pectoral fins banded. Males with a blackish spot on posterior spiny dorsal fin. Cirri small and obscure when present. Snout comparatively short, eyes comparatively large. Nasal spines long, sharp. Upper preopercular spine simple, upturned. Small scales present on snout, on top of head, on opercle, and as series above lateral line. Lateral line plates enlarged, with a spiny keel. Body otherwise scaleless. HABITAT: Amchitka Island, AK, and western Gulf of Alaska at Semidi Islands to nothern Baja California. Demersal over soft bottoms from about 30 to 930 ft.

Darter Sculpin - *Radulinus boleoides* Gilbert, 1898

FEATURES: Tannish to grayish, with about four dark brown saddles dorsally and brownish mottling that extends to lower sides. Anterior dark saddle under spiny dorsal fin merges with dark color of head. Ventral area whitish. Dorsal, caudal, and pectoral fins banded. Snout comparatively long, eyes comparatively small. Nasal spines short, sharp. Upper preopercular spine simple, upturned. Simple cirri present on posterior margin of eye and may be present on nape. Small scales present on snout, below eyes, on top of head, on opercle, and as series above lateral line. Lateral-line plates enlarged, with spiny margins. Body otherwise scaleless. HABITAT: East of Kodiak Island, AK, and Langara Island, British Columbia, to Santa Catalina Island, CA. Occur demersally over sandy and gravel bottoms from about 50 to 480 ft.

variants

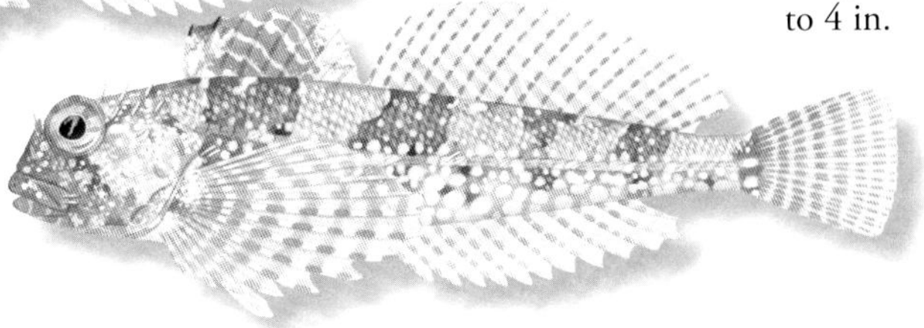

Snubnose Sculpin
to 4 in.

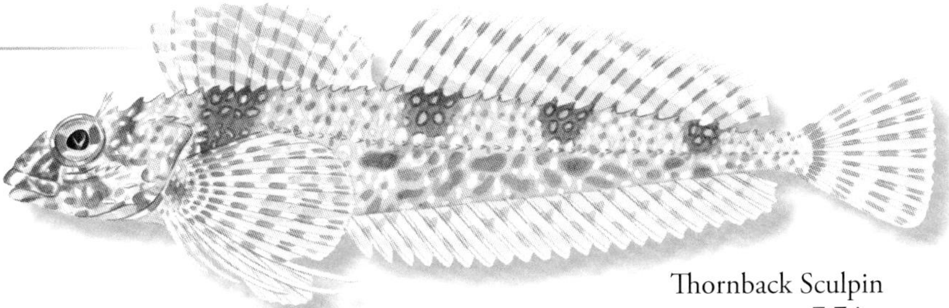

Thornback Sculpin
to 7.7 in.

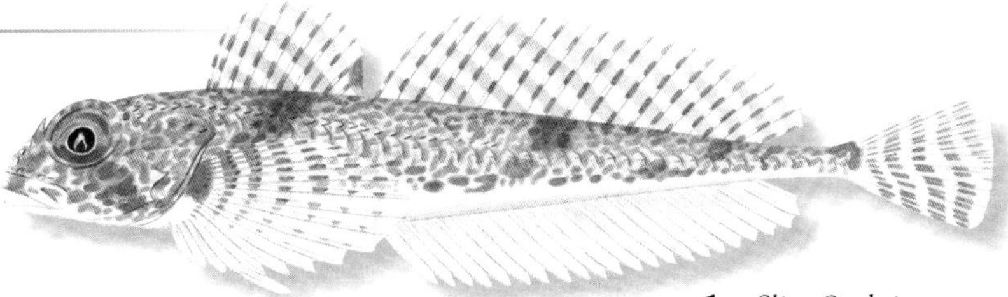

♂ Slim Sculpin
to 6 in.

Darter Sculpin
to 5 in.

Spinynose Sculpin - *Radulinus taylori* (Gilbert, 1912)

FEATURES: Color and pattern variable. Mottled in shades of pale tan, gray, or pinkish, with dark brown to dark gray saddles that extend onto lower sides. A distinct to broken pale tan to whitish band extends from below eyes. Dorsal, caudal, and pectoral fins finely spotted. Snout short, steep. Nasal spine short, sharp. Simple cirri present behind eyes, on nape, and on lower sides. Upper preopercular spine simple. Scales present above lateral line in one to four rows. Lateral-line scales enlarged. HABITAT: Southeastern AK off Strait Creek, Warm Chuck Inlet, and Sombrero Island to San Juan Islands, WA. Demersal over hard and mixed bottoms from about 15 to 160 ft. NOTE: Some authors place this species in the genus *Asemichthys*.

Roughskin Sculpin - *Rastrinus scutiger* (Bean, 1890)

FEATURES: Dark brownish black, with several obscure blackish blotches on back that extend into dorsal fins. Lower sides with irregular dark blotches. A blackish streak extends from upper pectoral-fin base and out along middle fin rays. Snout short, steep. Mouth and eyes large. Simple cirri present behind eyes and on top and sides of head. Upper preopercular spine sharp, upturned. Spiny dorsal fin tall. Head with minute, spiny scales. Body above and below lateral line with dense covering of small, spiny scales of differing sizes. HABITAT: Aleutian Islands and southern Bering Sea to Trinity Islands, AK. Also to Commander Islands. Demersal from about 385 to 1,680 ft.

Roughcheek Sculpin - *Ruscarius creaseri* (Hubbs, 1926)

FEATURES: Tannish to olive, with pale bluish variegations and dark brown to blackish saddles dorsally. Ventral area pinkish with pale spots. Dark bands below eyes. Pelvic and anal fins pale in females, blackish in males. Simple cirri at nasal spine, in front of eyes, at maxilla margin, on cheeks, on lower preopercle, and on spiny dorsal-fin tips. A multi-branched, flap-like cirrus present behind eyes, followed by clusters of cirri on head, at upper preopercular spine, on upper preopercle, and along lateral line. Scales present on head and cheeks, and nearly covering the back. Lateral line scales large and serrated. Body robust. HABITAT: Carmel Bay, CA, to central Baja California. Found demersally from intertidal zone to about 90 ft. NOTE: Previously known as *Artedius creaseri*.

Puget Sound Sculpin - *Ruscarius meanyi* Jordan & Starks, 1895

FEATURES: Olive gray to creamy with about five broken, dark brown saddles dorsally. Ventral area pale, with wavy dark bars. Dark bands below eyes. Dorsal fins mottled, banded, or pale. Anal fin pale or dark—may be dark in males. Simple cirri present in front of eyes, on head, on spiny dorsal-fin tips, and along lateral line. A three-branched cirrus present behind eyes. Upper preopercular spine elongate, with two prongs. Scales present on head, top of eyes, and cheeks and nearly covering the back. Lateral line scales enlarged. Body robust. HABITAT: Fillmore Island, AK, to Arena Cove, CA. Found over rocky bottoms and vertical rock faces from intertidal zone to about 270 ft. Usually occur between 5 and 65 ft. NOTE: Previously known as *Artedius meanyi*.

Spinynose Sculpin
to 3 in.

variants

Roughskin Sculpin
to 3.4 in.

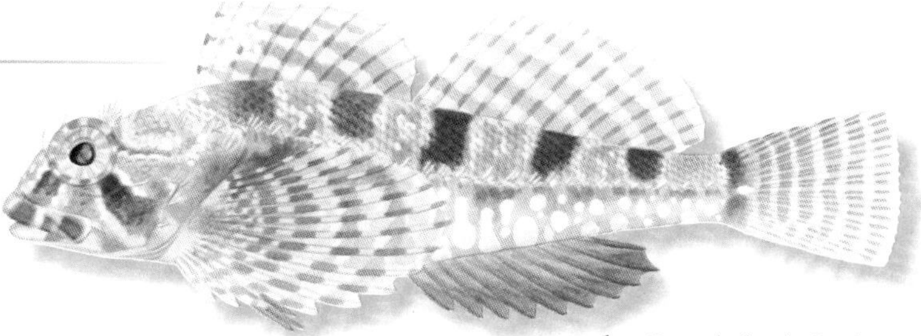

♂ Roughcheek Sculpin
to 3 in.

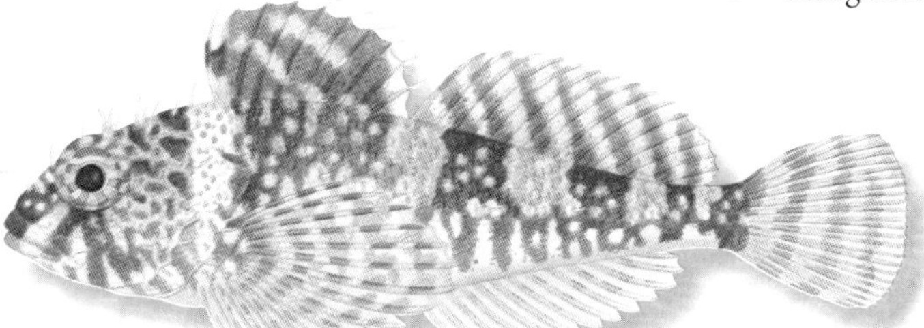

Puget Sound Sculpin
to 2.3 in.

Cabezon - *Scorpaenichthys marmoratus* (Ayres, 1854)

FEATURES: Color and pattern highly variable. Mottled in shades of brown, olive, gray, or lavender, with variable large and small pale blotches dorsally and on sides. Fins variably spotted and banded. Juveniles shades of bright red to brown, with few to many variable whitish blotches. All with a triangular, fleshy cirrus at tip of snout. A tall, fringed cirrus present behind eyes, and a fleshy tab present at maxilla margin. Upper preopercular spine simple. Spiny dorsal fin with 8–12 spines, middle margin depressed. Scales along lateral line small and often deeply embedded. Body otherwise scaleless. HABITAT: Southeastern AK near Sitka to central Baja California. Found demersally over hard bottoms from intertidal zone to about 360 ft. Small specimens often found in tidepools; larger specimens in deeper waters. BIOLOGY: Adults move inshore during high tide to feed. Flesh is palatable but eggs are poisonous.

Manacled Sculpin - *Synchirus gilli* Bean, 1890

FEATURES: Shades of translucent ochre, green, brown, or reddish dorsally. May be almost uniformly colored or with scattered large and small pale spots on back. Abdomen pale to whitish. Males with a silvery lateral stripe that may be continuous or broken. Snout long and pointed. Mouth large. Small, simple cirri sometimes present at nasal spine, behind eyes, and at preopercular margin. Scales sometimes present along dorsal profile. Pectoral fins united ventrally. Pelvic fins long in males, short in females. HABITAT: Unalaska Island and western Gulf of Alaska at Kodiak Island to San Miguel Island, CA. Occur near bottom and in kelp beds from intertidal zone and tidepools to about 45 ft. Often found clinging to pilings, seaweeds, and other substrates.

Scissortail Sculpin - *Triglops forficatus* (Gilbert, 1896)

FEATURES: Brownish, with five vague dark saddles dorsally. Lower sides with brownish blotches and several silvery spots anteriorly. Ventral area whitish. Opercular lining and upper gill membranes black. Spots on lower pectoral-fin rays faint in females, black in males. Tips of caudal-fin lobes with a large black spot. Snout moderately long. Lower pectoral-fin rays free in both sexes. Caudal fin forked in females, deeply forked in males. Lateral line with enlarged plates. Back covered with fine prickles. Ventral area with scales on 108–138 skin folds. HABITAT: Bering Sea and Aleutian Islands to Cook Inlet and False Point Retreat, AK. Also to Commander and northern Kuril Islands. Demersal over sandy, pebble, gravel, and shelly bottoms from about 60 to 1,400 ft.

Roughspine Sculpin - *Triglops macellus* (Bean, 1884)

FEATURES: Grayish with irregular brownish blotches and four blackish blotches that are sometimes bordered with white spots dorsally. A wavy, often broken blackish stripe runs along midline. Ventral area whitish. Dark streaks below eyes. Dorsal, caudal, and upper pectoral fin with spots that form bands. Snout long. Lower pectoral-fin rays free, elongate. Lateral line with plate-like scales. Back covered with fine prickles. Ventral area with scales on 122–162 skin folds. HABITAT: Eastern Bering Sea and Aleutian Islands to WA. Demersal over flat to sloping bottoms from about 60 to 900 ft.

Cabezon
to 3.2 ft.

juvenile

♀

Manacled Sculpin
to 2.7 in.

variants

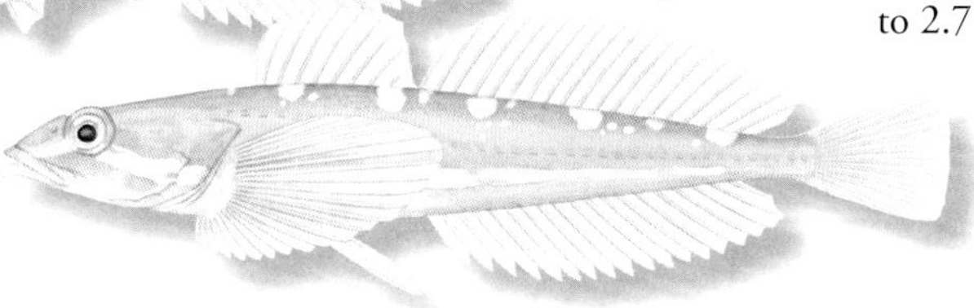

♂

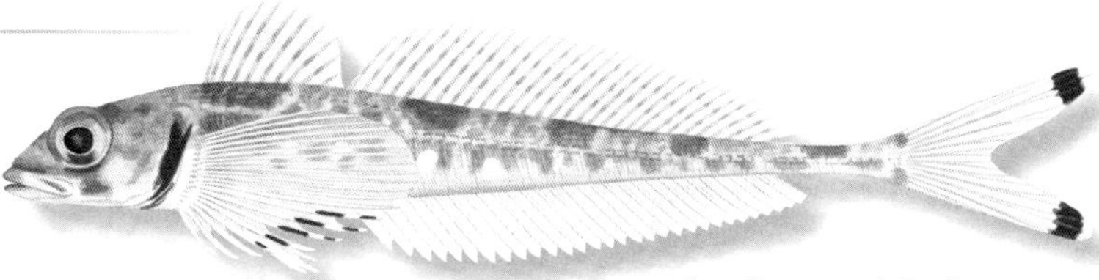

♂ Scissortail Sculpin
to 10.8 in.

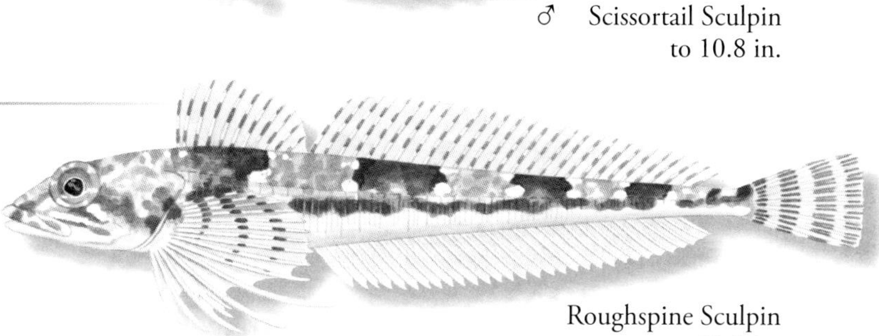

Roughspine Sculpin
to 11.8 in.

Cottidae - Sculpins, *cont.*

Ribbed Sculpin - *Triglops pingelii* Reinhardt, 1837

FEATURES: Finely speckled and mottled in shades of brown, with four darker saddles dorsally. Females overall paler, with brownish blotches on lower sides; males overall darker, with a continuous or broken blackish stripe on lower sides, especially during breeding. Upper and lower tips of caudal fin with a brown to blackish spot. Snout long. Lower pectoral-fin rays not elongate. Lateral line with plate-like scales. Back covered with fine prickles and a row of small scales. Ventral area with scales on 49–69 folds of skin. HABITAT: Circumpolar. Beaufort, Chukchi, Bering seas and Aleutian Islands to Puget Sound, WA. Also to Commander Islands and Sea of Japan to North Korea. Demersal over sandy, pebble, gravel, and rocky bottoms from about 15 to 1,580 ft.

Spectacled Sculpin - *Triglops scepticus* Gilbert, 1896

FEATURES: Brownish to olivaceous, with four faint, darker saddles dorsally and a few dark blotches on lower sides. Dorsal, caudal, and pectoral fins with dark blotches. Gill cavity and membranes blackish. Snout moderately long. Lower pectoral-fin rays incised. Lateral line with plate-like scales that become minute posteriorly. Back covered with large prickles and a row of well-developed scales. Ventral area with rough scales on about 180 skin folds. HABITAT: Bering Sea and Aleutian Islands to Cape Ommaney, AK. Also to Commander Islands and Sea of Japan off North Korea. Occur demersally over muddy, sandy, pebble, gravel, and shelly bottoms from about 80 to 3,000 ft.

NOTE: There are 18 other uncommon to rare Sculpins in the area.

Hemitripteridae - Sea Ravens

Crested Sculpin - *Blepsias bilobus* Cuvier, 1829

FEATURES: Reddish brown to olivaceous, with scattered darker and paler blotches on head and back. Others pale, with dark mottling. Fins irregularly banded. Pairs of prominent simple cirri present on tip of snout and under lower jaw. Head with bony ridges. Spiny dorsal fin short-based and without a deep notch. Head, body, and lower bases of fins covered with dense prickles. Pelvic fins small. Body comparatively deep. HABITAT: Chukchi and Bering seas and Aleutian Islands to northern British Columbia. Also to Sea of Japan. Occur near bottom from near shore to about 740 ft. Usually found above 390 ft. Juveniles found in shallower water and near surface.

Silverspotted Sculpin - *Blepsias cirrhosus* (Pallas, 1814)

FEATURES: Shades of reddish brown, olivaceous, or golden brown, with large silvery spots on lower sides behind pectoral fins. Dorsal area may have dark spots and/or pale wavy lines. Fins variably spotted. Pairs of prominent simple cirri present on tip of snout and under lower jaw. Top of head with distinct ridges. Spiny dorsal fin tall and deeply notched. Body and bases of fins almost entirely covered with dense prickles. HABITAT: Southwest Bering Sea and Aleutian Islands to San Simeon, CA. Also to Sea of Japan. Occur over a variety of bottoms and often in grass and kelp beds from near shore and brackish waters to about 310 ft. BIOLOGY: Reported to lay eggs in tissue of sponges.

Sculpins

Ribbed Sculpin
to 8.1in.

Spectacled Sculpin
to 12 in.

Sea Ravens

Crested Sculpin
to 10.6 in.

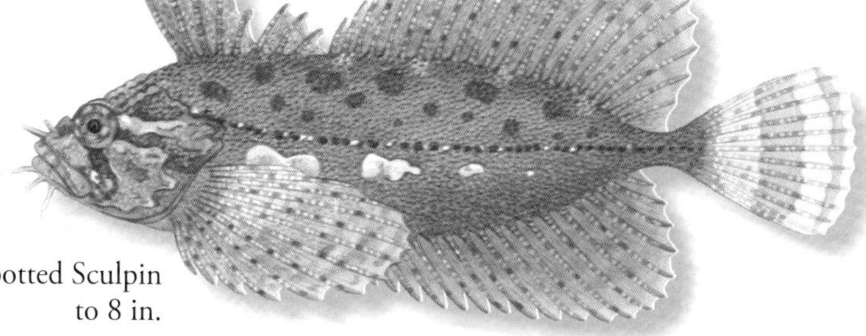

Silverspotted Sculpin
to 8 in.

Bigmouth Sculpin - *Hemitripterus bolini* (Myers, 1934)

FEATURES: Mottled and spotted in shades of brown, gray, red, or orange, with irregular darker saddles. Fins variably spotted and banded. Head very large, broad, and flattened, with numerous blunt spines and knobs. Mouth very large and broad. Numerous simple and complex cirri present on snout, lower jaw, maxilla margin, and head. Upper two preopercular spines pointed. Spiny dorsal fin single, long-based, and incised, with long filaments at spine tips. Head and body covered in dense prickles that extend onto fins. HABITAT: Bering Sea and Aleutian Islands to Eureka, CA. Also to Kuril Islands. Occur demersally over soft bottoms from about 80 to 3,000 ft. Juveniles sometimes migrate into water column at night.

Sailfin Sculpin - *Nautichthys oculofasciatus* (Girard, 1858)

FEATURES: Shades of tan, brown, or gray, with a distinct blackish bar below eyes. May have faint to distinct dark or pale spots and/or bars on body. Fins variably spotted and banded. Dark, flap-like cirri present above eyes. Head with knobby spines bearing small cirri. Spiny dorsal-fin base enlarged and greatly elevated. Spiny dorsal fin very tall and sail-like and barely connected to soft dorsal fin. Soft dorsal fin with 27–30 rays. Head and body almost completely covered with tiny prickles. HABITAT: Western Gulf of Alaska to San Miguel Island, CA. Found demersally over hard and mixed bottoms from intertidal zone and tidepools to about 360 ft. Often associated with vegetation. May also inhabit caves and crevices with octopi or wolf-eels. BIOLOGY: Nocturnal. Some females lay eggs between mussels in intertidal zones.

Eyeshade Sculpin - *Nautichthys pribilovius* (Jordan & Gilbert, 1898)

FEATURES: Shades of pale brown to gray with a distinct blackish bar below eyes and several faint to distinct dark spots on back. Fins variably spotted and banded. Dark, flap-like cirri present above eyes. Head with knobby spines bearing small cirri. Spiny dorsal-fin base moderately enlarged and elevated. Spiny dorsal fin tall, with spines descending in size and barely connected to soft dorsal fin. Soft dorsal fin with 22–26 rays. Head and body almost entirely covered with tiny prickles. HABITAT: Chukchi Sea off Wainwright, AK, Bering Sea, and Aleutian Islands to Steamer Bay, AK. Also to Sea of Japan. Occur demersally from near shore, from about 50 to 1,400 ft. Usually found above 420 ft in eastern Bering Sea and Gulf of Alaska.

Shortmast Sculpin - *Nautichthys robustus* Peden, 1970

FEATURES: Shades of tan to brown or gray with a distinct blackish bar below eyes and several faint dark spots on back. Fins variably spotted and banded. Dark, flap-like cirri present above eyes. Head with sharply pointed spines bearing small cirri. Spiny dorsal-fin base moderately enlarged and elevated. Spiny dorsal fin short and separated at base from soft dorsal. Soft dorsal fin with 19–21 rays. Head and body almost entirely covered with tiny prickles. HABITAT: Recorded from scattered locations from Bering Sea and Aleutian Islands to northern WA. May be more widely distributed. Found demersally over sandy, pebble, and rocky bottoms from near shore to about 320 ft.

Bigmouth Sculpin
to 2.4 ft.

Sailfin Sculpin
to 8 in.

Eyeshade Sculpin
to 4.3 in.

Shortmast Sculpin
to 3.2 in.

Kelp Poacher - *Agonomalus mozinoi* Wilimovsky & Wilson, 1979

FEATURES: Color and pattern highly variable. Mottled, blotched, barred, and spotted in mixed shades of red, brown, yellow, orange, white, olive, and black. Fins irregularly banded. Skin often covered with sponges and algae. Long, fleshy barbel at tip of snout. Head with numerous raised knobs. Spiny dorsal fin tall, elevated at base, incised, with fleshy tabs, and often oriented forward. Lower four or five pectoral-fin rays deeply incised. Body covered with spiny plates. Lateral line not enlarged anteriorly. HABITAT: Northern British Columbia at Dixon Entrance to San Simeon Point, CA. Found demersally over shallow rocky, coral, and vegetated bottoms to about 35 ft. Also in tidepools.

Southern Spearnose Poacher - *Agonopsis sterletus* (Gilbert, 1898)

FEATURES: Mottled in shades of tan to brown or gray, with seven or eight irregular darker bars dorsally. Pale bar usually present on head. Fins finely speckled and banded. Pelvic fins white. Snout comparatively short, with two pairs of spines at tip. Head with sharp spines dorsally and at sides. Several pairs of barbels under mouth, along throat, and along opercular margin. Barbels absent under snout tip. Anal fin with nine rays. HABITAT: San Simeon Point, CA, to southern tip of Baja California. Occur over soft and mixed bottoms from near shore to about 330 ft.

Northern Spearnose Poacher - *Agonopsis vulsa* (Jordan & Gilbert, 1880)

FEATURES: Mottled in shades of tan, brown, gray or black, with six or seven irregular dark bars dorsally. Some specimens uniformly dark. Fins finely speckled and banded. Pelvic fins dark brown with pale tips. Caudal fin usually with a semi-transparent blotch. Snout comparatively long with two pairs of spines at tip. Head with sharp spines dorsally and at sides. Numerous barbels under chin and mouth and along throat and opercular margin. Barbels usually present under tip of snout. Anal fin with 10–12 rays. HABITAT: Northern Gulf of Alaska at Katchemak Bay to Point Loma, CA. Occur demersally over soft and mixed bottoms from near shore to about 1,300 ft. Sometimes in tidepools.

Smooth Alligatorfish - *Anoplagonus inermis* (Günther, 1860)

FEATURES: Color varies: tannish dorsally, dark brown below; gray dorsally, blackish below; or almost uniformly orange. Usually with a dark stripe below eyes. May have vague darker body bars. Snout moderately long. Lower jaw protrudes. Spines on snout, barbels on head, and spiny dorsal fin absent. Body with smooth plates and a broadly concave dorsal surface. Anal fin with four to five rays. HABITAT: Aleutian Islands to Point Arena, CA. Found demersally from about 15 to 375 ft. Often occur around rocks.

Alligatorfish - *Aspidophoroides monopterygius* (Bloch, 1786)

FEATURES: Shades of tan to brown with about eight darker bars dorsally. Pale below. Dorsal and pectoral fins banded. Snout very short, with two recurved spines. Barbels absent on head. Spiny dorsal fin absent. Rows of smooth plates cover body. Body low and slender. HABITAT: Eastern Chukchi Sea to Aleutian Islands and western Gulf of Alaska. Also to Sea of Japan and western Atlantic Ocean off New Jersey. Occur demersally over a variety of grounds from about 25 to 1,155 ft.

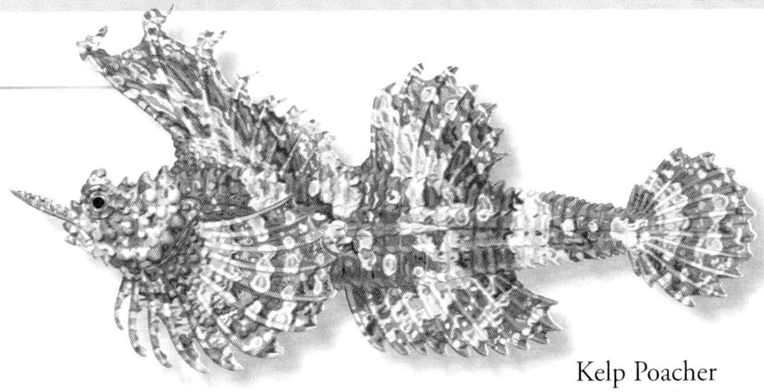

Kelp Poacher
to 3.5 in.

Southern Spearnose Poacher
to 5.8 in.

Northern Spearnose Poacher
to 8 in.

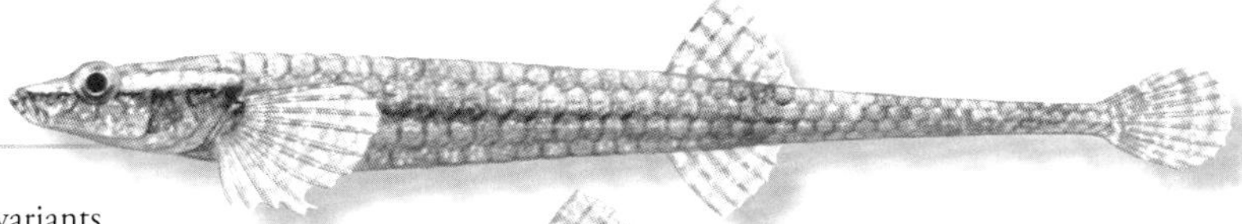

variants

Smooth Alligatorfish
to 6 in.

Alligatorfish
to 7 in.

Arctic Alligatorfish - *Aspidophoroides olrikii* Lütken, 1877

FEATURES: Tan to grayish with several darker bands dorsally. Pale below. Single dorsal fin with two dark bands. Caudal fin with a dark band near base and margin. Pectoral fins spotted. Snout very short with a pair of spines. Maxilla margin with a short barbel. Head and anterior body robust. Body plates rounded. HABITAT: Beaufort Sea to northern Bering Sea. Also along the Canadian Arctic coast to Greenland and Newfoundland. Found demersally over sandy, muddy, and rocky bottoms from about 20 to 1,700 ft.

Gray Starsnout - *Bathyagonus alascanus* (Gilbert, 1896)

FEATURES: Grayish green to brownish, with darker blotches and irregular bars dorsally. Dorsal, caudal, and pectoral fins banded. Snout short with five or more spines at tip. A pair of barbels present at maxilla margin, and one or two barbels also at sides of mouth. Lower pectoral-fin rays incised. Anal fin with six to eight rays, its origin under dorsal-fin interspace. Head and body plates spiny. HABITAT: Southeastern Bering Sea west of Pribilof Islands, AK, to northern CA. Found demersally over sandy and muddy bottoms from about 60 to 975 ft.

Spinycheek Starsnout - *Bathyagonus infraspinatus* (Gilbert, 1904)

FEATURES: Tannish to olivaceous, with darker blotches and spots dorsally. Dorsal, caudal, and pectoral fins banded. Snout short with five or more spines at tip. A pair of barbels present under mouth near tip, and a pair at maxilla margin. Cheeks spiny. Lower pectoral-fin rays may be incised. Anal fin with five to eight rays, its origin under rear spiny dorsal fin. Head and body plates spiny. HABITAT: Southeast Bering Sea from Pribilof Islands, AK, to near Eureka, CA. Found demersally over sandy and muddy bottoms from about 20 to 600 ft.

Blackfin Poacher - *Bathyagonus nigripinnis* Gilbert, 1890

FEATURES: Reddish brown to brown dorsally, pale below. Gill membranes blackish. All fins uniformly bluish black to black. Snout short, with three small spines at tip. Lower jaw protrudes. Maxilla with two small barbels at margin. Upper pectoral-fin rays long. Pelvic fins short. Anal fin with 6–10 rays. Head and body plates spiny. HABITAT: Eastern Bering Sea and Aleutian Islands to Eureka, CA. Also to northern Japan. Occur demersally over fine sand and muddy bottoms from about 60 to 4,100 ft. Most often on outer continental shelves and upper slopes.

Bigeye Poacher - *Bathyagonus pentacanthus* (Gilbert, 1890)

FEATURES: Grayish brown to olive brown with four to six darker bars dorsally. Dorsal and pectoral fins with dusky areas. Caudal fin dusky to blackish. Snout short, with five spines at tip. Eyes comparatively large. Maxilla margin with one or two barbels. Barbels at sides of mouth small to absent. Anal fin with six to nine rays, its origin under dorsal-fin interspace. Head and body plates spiny. HABITAT: Western Gulf of Alaska to Cortes Bank, CA. Found demersally over sandy and muddy bottoms from about 330 to 3,760 ft.

Arctic Alligatorfish
to 3.4 in.

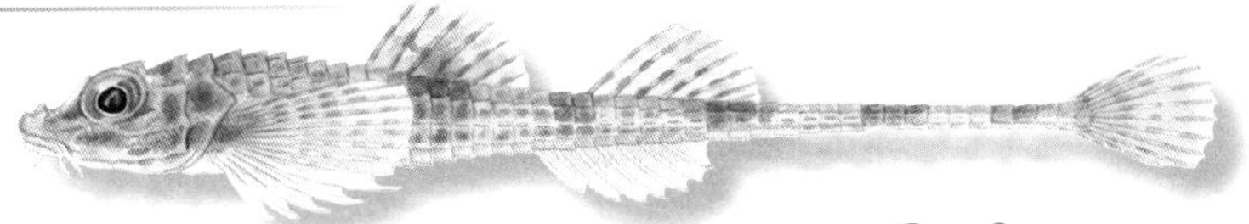

Gray Starsnout
to 5.6 in.

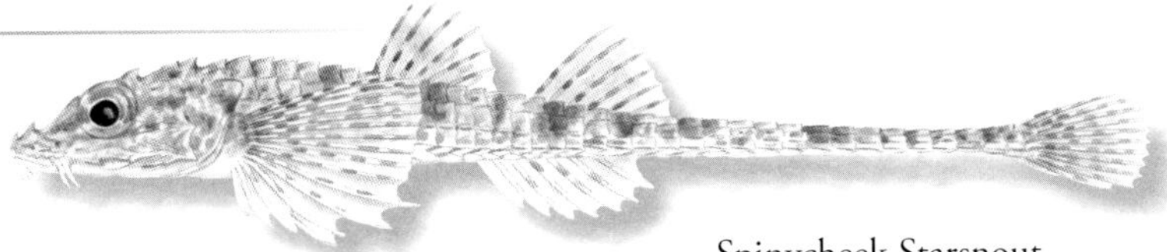

Spinycheek Starsnout
to 5.6 in.

Blackfin Poacher
to 11.4 in.

Bigeye Poacher
to 10.3 in.

Rockhead - *Bothragonus swanii* (Steindachner, 1876)

FEATURES: Shades of ochre, orange, red, brown, or white, with several brownish, blackish, or reddish bars on upper body. Others described as bluish. Pectoral fins darkly mottled, with a broad, pale central band. Other fins variably banded and speckled. Mouth small. Head very large and broad, with two large, rounded dorsal ridges and a deep dorsal pit. Dorsal and anal fins small. Body plates large and rounded. HABITAT: Kodiak Island, AK, and eastern Gulf of Alaska to San Luis Obispo County, CA. Found demersally along exposed coastal areas from intertidal zone to about 60 ft.

Warty Poacher - *Chesnonia verrucosa* (Lockington, 1880)

FEATURES: Mottled brownish to grayish dorsally, with variable darker bars on body and fins that may fade with age. Males with large brownish red areas on pectoral fins, orange and black on pelvic fins, and blackish anal fin. Females with spotted pectoral fins and pale pelvic and anal fins. Lower jaw protrudes. Head broad and depressed. Small barbel present at maxilla margin. Plates on abdomen with blunt spines. Anal fin with 10–12 rays. HABITAT: Bristol Bay and Shelikov Bay, AK, and Vancouver Island to Point Montara, CA. Found demersally over soft bottoms from about 65 to 900 ft.

Fourhorn Poacher - *Hypsagonus quadricornis* (Valenciennes, 1829)

FEATURES: Shades of white to yellow or brown, with several dark brownish, reddish, or blackish bars on body. Fins variably speckled and banded. Caudal fin with a dark margin. Skin sometimes covered with hydroids and seaweeds. Snout with a long barbel at tip. Spiny dorsal-fin base elevated. Lower pectoral-fin rays free. Body covered with spiny plates. First two lateral-line plates enlarged. HABITAT: Bering Sea and Aleutian Islands to Puget Sound, WA. Also to northern Sea of Japan. Occur demersally over rocky sand, gravel, or pebble bottoms from about 50 to 1,500 ft. Usually above 410 ft.

Atlantic Poacher - *Leptagonus decagonus* (Bloch & Schneider, 1801)

FEATURES: Tan to yellowish gray, with irregular brown spots and blotches dorsally. Dorsal fins banded. Caudal fin dark brown, usually blackish toward margin. Pectoral fins spotted with black at margin. Snout relatively short, with two spines at tip and sometimes with one or two short barbels underneath. A pair of forked barbels present under chin and four pairs at maxilla margin. Head and body plates spiny. HABITAT: Bering and Chukchi seas. Also Sea of Okhotsk, Gulf of Anadyr, and Arctic to North Atlantic and eastern Canada. Demersal over soft bottoms from about 80 to 3,070 ft.

Bering Poacher - *Occella dodecaedron* (Tilesius, 1813)

FEATURES: Olive gray to olive brown with darker speckles dorsally. White below. Males with yellowish pectoral fins spotted with red and black, white dorsal fins with two black bands, and white pelvic fins with a black blotch. Females more drably colored; pelvic fins white. Mouth upturned, lower jaw protruding. Small barbel at maxilla margin. Dorsal fins closely set. Anal fin long-based. HABITAT: Kotzebue Sound and eastern Bering Sea to western Gulf of Alaska. Also to northern Sea of Japan. Demersal over sandy and muddy bottoms from near shore to about 300 ft. Sometimes in kelp beds.

Rockhead
to 3.5 in.

♂ Warty Poacher
to 8 in.

Fourhorn Poacher
to 4.7 in.

Atlantic Poacher
to 9 in.

♂ Bering Poacher
to 9 in.

Pygmy Poacher - *Odontopyxis trispinosa* Lockington, 1880

FEATURES: Gray to tan, with six or seven darker bars dorsally. Sides with spots and blotches. Dorsal fins speckled and banded. Snout relatively short, with three spines at tip. Maxilla with a single barbel at margin. Top of head with two pits or one heart-shaped pit. Lower pectoral-fin rays elongate, incised. Anal fin with seven rays. Head and body plates spiny. HABITAT: Prince William Sound, AK, to central Baja California. Found demersally over sandy and muddy bottoms from near shore to about 1,200 ft.

Tubenose Poacher - *Pallasina barbata* (Steindachner, 1876)

FEATURES: Shades of brown to green or gray dorsally. Ventral area white to golden. Often with a dark stripe below eyes. Dorsal and pectoral fins spotted. Snout very long and tube-like. Lower jaw protrudes with a very short to very long barbel at tip. Maxilla margin without barbels. Dorsal fins touching at base or slightly separated. Body very long with smooth plates. HABITAT: Eastern Chukchi and Bering seas, St. Lawrence and Aleutian Islands to Bodega Bay, CA. Also to Wonsan, North Korea. Found demersally, in the water column, around eelgrass and algae, and over sandy and gravel bottoms from intertidal zone to about 420 ft.

Dragon Poacher - *Percis japonica* (Pallas, 1769)

FEATURES: Shades of tan to olive with four or five variable darker saddles dorsally. Sides and ventral area may have several dark blotches. Dark bands radiate from eyes. Fins marbled and banded. Snout short, barbel at tip absent. Spiny dorsal-fin base elevated. Dorsal fins well separated. Body plates with large spines. Head and body comparatively long and low. HABITAT: Occur at scattered locations from Bering Sea to Gulf of Alaska. Also to Gulf of Anadyr, Sea of Okhotsk, and Sea of Japan. Found demersally over hard and soft bottoms from about 65 to 2,500 ft.

Sturgeon Poacher - *Podothecus accipenserinus* (Tilesius, 1813)

FEATURES: Shades of brown, with irregular dark brown to blackish bars dorsally. Ventral area white to orange. Dorsal and caudal fins may have blackish margins. Snout long with two pairs of spines at tip. Dense clusters of 13–19 yellow barbels under snout and clusters of white barbels at corners of mouth. Upper preopercular spine not laterally expanded. Pectoral fins with 17–19 rays. Head spiny. Body plate spines low. HABITAT: Bering Sea and Aleutian Islands to Point Reyes, CA. Also from Cape Navarin to northern Kuril Islands. Occur demersally over soft bottoms from about 7 to 1,640 ft.

Veteran Poacher - *Podothecus veternus* Jordan & Starks, 1895

FEATURES: Brown dorsally with numerous darker bars and blotches. Pale below. Fins with dark margins. Snout long with two pairs of spines. Clusters of 6–11 white barbels under snout and clusters of white barbels at corners of mouth. Upper preopercular spine laterally expanded. Pectoral fins with 15–17 rays. Head spiny. Body plate spines low. HABITAT: Found from scattered locations of Beaufort and Chukchi seas to eastern Bering Sea. Also to Sea of Japan. Occur demersally over soft bottoms from about 30 to 790 ft.

Pygmy Poacher
to 5 in.

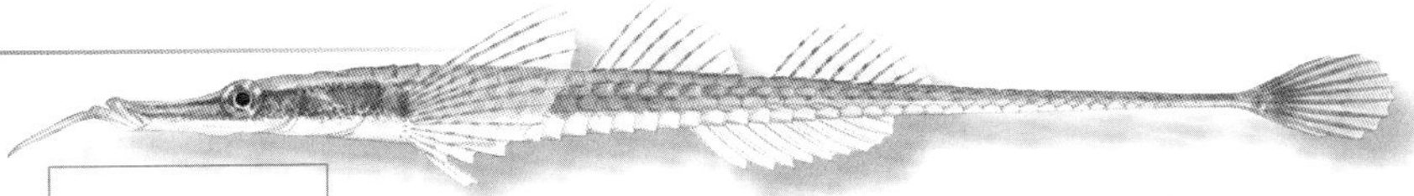

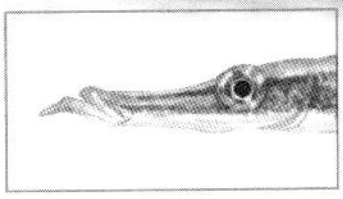

barbel variation

Tubenose Poacher
to 8.2 in.

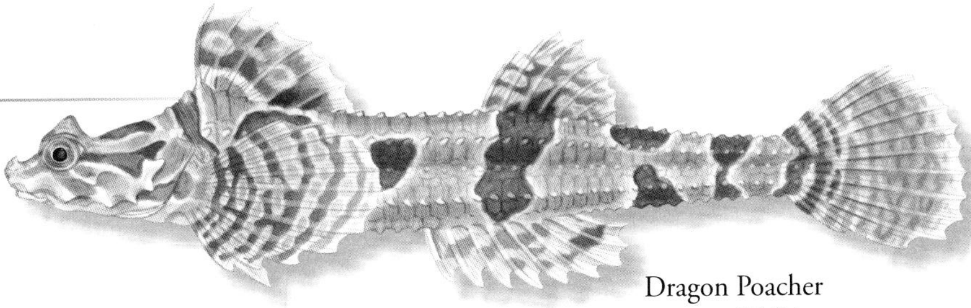

Dragon Poacher
to 17.3 in.

Sturgeon Poacher
to 13 in.

Veteran Poacher
to 11.2 in.

Sawback Poacher - *Sarritor frenatus* (Gilbert, 1896)

FEATURES: Shades of brown to gray dorsally with several dark blotches on back. Head with small, dark spots. Dorsal and pectoral fin brownish to blackish, with faint, pale spots. Snout blunt with small spines at tip and a pair of barbels underneath. Chin and maxilla with five pairs of barbels. Head spiny. Body plates with strong, sharp spines. Anal fin with six to eight rays. HABITAT: Norton Sound, Bering Sea, and Aleutian Islands to Observatory Inlet, British Columbia. Also to Hokkaido, Japan. Occur demersally from about 60 to 3,200 ft. Usually over continental shelf from about 60 to 3,200 ft.

Longnose Poacher - *Sarritor leptorhynchus* (Gilbert, 1896)

FEATURES: Dark brown with darker bars and blotches dorsally. White below. First dorsal fin with a white and dark brown to blackish band. Second dorsal fin spotted. Upper pectoral fins dark, lower rays white. Caudal fin dark brown. Snout long with small spines at tip and a pair of barbels underneath. Lower jaw with a pair of barbels, and maxilla with four pairs of barbels. Head spiny. Body plates comparatively low. Anal fin with five to eight rays. HABITAT: Southeastern Bering Sea and Aleutian Islands to Prince William Sound, AK. Found demersally over a variety of bottoms from about 50 to 1,510 ft.

Pricklebreast Poacher - *Stellerina xyosterna* (Jordan & Gilbert, 1880)

FEATURES: Olivaceous with darker speckles dorsally. White below. Pectoral fins with a broad, dark, inner band. Dorsal fins with blackish margins. Caudal fin blackish. Posterior anal-fin margin black. Mouth upturned, with protruding lower jaw. A single barbel present at maxilla margin. Head large and depressed. Chest with numerous small prickles. HABITAT: Northern Gulf of Alaska to northern Baja California. Demersal over muddy and sandy bottoms from near shore to about 330 ft. Usually above 250 ft.

Blacktip Poacher - *Xeneretmus latifrons* (Gilbert, 1890)

FEATURES: Tan to brown, with five to eight dark blotches dorsally. Ventral area pale. First dorsal fin with black margin of uniform width. Snout short, with three spines. Lower jaw with two pairs of small barbels. Maxilla margin with a single barbel. Lower four to five pectoral-fin rays moderately free. Anal fin with six to nine rays; originates under dorsal-fin interspace. HABITAT: Rennell Sound and Skidgate Channel, British Columbia, to northern Baja California. Demersal over sandy and mud bottoms from near shore to about 4,235 ft.

Bluespotted Poacher - *Xeneretmus triacanthus* (Gilbert, 1890)

FEATURES: Olive brown, with about six faint blotches dorsally. Pale below. Pale blue spots on and behind head. Dorsal fins dusky to clear, with darker rays. Snout short, with three spines. Lower jaw rarely with three pairs of short barbels. Maxilla margin with two barbels. Anal fin with 5–7 rays; originates under soft dorsal-fin origin. HABITAT: Kwatna Inlet, British Columbia, to northern Baja California. Demersally over sandy and muddy bottoms from about 50 to 1,400 ft.

ALSO IN THE AREA: Smootheye Poacher, *Xeneretmus leiops*, page 349.

Sawback Poacher
to 11.4 in.

Longnose Poacher
to 7.9 in.

Pricklebreast Poacher
to 6.5 in.

Blacktip Poacher
to 7.5 in.

Bluespotted Poacher
to 7 in.

Spinyhead Sculpin - *Dasycottus setiger* Bean, 1890

FEATURES: Pinkish gray, with irregular gray to dark brown spots and marbling that form lose bands and bars on head and body. Fins irregularly spotted and banded. Head large. Mouth large and broad. Four strong and variably directed spines above eyes. Head with large single or double spines. Preopercle with four spines. Numerous cirri on head and body. Dorsal fins separate or slightly connected. Spiny scales below first dorsal fin, and spiny plates scattered below lateral line. HABITAT: Bering Sea and Aleutian Islands to WA. Also to Sea of Japan. Demersal over soft bottoms from about 50 to 2,800 ft.

Smoothcheek Sculpin - *Eurymen gyrinus* Gilbert & Burke, 1912

FEATURES: Color highly variable. Observed specimens mottled in mixed shades of brown, gray, red, and tan. Some with scattered white spots. Caudal fin with a large pale area at center. Other fins variably banded and blotched. Mouth large, with protruding lower jaw. Thick, branched cirri present under chin and at maxilla margin. Head with numerous fleshy cirri and raised pores. Spines on head and preopercle absent. Dorsal fin continuous but may be notched. HABITAT: Chukchi and Bering seas, Aleutian Islands, and western Gulf of Alaska. Also to South Korea. Demersal from about 7 to 1,300 ft.

Blackfin Sculpin - *Malacocottus kincaidi* Gilbert & Thompson, 1905

FEATURES: Tannish to grayish, with about three broad, broken darker saddles dorsally. Ventral area with pale blotches. Dorsal, caudal, and pectoral fins with dark brown to blackish inner margins. Head large and rounded, with cirri absent or with a few minute cirri. Preopercle with three spines and a fourth spine that is small to obscure. Anterior accessory preopercular spine absent. Anal fin with 11–12 rays. Lateral line with 14–17 pores. HABITAT: British Columbia to Puget Sound, WA. Occur demersally on soft, silty, and shelly bottoms in shallow water.

Darkfin Sculpin - *Malacocottus zonurus* Bean, 1890

FEATURES: Tannish to grayish with three to four vague, darker saddles dorsally. Ventral area with irregular pale blotches. Dorsal, caudal, and pectoral fins with dark brown to blackish inner margins. Head large and rounded with cirri above eyes and numerous cirri on top and sides. Preopercle with four spines and an anterior accessory spine. Anal fin with 9–12 rays. Lateral line with 12–20 pores. HABITAT: Bering and Aleutian Islands to WA. Reported to OR. Also to South Korea. Demersal over soft and hard bottoms from about 90 to 6,500 ft. Usually between about 330 to 1,640 ft.

Tadpole Sculpin - *Psychrolutes paradoxus* Günther, 1861

FEATURES: Creamy to pale brown with fine stippling and distinct dark brown bars and blotches on body. Ventral area pale. A dark V shape usually present at caudal-fin base. Fins loosely banded. Spines absent from head and preopercle. Cirri absent. Dorsal fin continuous, with spiny portion low to mostly buried in skin. Head, body, and fins with numerous wart-like papillae. HABITAT: Norton Sound and Bering Sea to Puget Sound, WA. Also to Sea of Japan. Demersal over soft and hard bottoms from about 20 to 700 ft.

Spinyhead Sculpin
to 17.7 in.

Smoothcheek Sculpin
to 15.3 in.

Blackfin Sculpin
to 4.2 in.

Darkfin Sculpin
to 13.8 in.

Tadpole Sculpin
to 2.6 in.

Psychrolutidae - Fathead Sculpins, *cont.*

Soft Sculpin - *Psychrolutes sigalutes* (Jordan & Starks, 1895)

FEATURES: Brownish to pinkish or gray, with fine stippling. May have vague darker marks on body. Capable of rapid color change. Fins vaguely banded. Spines on head and preopercle absent. Dorsal fin continuous. Bases of dorsal, anal, pectoral, and pelvic fins buried in skin. Pectoral fins with 14–18 rays. Anal fin with 12–15 rays. Skin covered with minute wart-like papillae. HABITAT: Adak Island, AK, to Puget Sound, WA. Also to Commander Islands. Occur around rocky areas and sponges and over silty bottoms from surface to about 740 ft.

Cyclopteridae - Lumpfishes

Smooth Lumpsucker - *Aptocyclus ventricosus* (Jordan & Starks, 1895)

FEATURES: Shades of blue gray to gray dorsally. May be almost uniformly colored or with numerous, small, blackish spots. Fins dusky. Head and body round. First dorsal fin completely embedded in skin. Pectoral fins broad and nearly unified at throat. Pelvic fins modified to form a strong sucking disk. Skin flabby and wrinkled when body is deflated, appears smooth when body is swollen. Spines and tubercles absent. HABITAT: Bering Sea and Aleutian Islands to Mathieson Channel, British Columbia. Also to South Korea. Occur at surface, pelagically, and demersally to about 1,650 ft. BIOLOGY: After spawning, most females die. Males guard and moisten eggs by spitting water on nests that are above high-tide line.

Pacific Spiny Lumpsucker - *Eumicrotremus orbis* (Günther, 1861)

FEATURES: Color varies. Observed specimens shades of gray, white, yellow, pink, or purplish. Females described as shades of pale to dark green; males described as shades of orange to red or reddish brown. Head and body round and covered with numerous large and small rough, conical tubercles. Tubercles generally larger and more numerous in females. Several cirri present on lower jaw. Spiny dorsal fin covered with thin skin and with small tubercles on spines. Pectoral fins broad and nearly unified at throat. Pelvic fins form a strong sucking disk. HABITAT: Bering Sea and Aleutian Islands to Puget Sound, WA. Also to Sea of Okhotsk and off Hokkaido, Japan. Found pelagically and demersally over sandy, pebble, and rocky bottoms from near shore to about 1,180 ft.

Toad Lumpsucker - *Eumicrotremus phrynoides* Gilbert & Burke, 1912

FEATURES: Shades of tan to brown or golden, with dense covering of minute darker flecks and scattered spots. Head and body round and covered with numerous wart-like tubercles. Some tubercles with rosettes of short spines, others concealed. Pectoral fins broad and nearly unified at throat. Pelvic fins form a strong sucking disk. Skin appears mostly smooth when the body is swollen. HABITAT: Aleutian Islands and western Gulf of Alaska. Occur demersally over sandy, shelly, and gravel bottoms from about 230 to 770 ft. NOTE: May be a junior synonym of *Eumicrotremus gyrinops*.

NOTE: There are six or seven other rare Lumpfishes in the area.

Fathead Sculpins

Soft Sculpin
to 3.3 in.

Lumpfishes

Smooth Lumpsucker
to 16.5 in.

♀

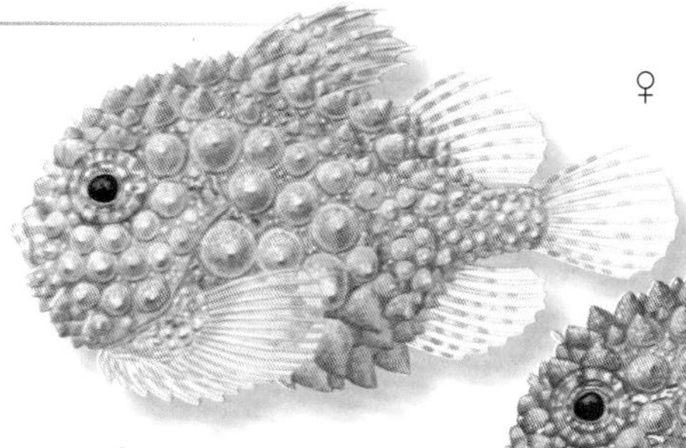

♂

Pacific Spiny Lumpsucker
to 5 in.

Toad Lumpsucker
to 3 in.

Gray Snailfish - *Careproctus canus* Kido, 1985

FEATURES: Head and body pale to dark grayish to brownish gray. Caudal fin and posterior portions of dorsal and anal fins dark brown to blackish. Peritoneum pale. Snout slightly projecting. Opercular flap absent. Gill opening entirely above pectoral fin. Pectoral fins shallowly notched. Pelvic fins form a sucking disk. Posterior dorsal- and anal-fin rays extend over two-thirds of caudal fin. HABITAT: Known from western to central Aleutian Islands. Demersal from about 160 to 1,400 ft.

Alaska Snailfish - *Careproctus colletti* Gilbert, 1896

FEATURES: Reddish to reddish black and finely dotted with black. Lips, mouth, and gill cavity dusky. Peritoneum black. Caudal fin and posterior portions of dorsal and anal fins black. Snout slightly projecting. Gill opening entirely above pectoral fin. Pectoral fins notched, with lower rays very long, slender, and free. Pelvic disk small, triangular. HABITAT: Bering Sea to northern Gulf of Alaska. Also to Sea of Japan and Sea of Okhotsk. Demersal over muddy and sandy bottoms from about 210 to 5,105 ft.

Emarginate Snailfish - *Careproctus furcellus* Gilbert & Burke, 1912

FEATURES: Head and body pinkish. Caudal fin and posterior margins of pectoral, dorsal, and anal fins black. Gill cavity dusky. Peritoneum black. Snout projects slightly. Gill opening entirely above pectoral fin. Pectoral fins unnotched, with a nearly horizontal base, and with lower rays free at tips. Pelvic disk small, triangular. Caudal-fin margin nearly straight. HABITAT: Bering Sea and in Pacific Ocean off Aleutian Islands to Gulf of Alaska. Reported to northern CA. Also to Sea of Okhotsk and Hokkaido, Japan. Occur from about 320 to 4,200 ft.

Blacktail Snailfish - *Careproctus melanurus* Gilbert, 1892

FEATURES: Head and body shades of pale pink to deep rosy red. Caudal fin and posterior margins of dorsal and anal fins black. Underside of pectoral fins blackish. Lips and front of snout often blackish. Gill cavity and peritoneum black. Snout blunt, not projecting. Gill opening entirely above pectoral fin. Pectoral fins with a horizontal base and free lower rays. Pelvic disk small, oval. HABITAT: Southern Bering Sea and in Pacific Ocean off Aleutian Islands to Baja California. Also off Honshu, Japan. Demersal from about 200 to 7,500 ft.

Salmon Snailfish - *Careproctus rastrinus* Gilbert & Burke, 1912

FEATURES: Head and upper body shades of deep pinkish red to yellowish pink or salmon. Abdomen pale to white. Peritoneum pale. Head round and steeply sloping. Gill opening extends over 1st to 6th pectoral-fin rays; rarely over entire fin. Pectoral fins notched with lower rays long and mostly free. Pelvic disk small, triangular. Caudal-fin margin straight. HABITAT: Bering Sea and Aleutian Islands to southeastern AK. Also to Sea of Okhotsk and Sea of Japan. Found over sandy, silty, and muddy bottoms from about 180 to 2,995 ft.

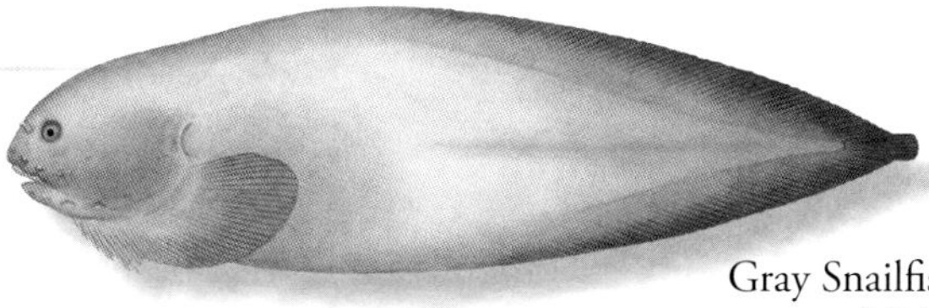

Gray Snailfish
to 6.3 in.

Alaska Snailfish
to 15.6 in.

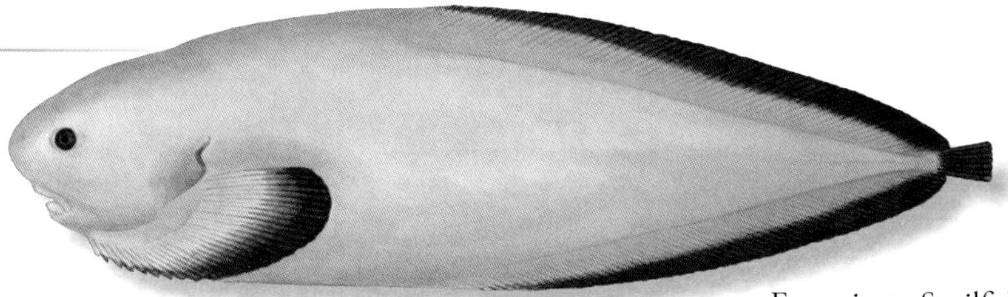

Emarginate Snailfish
to 21.3 in.

Blacktail Snailfish
to 13.8 in.

Salmon Snailfish
to 19.3 in.

Blotched Snailfish - *Crystallichthys cyclospilus* Gilbert & Burke, 1912

FEATURES: Pinkish, yellowish, or white, with numerous round and oval spots in shades of pink, orange, red, or brown. Spots may be ocellated. Snout short and steep. Opercular flap absent. Gill opening entirely above pectoral fin. Pectoral fins notched, with lower rays thick and free at tips. Caudal-fin margin nearly straight. HABITAT: Bering Sea and Aleutian Islands to south of Shumagin Islands, Gulf of Alaska. Also to Sea of Okhotsk. Occur over sand, mud, and gravel bottoms from about 180 to 1,900 ft.

Dimdisk Snailfish - *Elassodiscus tremebundus* Gilbert & Burke, 1912

FEATURES: Head and body translucent rosy red. Mouth, gill cavity, peritoneum, and stomach blackish. Caudal fin and posterior portions of dorsal and anal fins black. Snout short and steep. Gill opening above pectoral fin, rarely extending to 1st or 4th ray. Pectoral fins notched, with lower rays relatively short. Pelvic disk minute. HABITAT: Eastern Bering Sea and Aleutian Islands. Also to Hokkaido, Japan. Demersal over continental slopes from about 425 to 4,220 ft.

Spotted Snailfish - *Liparis callyodon* (Pallas, 1814)

FEATURES: Dark to pale olive brown, reddish brown, or brown, with minute flecks and irregular dark spots that may form loose rows. Fins irregularly spotted and banded. Gill opening above pectoral fin or extending over 1st to 3rd ray. Pectoral fins short, broad, and notched. Pelvic disk round and moderately large. Dorsal fin notched with blunt anterior lobe. Dorsal and anal fins barely reach caudal fin. HABITAT: Bering Sea and Aleutian Islands to OR. Also to Kamchatka. Occur demersally over rocky intertidal areas and in tidepools.

Ribbon Snailfish - *Liparis cyclopus* Günther, 1861

FEATURES: Shades of olive green to grayish green, with darker and lighter fine speckles and flecks. Abdomen sometimes white. Sometimes with four pale stripes on sides. Gill opening extends over 5th to 10th pectoral-fin rays—usually to 6th or 8th ray. Pectoral fins short, notched. Pelvic disk moderately large. Dorsal fin unnotched or barely notched. Dorsal and anal fins barely reach caudal fin. HABITAT: Southeastern Bering Sea and Aleutian Islands to OR. Also Commander Islands to Petropavlovsk, Kamchatka. Demersal in shallow water and rocky tidepools to about 600 ft.

Marbled Snailfish - *Liparis dennyi* Jordan & Starks, 1895

FEATURES: Color and pattern highly variable. Three described forms: tan, olive, or brown to blackish with dense marbling; pale with dark wavy stripes; tannish with whitish wavy lines. All with a short, skin-covered spine on opercle. Gill opening extends over 10th to 20th pectoral-fin rays. Pectoral fins large, broad, and notched. Pelvic disk moderately large. Dorsal fin relatively high anteriorly but without a notch; posterior rays extend over one-third of caudal fin. Anal fin with 29–34 rays; extends to middle of caudal fin. HABITAT: Eastern Aleutian Islands to Puget Sound, WA. Not confirmed from Bering Sea. Demersal from near shore to about 740 ft. Rarely in tidepools.

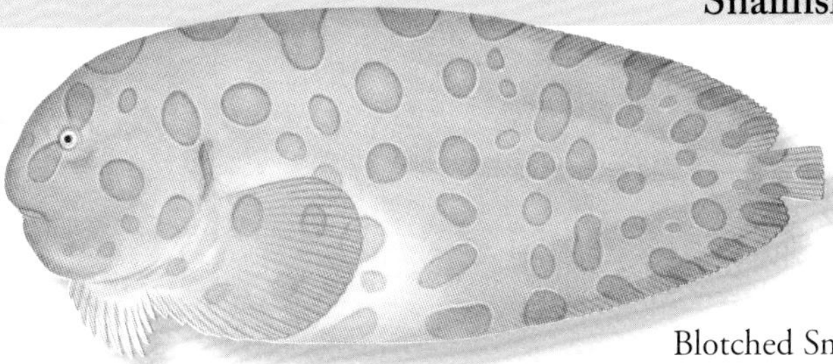

Blotched Snailfish
to 11.3 in.

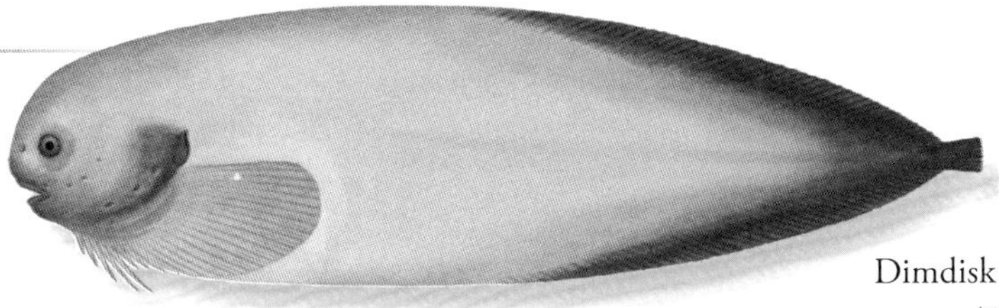

Dimdisk Snailfish
to 15.4 in.

Spotted Snailfish
to 5 in.

Ribbon Snailfish
to 4.5 in.

variants

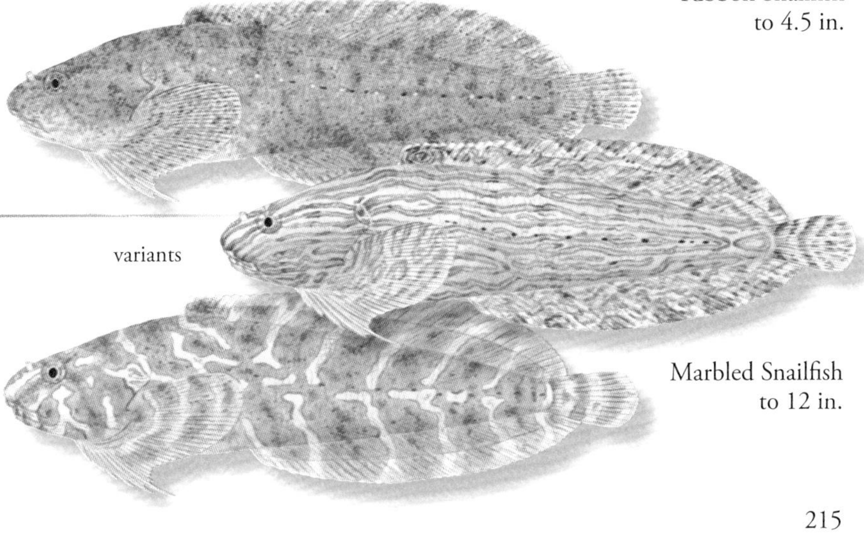

Marbled Snailfish
to 12 in.

Gelatinous Seasnail - *Liparis fabricii* Krøyer, 1847

FEATURES: Head and body uniformly dark gray to dark brown or black, or with variable dark blotches. Dorsal and anal fins dark or with dark bands. Snout steep. Gill opening extends over 3rd to 16th pectoral ray. Pectoral fins notched with lower rays free at tips. Dorsal fin unnotched or with a slight notch. Dorsal and anal fins extend to about middle of caudal fin. HABITAT: Circumpolar. Arctic Alaska. Pelagic and demersal from near shore to beneath pack ice over bottom depths of more than 6,560 ft.

Tidepool Snailfish - *Liparis florae* (Jordan & Starks, 1895)

FEATURES: Generally uniformly colored in shades of olive brown, yellow brown, reddish brown, to purplish. Gill opening extends over 2nd to 5th pectoral rays. Pectoral fins notched with lower rays free at tips. Pelvic disk moderately large. Dorsal fin with anterior lobe. Dorsal and anal fins barely extend over caudal fin. Anus closer to disk than to anal fin. Body deepest at anal-fin origin. HABITAT: Kodiak Island, AK, to Point Conception, CA. Demersal in intertidal zone and in exposed coastal tidepools.

Slipskin Snailfish - *Liparis fucensis* Gilbert, 1896

FEATURES: Head and body shades of pale olive brown to dark brown, with faint mottling and often tinged with pink. Fins irregularly banded. Caudal fin sometimes with a pale diagonal band. Gill opening extends over 12th to 16th pectoral rays. Pectoral fins notched, with 38–43 rays, and with lower rays elongate and free at tips. Pelvic disk moderately large. Dorsal fin usually notched. Dorsal and anal fins barely attached to caudal fin. HABITAT: SE Bering Sea and Unimak Pass, AK, to San Simeon Point, CA. Demersal from near shore to about 1,270 ft. Juveniles often occur in rocky tidepools.

Variegated Snailfish - *Liparis gibbus* Bean, 1881

FEATURES: Color and pattern highly variable. Pale, with dense brownish to blackish mottling; pink to brown with tan stripes, or brownish with large blackish blotches. Peritoneum pale. Gill opening extends over 8th to 16th pectoral rays—usually to the 14th. Pectoral fins broad, notched, with lower rays elongate and free at tips. Dorsal fin unnotched. Dorsal and anal fins extend well onto caudal fin. Head very broad. Body deepest under dorsal-fin origin. HABITAT: Circumpolar. Arctic Alaska to northern British Columbia. Demersal over hard and soft bottoms on continental shelves and upper slopes from about 80 to 1,400 ft.

Lobefin Snailfish - *Liparis greeni* (Jordan & Starks, 1895)

FEATURES: Head and body olivaceous, reddish, brown, gray, or purplish brown. Lower jaw and chest paler. Head very broad. Eyes comparatively small. Opercle with skin-covered spine. Gill opening extends over 1st to 3rd pectoral rays, sometimes located above fin. Pectoral fins notched, lower rays free at tips. Pelvic disk comparatively small. Dorsal fin notched with a distinct anterior lobe. Dorsal and anal fins barely reach caudal fin. HABITAT: Pribilof and Aleutian Islands to WA. Also Commander Islands. Demersal in tidepools and nearshore shallow waters. Also known as *Polypera greeni*.

Gelatinous Seasnail
to 7.6 in.

Tidepool Snailfish
to 7.2 in.

Slipskin Snailfish
to 7 in.

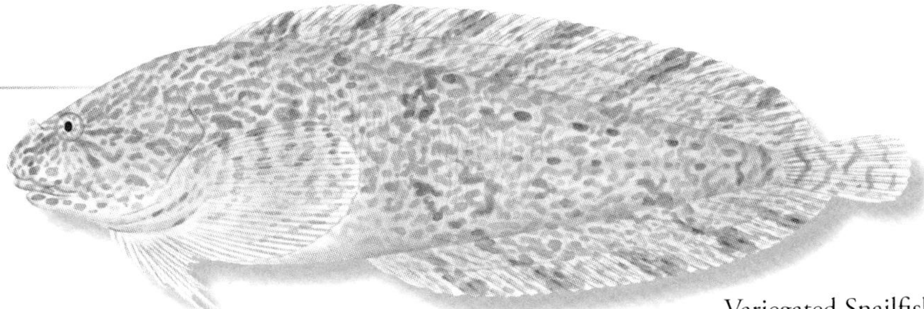

Variegated Snailfish
to 20.6 in.

Lobefin Snailfish
to 12.1 in.

Thumbtack Snailfish - *Liparis micraspidophorus* (Gilbert & Burke, 1912)

FEATURES: Head and body reddish brown to olive gray. Dorsal, caudal, and anal fins indistinctly speckled and barred. Gill opening extends over 4th to 5th pectoral rays. Pectoral fins notched, with lower rays comparatively short. Pelvic disk relatively large. Dorsal fin notched, with distinct anterior lobe. Dorsal and anal fins do not reach caudal-fin rays. Body with or without scattered thumbtack-shaped prickles. HABITAT: Aleutian Islands to Cold Bay, AK. Also Commander Islands. Demersal in tidepools and intertidal areas.

Slimy Snailfish - *Liparis mucosus* Ayres, 1855

FEATURES: Color and pattern highly variable. Uniformly colored in yellow, dark brown, reddish, olive, gray, or pink, to speckled and reticulated, or with dark stripes. All usually with a dark bar at caudal-fin base. Gill opening above pectoral fins or over 1st to 6th rays; usually to below 3rd ray. Pectoral fins shallowly notched. Pelvic disk very large. Dorsal and anal fins barely extend onto caudal fin, if at all. HABITAT: Samsing Cove, AK, to Baja California. Demersal from intertidal zone to about 50 ft. Not usually found on tidepools.

Showy Snailfish - *Liparis pulchellus* Ayres, 1855

FEATURES: Color and pattern variable. Uniformly colored in shades of pale to dark or reddish brown, to mottled, or with dark stripes. Fins may be spotted. Gill opening above pectoral fins or extends over 1st to 7th rays; usually only to 4th ray. Pectoral fins notched. Pelvic disk moderately large. Dorsal fin unnotched. Dorsal and anal fins extend almost to caudal-fin margin. HABITAT: Southeastern Bering Sea and Aleutian Islands to Monterey Bay, CA. Demersal over soft bottoms from intertidal zone to about 600 ft.

Ringtail Snailfish - *Liparis rutteri* (Jordan & Starks, 1898)

FEATURES: Head and body bluish black to brown; may have wavy streaks. Usually with whitish margins on pectoral, dorsal, and anal fins and with a whitish bar at caudal-fin base. Gill opening entirely above pectoral fin. Pelvic disk large. Anterior dorsal fin in females low or slightly notched; in males, anterior rays may be elongate and separate. Dorsal and anal fins barely extend onto caudal fin. Skin loose and wrinkled. HABITAT: Aleutian Islands to British Columbia. Demersal in tidepools and shallow intertidal zones.

Kelp Snailfish - *Liparis tunicatus* Reinhardt, 1836

FEATURES: Color and pattern variable. Uniformly yellowish brown, with dark fin margins; boldly banded and blotched with dark brown, or with creamy stripes. Peritoneum pale. Gill opening above pectoral fin or extends to 12th ray; usually to 4th or 6th ray. Pelvic disk moderately large. Dorsal fin unnotched and extends onto first quarter of caudal fin. Anal fin extends sightly further than dorsal. HABITAT: Circumpolar. Arctic Alaska to south of St. Lawrence Island, AK. Demersal in tidepools and coastal waters to about 490 ft. Commonly found clinging to kelp.

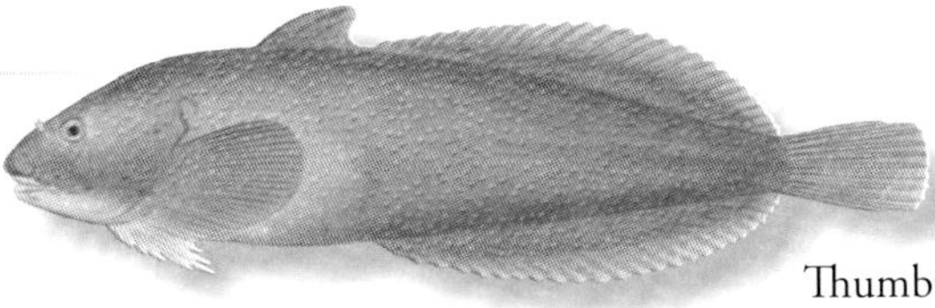

Thumbtack Snailfish
to 3.9 in.

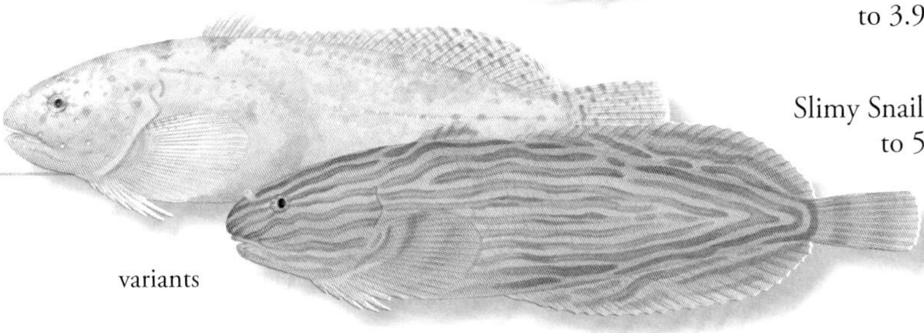

Slimy Snailfish
to 5 in.

variants

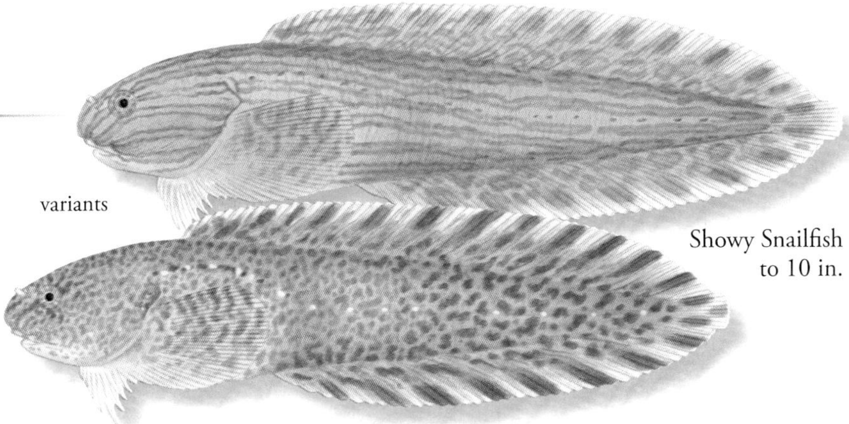

variants

Showy Snailfish
to 10 in.

Ringtail Snailfish
to 2.8 in.

Kelp Snailfish
to 6.6 in.

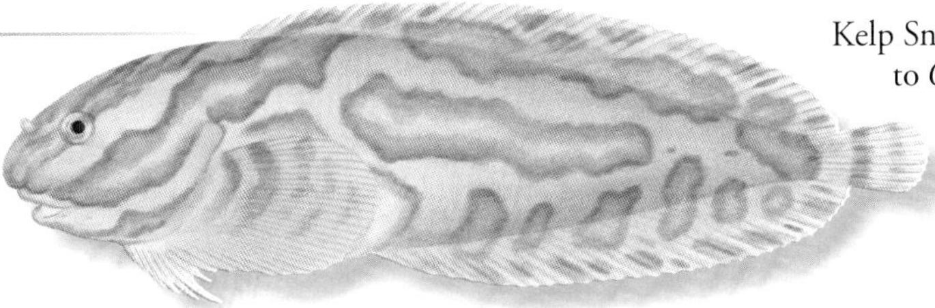

Liparidae - Snailfishes, *cont.*

Pygmy Snailfish - *Lipariscus nanus* Gilbert, 1915

FEATURES: Mouth area, gills, and abdomen black. Snout, body, and tail dusky. Small blackish flecks scattered on body and fins, very dense posteriorly. Skin translucent. Gill opening above pectoral fins. Pectoral fins notched and with lower two rays free. Pelvic disk absent. Prickles present on body and fins. HABITAT: Eastern Bering Sea to Monterey Bay, CA. Also to Hokkaido, Japan. Pelagic at depths between about 190 and 3,000 ft.

Tadpole Snailfish - *Nectoliparis pelagicus* Gilbert & Burke, 1912

FEATURES: Lower head and abdomen silvery. Body and tail tannish to brownish. Small blackish flecks scattered on body and fins. Skin translucent, naked. Gill opening in front of pectoral fins. Pectoral fins with two distinct lobes; lower rays short to long and free. Pelvic disk absent. HABITAT: Bering Sea to Santa Barbara, CA. Also to Sea of Okhotsk. Pelagic from surface to about 11,200 ft. Usually in mid-water.

NOTE: There are about eleven other poorly recorded, shallow-water Snailfishes in the area.

Moronidae - Temperate Basses

Striped Bass - *Morone saxatilis* (Walbaum, 1792)

FEATURES: Grayish green to steel blue, with brassy highlights dorsally. Sides silvery, with about seven black to dark olive stripes along scale rows. Pelvic fins white, other fins dusky. Head profile relatively straight. Body oblong, slightly compressed. HABITAT: Barkley Sound, British Columbia, to just south of the US-Mexico border. Also in Atlantic from Canada to east TX. Occur in rivers, estuaries, and nearshore waters. Also in shallow bays, along beaches, and over rocky areas. Some populations are landlocked. BIOLOGY: Anadromous and seasonally migratory. A voracious predator. Sought commercially and for sport. Introduced into the area. Other name: Striper.

Polyprionidae - Wreckfishes

Giant Sea Bass - *Stereolepis gigas* Ayres, 1859

FEATURES: Adults shades of gray to dark brown or silvery, with large black spots and blotches dorsally and on sides. Ventral area usually whitish. May rapidly change color. Mouth large, with maxilla extending to at least middle of eye. Two spines present on opercle; may be embedded in skin. Head large and broad. Spiny dorsal fin low. Soft dorsal fin with 10–11 rays. Body very robust. Juveniles reddish to golden, with black and white spots. Dorsal fin tall. Pelvie fins fan-like. Body deep. HABITAT: Humboldt Bay, CA, to Oaxaca, Mexico, including Gulf of California. Reports from Sea of Japan are erroneous. Found over rocky and sandy bottoms, around kelp beds, and along drop-offs from near shore to about 150 ft. BIOLOGY: Known to migrate inshore in spring and offshore in fall. Visit a variety of fish-cleaning stations. May grow to at least 75 years. Other name: Black Seabass. Listed as Critically Endangered.

Snailfishes

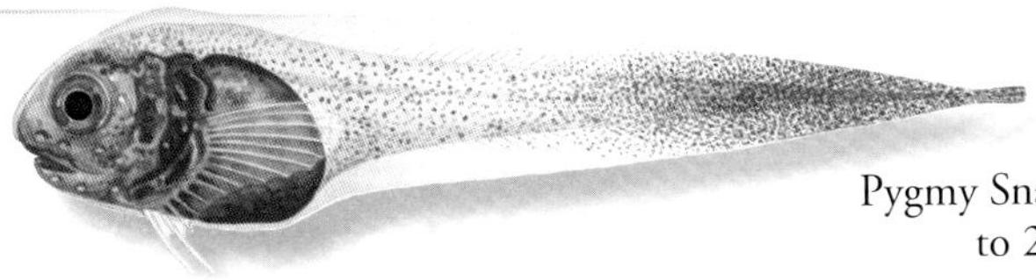

Pygmy Snailfish
to 2.8 in.

Tadpole Snailfish
to 2.5 in.

Temperate Basses

Striped Bass
to 6 ft.

Wreckfishes

Giant Sea Bass
to 7.4 ft.

juvenile

Splittail Bass - *Hemanthias peruanus* (Steindachner, 1875)

FEATURES: Rosy red dorsally. Sides pinkish and washed with yellow blotches and spots. Ventral area white. A yellow streak present below eyes. Fins rosy with yellow spots and margins. Mouth upturned, with lower jaw protruding. Third dorsal-fin spine long and filamentous. Caudal fin deeply forked, with middle rays of upper and lower lobes longest. Pectoral fins with 17–18 rays (rarely 16). HABITAT: Redondo Beach, CA, to Peru, including the Gulf of California. Found near bottom over rocky reefs and around reef edges from about 65 to 400 ft. Rare in the area.

Hookthroat Bass - *Hemanthias signifer* (Garman, 1899)

FEATURES: Rosy red dorsally. Sides pinkish and washed with yellow blotches and spots. Ventral area white. A yellow streak present below eyes. Fins rosy, with yellow spots and margins. Mouth upturned, with lower jaw projecting. A sharp, projecting, bony spine present at throat where lower edges of gill covers meet. Third dorsal-fin spine long and filamentous. Caudal fin deeply forked, with uppermost and lowermost rays longest. Pectoral fins usually with 19 rays. HABITAT: Playa del Rey, CA, to Talara, Peru, including Gulf of California. Demersal over rocky reefs and reef edges from about 75 to 1,000 ft.

Kelp Bass - *Paralabrax clathratus* (Girard, 1854)

FEATURES: Color varies with habitat and season. Brown to gray brown or olivaceous with white oblong blotches and pale spots dorsally. Ventral area silvery to golden. Fins dusky to yellowish or olivaceous. Can quickly change color. Breeding males usually have bright orange on the snout. Dorsal fin with 10–11 spines; 3rd spine about as long as 4th and 5th. Caudal-fin margin straight. HABITAT: Columbia River, WA, to southern Baja California. More common in southern range. Occur from near bottom to surface around a variety of hard structures, reefs, and kelp beds from near shore to about 150 ft.

Spotted Sand Bass - *Paralabrax maculatofasciatus* (Steindachner, 1868)

FEATURES: Females generally washed with pale to bright yellow and with yellow on chin and jaws. Background color of males usually whitish to gray. All with about seven dark bars and covered with dense blackish to brownish spots dorsally and on sides. Fins spotted. Juveniles with a broad dark stripe on sides. Dorsal fin with ten spines; 3rd spine tallest. Caudal-fin margin straight. HABITAT: Monterey, CA, to central Mexico, including Gulf of California. More common in southern embayments. Occur from interitdal zone to about 200 ft. Tolerant of extreme temperatures.

Barred Sand Bass - *Paralabrax nebulifer* (Girard, 1854)

FEATURES: Gray to olivaceous with irregular, oblique dusky to blackish bars dorsally. Most with orange brown spots on snout. Heavily blotched with brown, yellow, and silver at night. Juveniles with large spots forming bars and stripes. Third dorsal-fin spine longest. Caudal-fin margin straight. HABITAT: Santa Cruz, CA, to southern Baja California. Possibly also off Acapulco, Mexico. Occur in embayments and along coast over rocks and soft bottoms and around structures and kelp beds from shore and surface to about 200 ft. Other name: Sand Bass.

Splittail Bass
to 17.7 in.

Hookthroat Bass
to 16.5 in.

Kelp Bass
to 2.3 ft.

♀ Spotted Sand Bass
to 15 in.

juvenile

Barred Sand Bass
to 2.1 ft.

Serranidae - Sea Basses, *cont.*

Threadfin Bass - *Pronotogrammus multifasciatus* Gill, 1863

FEATURES: Rosy red, with dark to faint bronze bars dorsally. Fading to pink below, with yellow bars on lower sides. Dorsal fin rosy, with yellow margin and spots on soft portion. Caudal fin rosy with yellow margin. Mouth upturned. Caudal-fin margin with short to very long filaments. Pelvic fins may be very long and filamentous. HABITAT: Off eastern Anacapa Island and south of Point Muga, CA, to Peru, including Gulf of California. Over rocky reefs and around boulders and ridges from about 45 to 1,000 ft.

ALSO IN THE AREA: Greater Sand Perch, *Diplectrum maximum*; Deepwater Serrano, *Serranus aequidens*, page 350.

Epinephelidae - Groupers

Leather Bass - *Dermatolepis dermatolepis* (Boulenger, 1895)

FEATURES: Color and pattern somewhat variable. Shades of brown to grayish, with distinct to vague alternating dark and pale bars overlaid with irregular paler blotches and blackish spots. Head may be washed with yellow. Markings may be muted or may be very dark overall. Fin margins yellow. Juveniles with distinct blackish or brown and white bars. Snout bluntly pointed, head profile steep. Dorsal fin continuous and slightly notched. Body very deep. HABITAT: El Segundo, CA, to northern Peru, including the Galápagos Islands. Found over rocky reeks from about 15 to 130 ft. More common in southern portion of range. BIOLOGY: Feed on fishes and decapods. Form schools and synchronized spawning aggregations.

Spotted Cabrilla - *Epinephelus analogus* Gill, 1863

FEATURES: Pale tan to gray, with five or six diffuse to distinct and irregular reddish to dark brown bars dorsally that extend onto fins. Head and body covered with brown to blackish spots. Often with a dark stripe along midline formed by close-set spots. Spots on juveniles large, spots on adults very small. Caudal, anal, pelvic, and pectoral fins mostly dark, spotted at bases. Dorsal fin continuous and slightly notched, with second through fourth spines about the same height. Caudal-fin margin broadly rounded. HABITAT: San Pedro, CA, to Peru, including Gulf of California. More common in southern portion of range. Juveniles in tidepools, estuaries, and lagoons. Adults occur over reefs. Found from near shore to about 350 ft.

Flag Cabrilla - *Epinephelus labriformis* (Jenyns, 1840)

FEATURES: Color and pattern vary. Uniformly olivaceous, grayish, or ochre to brown; or with distinct, irregular dark brownish bars on back and sides. All with large and small white spots generally arranged in rows on head and body, and a black blotch on upper caudal peduncle. Fins often with whitish spots and red margins. Spiny dorsal fin with white-tipped membranes. Dorsal fin with 11 spines and 15–16 rays. HABITAT: Southern California to Chile, including Gulf of California and Galápagos Islands. Occur over rocky bottoms and reefs from near shore and in tidepools to about 165 ft.

Sea Basses

Threadfin Bass
to 10.3 in

Groupers

Leather Bass
to 3.3 ft.

juvenile

Spotted Cabrilla
to 2.8 ft.

Flag Cabrilla
to 20.3 in.

Gulf Coney - *Hyporthodus acanthistius* (Gilbert, 1892)

FEATURES: Uniformly brick red to brown or grayish brown, with a narrow dark "mustache" above upper jaw. Juveniles with pale flecks and enlarged soft dorsal and anal fins. Preopercular corner serrated. Spiny dorsal fin with nine deeply-incised spines—third spine very tall. Anal fin with nine rays. Caudal fin broadly rounded. Body deep and robust. HABITAT: North of Santa Barbara, CA, to Peru, including Gulf of California. Occur over rocky reefs and adjacent sandy bottoms from near shore to about 365 ft. BIOLOGY: Overfished in the Gulf of California. Listed as Endangered.

Star-studded Grouper - *Hyporthodus niphobles* (Gilbert & Starks, 1897)

FEATURES: Brown to gray brown, with four rows of bright to diffuse and faded spots on upper sides. Large adults often lack spots. Dark "mustache" above upper jaw and a black blotch present on upper caudal peduncle. Lower preopercular corner serrated and angular. Spiny dorsal fin with 11 spines, tall anteriorly, and deeply incised. Soft dorsal fin with 13–15 rays. Anal fin with nine rays. Pectoral fins with 17–21 rays. Caudal-fin margin broadly rounded. Body deep and robust. HABITAT: Piedras Blancas, CA, to Peru, including Gulf of California and Galápagos Islands. Occur over rocky reefs and adjacent sandy bottoms from near shore to about 1,500 ft.

Gulf Grouper - *Mycteroperca jordani* (Jenkins & Evermann, 1889)

FEATURES: Greenish gray to greenish brown, with irregular rows of dark brown to olive brown blotches and spots on back and sides. Dark bars radiate from eyes. Fins dark, with very narrow white margins. Lower preopercular corner rounded. Dorsal fin with 11 spines and 15–17 rays. Anal fin with 10–11 rays and with lower corner squared to pointed. Pectoral fins usually with 17 rays. Caudal-fin margin nearly straight. HABITAT: La Jolla, CA, to Mazatlán, Mexico, including Gulf of California. Found over rocky reefs from near shore to about 165 ft. Usually in deep water during summer.

Broomtail Grouper - *Mycteroperca xenarcha* Jordan, 1888

FEATURES: Grayish to pale brown or grayish green, with irregular dark brown to olive brown blotches loosely arranged in vermiculations and broken rows. Dark bands radiate from eyes. Fins dark with spots at bases. Preopercular corner serrated. Soft dorsal and anal fins with elongate middle rays. Caudal-fin margins with short to very long and trailing filaments. Body robust. HABITAT: San Francisco, CA, to Peru, including Gulf of California. Rare in northern range. Juveniles in mangroves and embayments. Adults over rocky bottoms and near caves and crevices. From intertidal zone to about 230 ft.

Pacific Creolefish - *Paranthias colonus* (Valenciennes, 1846)

FEATURES: Greenish brown to grayish dorsally, fading to pinkish below. Five small white spots usually present on posterior body. Dark red blotch present at upper pectoral-fin base. Juveniles lemon yellow, with blackish spots posteriorly. Dorsal fin continuous. Pectoral fins long. Caudal fin deeply forked. HABITAT: San Diego, CA, to Peru, including Gulf of California. Found around rocky reefs and in water column from surface to about 330 ft. NOTE: Classified by some authors as *Cephalopholis colonus*.

Gulf Coney
to 4.3 ft.

Star-studded Grouper
to 3.6 ft.

Gulf Grouper
to 6.5 ft.

Broomtail Grouper
to 5 ft.

Pacific Creolefish
to 14 in.

juvenile

Priacanthidae - Bigeyes

Popeye Catalufa - *Pristigenys serrula* (Gilbert, 1891)

FEATURES: Irises, head, and body bright red. May display five faint to distinct pale saddles or bars on body. Tips of spiny dorsal fin white. Soft dorsal, caudal, and anal fins with deep red to black posterior margins. Mouth large and oblique. Eyes large. Spiny dorsal fin very tall and incised. Pelvic fins broad, with inner rays attached to abdomen by membrane. Caudal-fin margin rounded. Body very deep and laterally compressed. HABITAT: Newport, OR, to Chile, including Gulf of California and Galápagos Islands. More common in southern portion of range. Occur over rocky bottoms and high-relief reefs and in crevices from about 10 to 655 ft. Often found in caves and crevices. BIOLOGY: Nocturnal. Feed on small fishes and crustaceans.

Apogonidae - Cardinalfishes

Plain Cardinalfish - *Apogon atricaudus* Jordan & McGregor, 1898

FEATURES: Lower head and abdomen yellowish orange, back and upper peduncle violet to reddish blue. Tiny dark flecks cover head and body. Middle portion of spiny dorsal fin with a blackish blotch. Caudal fin violet to reddish blue and with dark flecks that may concentrate at tip of upper lobe. Mouth and eyes very large. Body and caudal peduncle comparatively elongate. HABITAT: San Clemente Island, CA, and Isla Guadalupe, Baja California, to Gulf of California and islands of eastern Pacific. Occur over rocky reefs from about 10 to 165 ft. BIOLOGY: Nocturnal. Feed on small crustaceans, mollusks, and other invertebrates. Mouth-brood eggs.

Guadalupe Cardinalfish - *Apogon guadalupensis* (Osburn & Nichols, 1916)

FEATURES: Lower head and abdomen bright yellow to deep orange; back and upper peduncle violet to bluish gray. Tiny dark flecks cover head and body. Fins unmarked, with membranes comparatively clear. Small specimens reported to have a dark blotch on spiny dorsal fin. Mouth and eyes very large. Body and caudal peduncle comparatively deep. HABITAT: Anacapa Island and Santa Barbara, CA, to tip of Baja California and Islas Revillagigedo. Found over rocky and reef bottoms and around crevices from about 30 to 100 ft. BIOLOGY: Nocturnal. Feed on planktonic crustaceans and small fishes. Mouth-brood eggs.

Pink Cardinalfish - *Apogon pacificus* (Herre, 1935)

FEATURES: Head and abdomen washed with yellow to orange. Body rosy to peach with darker speckles. A black stripe runs from snout through eyes. Pupils bordered above and below with bright golden. A black bar present below soft dorsal fin. Some also with a faint to dark saddle on upper peduncle. Fins unmarked. Mouth and eyes very large. HABITAT: San Clemete Island, CA, and mainland off Arrecife Sacramento, Baja California to southern Baja California and Peru, including Gulf of California. Found over high-relief rocky and reef bottoms and around crevices from about 3 to 315 ft. BIOLOGY: Nocturnal. Feed on small crustaceans and fishes. Mouth-brood eggs.

Bigeyes

Popeye Catalufa
to 13 in.

Cardinalfishes

Plain Cardinalfish
to 3.5 in.

Guadalupe Cardinalfish
to 5 in.

variants

Pink Cardinalfish
to 3.9 in.

Malacanthidae - Tilefishes

Pacific Golden-eyed Tilefish - *Caulolatilus affinis* Gill, 1865

FEATURES: Irridescent bluish gray to olive gray dorsally. Silvery below. Dorsal, caudal, and anal fins dusky with some yellow at margins. Pectoral fins dusky with blue at center and a black blotch at upper base. A bright yellow and bright blue band run from eyes to snout. Head profile comparatively steep, body comparatively deep. Pored lateral-line scales number 80–91. Caudal-fin margin nearly straight. HABITAT: Off Point Loma, CA, to Peru, including Gulf of California and Galápagos. Found over rocky and sandy bottoms from about 100 to 655 ft.

Ocean Whitefish - *Caulolatilus princeps* (Jenyns, 1840)

FEATURES: Pinkish tan to brownish dorsally and sometimes with a greenish luster. Whitish below. A pale white streak runs from eyes to snout. Dorsal and anal fins yellowish with a white band near margins. Caudal and pectoral fins yellowish. Head profile gently sloping, body comparatively slender. Pored lateral-line scales number 99–115. Caudal-fin margin concave in adults. HABITAT: Vancouver Island, WA, to Chile, including Gulf of California and Galápagos Islands. Occur over sandy bottoms, eelgrass beds, and reefs from about 10 to 490 ft. BIOLOGY: Feed on invertebrates and small fishes.

Nematistiidae - Roosterfish

Roosterfish - *Nematistius pectoralis* Gill, 1862

FEATURES: Irridescent blue to green or blue gray dorsally, with four to five distinct curved blackish bands on head and upper sides. Silvery on sides and below. Pectoral fins with a large blackish blotch on lower margin. Spiny dorsal fin very tall with trailing rays that are banded near base and black along tips. Caudal fin deeply forked. Pectoral fins very long and pointed. Body fusiform and robust. HABITAT: San Clemente, CA, to Isla San Lorenzo, Peru, including Gulf of California and Galápagos Islands. Found from near shore and surface to about 66 ft. Frequently occur along sandy beaches. Occasionally in estuaries. BIOLOGY: Sought as sport and food fish.

Coryphaenidae - Dolphinfishes

Dolphinfish - *Coryphaena hippurus* Linnaeus, 1758

FEATURES: Bright blue green above, golden to silvery below. Small dark spots on sides. Pectoral fins yellowish. Juveniles with bars that extend onto dorsal and anal fins. Capable of rapid color change. Dorsal fin tall, long-based. Caudal fin deeply forked. Dorsal and anal profiles nearly straight. Dorsal fin with 58–66 rays. Pectoral fins more than half of head length. Anal fin notched anteriorly. Males with an angular head profile, females with a rounded head profile. HABITAT: Circumglobal in tropical to warm temperate seas. In eastern Pacific from Grays Harbor, WA, to Chile, including Gulf of California and Galápagos Islands. Usually offshore and oceanic; may enter coastal waters. Associated with flotsam and floating seaweeds. BIOLOGY: Form small schools. Highly migratory. Feed on fishes, crustaceans, and squids. Other names: Dorado, Mahi-mahi.

Tilefishes

Pacific Golden-eyed Tilefish
to 20 in.

Ocean Whitefish
to 3.3 ft.

Roosterfish

Roosterfish
to 5.8 ft.

Dolphinfishes

♂

Dolphinfish
to 6.5 ft.

♀

Sharksucker - *Echeneis naucrates* Linnaeus, 1758

FEATURES: Dark bluish to brownish gray, with a dark stripe running from mouth to caudal-fin base that is bordered above and below by whitish stripes. White stripes may be obscure in larger specimens. Dorsal, caudal, and anal fins dark, with thin, pale to white margins. Cephalic disk with 21 to 28 laminae (usually 23). Caudal fin trilobed or with a nearly concave margin. HABITAT: Circumglobal in tropical to warm temperate seas. Southern CA to Chile, including Gulf of California and Galápagos Islands. BIOLOGY: Attach to a wide variety of hosts, including sharks, rays, bony fishes, sea turtles, and marine mammals. They are also free swimming.

Slender Suckerfish - *Phtheirichthys lineatus* (Menzies, 1791)

FEATURES: Dark brownish gray to black with a darker stripe running from mouth to caudal-fin base that is bordered above and below by whitish stripes. All fins with pale to white margins. Disk with 9 to 11 laminae (usually 10) and reaching only to about middle of pectoral fins. Head short, body very long. Caudal-fin margin slightly trilobed to rounded. HABITAT: Circumglobal in tropical to subtropical seas. Southern CA to Colombia, including Gulf of California and Galápagos Islands. BIOLOGY: Usually associated with barracudas but also with pufferfishes, other bony fishes, and sea turtles.

White Suckerfish - *Remora albescens* (Temminck & Schlegel, 1850)

FEATURES: Usually whitish. May be pale tannish or grayish. Cephalic disk wide and short, with 12 to 14 laminae and reaching to pectoral-fin tips. Dorsal-, anal-, and pectoral-fin margins rounded. Body robust. HABITAT: Circumglobal in tropical to warm temperate seas. San Francisco, CA, to Chile, including central and southern Gulf of California. BIOLOGY: Usually associated with Giant Manta and Black Marlin, with a few accounts of sharks as hosts. Previously known as *Remorina albescens*.

Whalesucker - *Remora australis* (Bennett, 1840)

FEATURES: Color varies. Uniform dark slate blue, bluish white, dark blue, grayish white, slate gray, to brown or almost black. Margins of dorsal, anal, and caudal fins may be white. Cephalic disk very large, about half as long as body and with 24 to 28 laminae (usually 26). Caudal-fin margin nearly straight to slightly lunate. Body robust. HABITAT: Worldwide in all seas (usually in warm seas). BIOLOGY: Associated with whales, porpoises, and dolphins. Previously known as *Remilegia australis*.

Spearfish Remora - *Remora brachyptera* (Lowe, 1839)

FEATURES: Color corresponds to host color. May be brownish or bluish brown to whitish. Dark specimens with pale dorsal- and anal-fin margins. Cephalic disk with 15 to 18 laminae, does not reach past pectoral-fin tips. Pectoral fins rounded. Caudal-fin margin slightly concave. Body robust. HABITAT: Circumglobal in tropical to warm seas. Southern CA to Chile, including Gulf of California and Galápagos Islands. Pelagic. BIOLOGY: Associated with billfishes but also with sharks and molas.

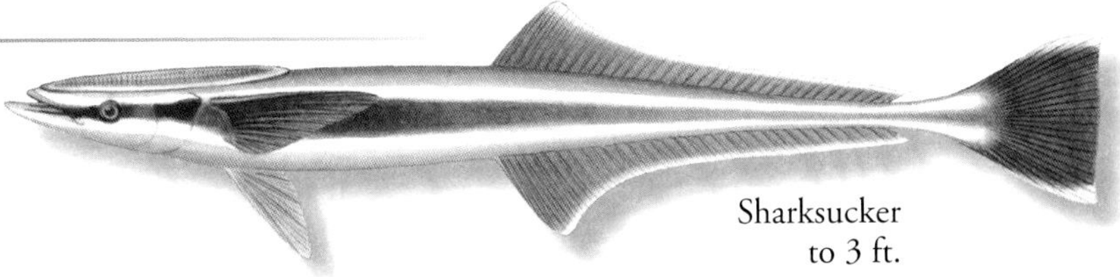

Sharksucker
to 3 ft.

Slender Suckerfish
to 17.6 in.

White Suckerfish
to 12 in.

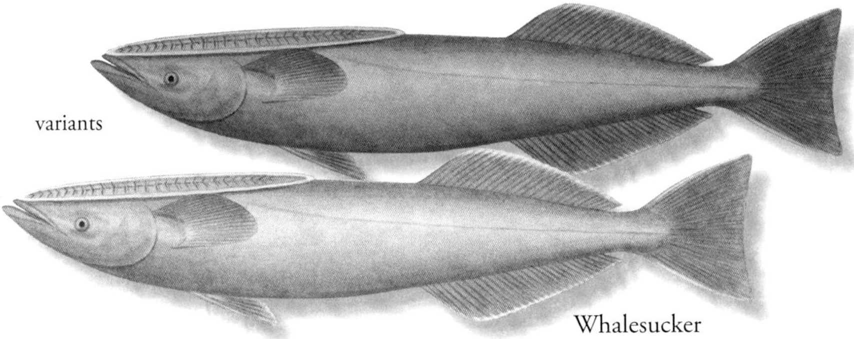

variants

Whalesucker
to 2.5 ft.

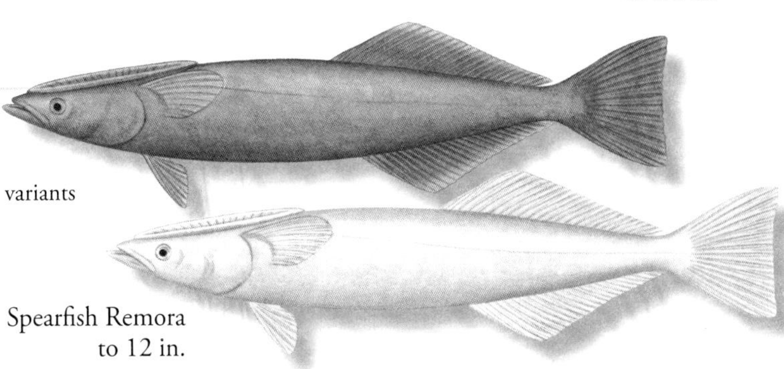

variants

Spearfish Remora
to 12 in.

Echeneidae - Remoras, *cont.*

Marlinsucker - *Remora osteochir* (Cuvier, 1829)

FEATURES: Brownish gray to grayish or blackish. Cephalic disk large, with 15 to 19 laminae and reaching well past pectoral-fin tips. Fin margins rounded. Caudal peduncle comparatively narrow. Caudal-fin margin concave. HABITAT: Circumglobal in tropical to warm temperate seas. Santa Catalina Island, CA, to northern Peru, including central and southern Gulf of California. Pelagic from surface to about 655 ft. BIOLOGY: Associated with billfishes, particularly White Marlin and Sailfish.

Remora - *Remora remora* (Linnaeus, 1758)

FEATURES: Brownish gray to sooty or blackish. Mottled or uniformly colored. Cephalic disk with 16 to 20 laminae and not reaching past pectoral-fin tips. Pectoral-fin tips blunt. Caudal-fin margin deeply concave. HABITAT: Circumglobal in tropical to warm temperate seas. San Francisco, CA, to Chile, including central and southern Gulf of California and Galápagos Islands. Pelagic. BIOLOGY: Usually associated with sharks but may attach to other large fishes and sea turtles. May also attach to ships.

Carangidae - Jacks and Pompanos

Green Jack - *Caranx caballus* Günther, 1868

FEATURES: Pale olive to dark greenish blue dorsally, fading to silver or golden below. Juveniles display seven narrow, pale bars on body. Black spot on rear opercular margin. Body otherwise unmarked. Eyes relatively large. Maxilla extends to below middle of pupil. Pectoral fins long and pointed. Lateral line arched anteriorly, with 45–56 scutes posteriorly. HABITAT: Monterey Bay, CA, to Chile, including Gulf of California, Galápagos Islands, and Hawaii. Pelagic from surf zone to about 330 ft.

Pacific Crevalle Jack - *Caranx caninus* Günther, 1867

FEATURES: Pale gray, bluish gray, or greenish gray, with silvery to golden highlights dorsally. Silver to golden below. Juveniles with narrow pale bars. Black spot at rear opercular margin. Blackish blotch at lower pectoral-fin edge. Caudal fin dusky yellow to yellow. Anal fin and lower caudal peduncle white or yellow. Eyes comparatively small. Lateral line arched anteriorly and with 35–42 scutes posteriorly. Breast naked, with a small patch of scales. Body robust. HABITAT: San Diego, CA, to northern Peru, including Gulf of California and Galápagos Islands. Pelagic in oceanic and coastal waters from surface to about 1,150 ft. Also in estuaries and brackish water. May ascend rivers.

Bigeye Trevally - *Caranx sexfasciatus* Quoy & Gaimard, 1825

FEATURES: Iridescent pale bluish green dorsally, silvery below. Small black spot at upper opercular margin. Tips of soft dorsal and anal fins white. Posterior lateral line usually blackish. Juveniles with yellow anal fin. Eyes comparatively large. Lateral line arched anteriorly and with 27–36 scutes posteriorly. HABITAT: San Diego Bay, CA, to Ecuador, including lower Gulf of California and Galápagos Islands. Occur over sandy bottoms and around reefs in shallow coastal waters. BIOLOGY: Form very large schools.

Remoras

Marlinsucker
to 15 in.

Remora
to 2.5 ft.

Jacks and Pompanos

Green Jack
to 2.3 ft.

Pacific Crevalle Jack
to 3.3 ft.

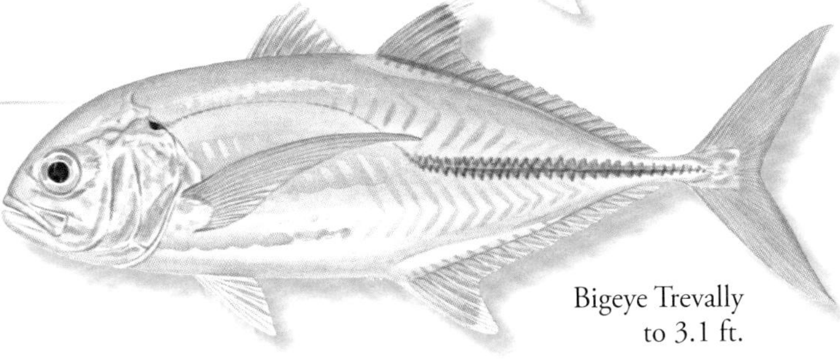

Bigeye Trevally
to 3.1 ft.

Pacific Bumper - *Chloroscombrus orqueta* Jordan & Gilbert, 1883

FEATURES: Iridescent blue dorsally, silvery below. A black blotch on upper opercular margin and a large black blotch on upper caudal peduncle. Caudal fin yellowish, other fins mostly yellowish. Mouth upturned. Body deep and laterally compressed, with ventral profile more convex than dorsal profile. Scutes on posterior lateral line very small to obsolete. HABITAT: San Pedro, CA, to Peru, including Gulf of California and Isla Malpelo. Found along shallow inshore areas and in estuaries. BIOLOGY: Form large schools. Make grunting noises when distressed. Sought commercially.

Amberstripe Scad - *Decapterus muroadsi* (Temminck & Schlegel, 1844)

FEATURES: Color varies. Blue green to olive dorsally. Silvery below. A golden to orange midlateral stripe may be present or absent. All with a large black blotch on rear opercular margin. Dorsal, caudal, and pectoral fins yellowish. Maxilla reaches to anterior portion of eye. Lateral line arched anteriorly with small scales; posterior lateral line with 28–41 scutes. Dorsal and anal fins followed by a single finlet. HABITAT: Nearly circumglobal. In eastern Pacific from Pacific Grove, CA, to Peru, including Galápagos Islands. From surface to about 100 ft. BIOLOGY: Schooling planktivore.

Pilotfish - *Naucrates ductor* (Linnaeus, 1758)

FEATURES: Bluish silver, with bluish black bars on body. Three bars extend into second dorsal fin. Caudal-fin lobes with white tips. Band from eyes to dorsal-fin origin absent. Eyes large. First dorsal fin with small, separate spines. Caudal peduncle with lateral keels. HABITAT: Circumglobal. In eastern Pacific from Vancouver Island, British Columbia, to Chile, including Gulf of California and offshore islands. Pelagic. BIOLOGY: Adults follow large sharks, rays, and sea turtles. Juveniles associate with seaweed and sea jellies. Feed on ectoparasites, host's prey scraps, and fishes.

Mexican Lookdown - *Selene brevoortii* (Gill, 1863)

FEATURES: Overall silvery; may have golden highlights. Caudal fin dusky to yellowish. Juveniles with filamentous first dorsal fin. Body laterally compressed, rhomboid in profile with a steeply sloping forehead. Soft dorsal fin with 20–24 rays, the first long and trailing. Anal fin with 17–18 rays, the first elongate. Gill rakers on lower limb of first arch number 29–34 (excluding rudimentary rakers). HABITAT: San Diego Bay, CA, to Isla San Lorenzo, Peru, including Gulf of California. Occur near bottom in shallow coastal waters and occasionally in estuaries. BIOLOGY: Form small schools. Popular in aquarium trade.

Pacific Moonfish - *Selene peruviana* (Guichenot, 1866)

FEATURES: Overall silvery; sometimes with a golden luster. Usually with a small black blotch on upper opercular margin. Upper caudal peduncle with a large black blotch. Caudal fin yellowish. Body very deep and laterally compressed. Head profile steeply sloping and humped over eyes. First dorsal- and anal-fin rays not elongate. HABITAT: Long Beach, CA, to Chile, including Gulf of California and Galápagos Islands. Occur near bottom in shallow coastal and inshore waters. BIOLOGY: Schooling. Feed on small fishes and a variety of invertebrates.

Pacific Bumper
to 12.2 in.

Amberstripe Shad
to 21.7 in.

Pilotfish
to 2.3 ft.

Mexican Lookdown
to 16.5 in.

Pacific Moonfish
to 2.8 ft.

Yellowtail Jack - *Seriola lalandi* Valenciennes, 1833

FEATURES: Shades of blue to green blue dorsally, silvery white below. Usually with a yellowish to bronze stripe from snout to caudal peduncle. Fins yellow to amber. Upper maxilla angular and broad posteriorly; reaches to below anterior margin of pupil. Second dorsal- and anal-fin lobes relatively tall. Body comparatively elongate. HABITAT: Nearly circumglobal in warm seas. In eastern Pacific from northern British Columbia to Chile, including Gulf of California. Pelagic around a variety of structures and kelp mats from surface to about 1,000 ft. BIOLOGY: Schooling. Adults usually migrate northward in winter to spring. Some may have home ranges. Sought commercially and for sport.

Almaco Jack - *Seriola rivoliana* Valenciennes, 1833

FEATURES: Brown, olivaceous, to bluish green dorsally, paler below. Dark band runs from eyes to first dorsal-fin origin. Faint amber stripe on sides from eyes to caudal fin. Upper maxilla angular and broad posteriorly; reaches below anterior margin of pupil. Second dorsal- and anal-fin lobes long, pointed. Body comparatively deep. HABITAT: Circumglobal in warm seas. In eastern Pacific from Oceanside, CA, to Cabo Blanco, Peru, including Gulf of California. Pelagic from near surface to about 820 ft.

Paloma Pompano - *Trachinotus paitensis* Cuvier, 1832

FEATURES: Pale metallic gray to pale bluish or greenish dorsally. Silvery white below. Head and body unmarked. Second dorsal and caudal fins grayish to blackish. Head, mouth, and eyes small. First dorsal fin reduced to small, separate, nearly embedded spines. Lateral line nearly straight. Second dorsal- and anal-fin lobes tall but not trailing. Body deep and laterally compressed. HABITAT: Redondo Beach, CA, to Callao, Peru, including central and southern Gulf of California. Occur near bottom, usually over shallow sandy inshore areas. BIOLOGY: Sought as food fish.

Gafftopsail Pompano - *Trachinotus rhodopus* Gill, 1863

FEATURES: Head and body silvery overall, with four to five narrow bars on upper sides. Second dorsal-, caudal-, and anal-fin lobes elongate and yellow to deep orange; usually dark posteriorly. Head and mouth small. First dorsal fin reduced to small, separate spines. Body deep and laterally compressed. HABITAT: Zuma Beach, CA, to Peru, including central and southern Gulf of California and Galápagos Islands. Found over sandy bottoms and around reefs and rocky areas from surface to about 100 ft.

Jack Mackerel - *Trachurus symmetricus* (Ayres, 1855)

FEATURES: Metallic greenish brown to bluish dorsally, silvery below. Opercular margin with a black blotch. Maxilla reaches past anterior rim of eye. Lateral line arched anteriorly, with moderately large scutes that increase in size posteriorly. Second dorsal and anal fins followed by a single finlet. HABITAT: South of Aleutian Islands to Gulf of California. Mostly pelagic from surface and surf zone to about 1,320 ft.

ALSO IN THE AREA: Cocinero, *Carangoides vinctus*; Whitemouth Jack, *Uraspis helvola*; Cottonmouth Jack, *Uraspis secunda*; page 350.

Yellowtail Jack
to 8.2 ft.

Almaco Jack
to 5 ft.

Paloma Pompano
to 20 in.

Gafftopsail Pompano
to 24 in.

Jack Mackerel
to 2.7 ft.

Pacific Pomfret - *Brama japonica* Hilgendorf, 1878

FEATURES: Bluish to bluish black on back, silvery on upper sides and below. Some overall silvery on head and body. Dorsal and caudal fins generally dark. Pectoral-fin margins transparent. Mouth moderately large, oblique, with protruding lower jaw. Snout very blunt and steep. Area between eyes arched. Dorsal, caudal, and anal fins scaled and rigid. Dorsal-fin lobe moderately tall. Caudal fin deeply forked. Anal-fin lobe small. HABITAT: South of the Aleutian Islands and southern Bering Sea to Chile. Also to Sea of Japan. Abundant from Gulf of Alaska to CA. Pelagic and oceanic. Usually found well offshore and near surface to about 2,000 ft. BIOLOGY: Form small schools. Adults generally migrate northward in spring, and southward in the fall to spawn. Feed on a wide variety of fishes and invertebrates. Taken as bycatch and by sport fishermen.

Pacific Fanfish - *Pteraclis aesticola* (Jordan & Snyder, 1901)

FEATURES: Head and body silvery. Dorsal and anal fins deep blue to black, greatly expanded, sail-like, and able to depress into scaled grooves. Snout short and blunt. Mouth large, oblique, with protruding lower jaw. Pectoral fins long and pointed. Size of dorsal and anal fins varies with age and conditions. Body long, slender, and laterally compressed. HABITAT: In northwest and southeast Pacific. In eastern Pacific from northern CA to Chile. Pelagic and oceanic from surface to about 330 ft. Juveniles possibly in shallow water. BIOLOGY: Rarely taken commercially, but considered a food fish. Preyed upon by predatory fishes such as tunas.

Rough Pomfret - *Taractes asper* Lowe, 1843

FEATURES: Overall blackish to dark brown with a silvery or coppery sheen. Margins of dorsal and anal fins black. Rear margin of caudal fin slightly paler. Mouth large and oblique, with protruding lower jaw. Snout, interorbital space, and lower jaw scaleless. Head profile straight to slightly concave. Dorsal- and anal-fin lobes scaled, stiff, long, and pointed. Caudal fin crescent shaped, with a longer upper lobe. Body comparatively shallow, somewhat compressed. Scales with a low central keel. HABITAT: Scattered circumglobablly. In eastern Pacific from Kodiak Island, AK, and off British Columbia to southern CA. Oceanic and primarily pelagic from surface to about 1,800 ft. BIOLOGY: Taken by gill nets, bottom trawls, and other fishing gear.

Sicklefin Pomfret - *Taractichthys steindachneri* (Döderlein, 1883)

FEATURES: Overall blackish, with a silvery to coppery sheen. Margins of dorsal, caudal, anal, and pelvic fins whitish. Mouth moderately large and oblique, with a protruding lower jaw. Snout short, steep, and scaleless in front of eyes. Head profile rounded. Interorbital space scaled. Dorsal- and anal-fin lobes scaled, stiff, long, and pointed. Caudal fin crescent shaped. Caudal peduncle with a midlateral groove. Body comparatively deep, somewhat compressed. Scales smooth. HABITAT: Indo-Pacific and eastern central Pacific. In eastern Pacific known from Point Conception, CA, and central Baja California. Pelagic over outer continental shelf to about 990 ft.

Pacific Pomfret
to 24 in.

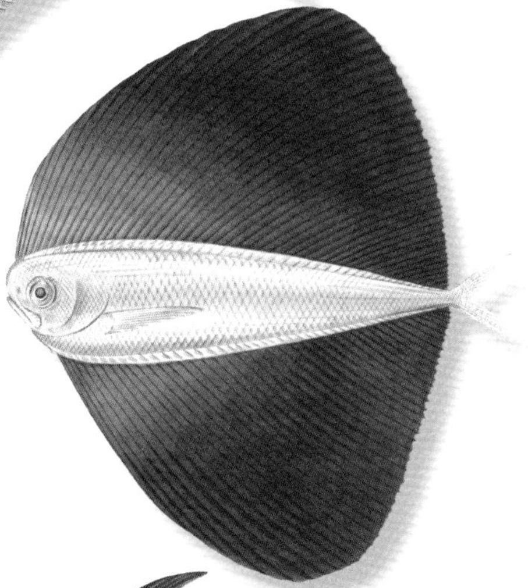

Pacific Fanfish
to 24 in.

Rough Pomfret
to 19.7 in.

Sicklefin Pomfret
to 3 ft.

Lutjanidae - Snappers

Amarillo Snapper - *Lutjanus argentiventris* (Peters, 1869)

FEATURES: Generally rosy anteriorly, fading to yellowish or greenish posteriorly. May be golden overall. A row of small blue spots or a blue streak present under eyes. Narrow pale bars on upper sides fade with age. May have a dark streak through eyes to nape. Teeth at tip of mouth enlarged and canine-like. Snout comparatively long. HABITAT: Oceanside, CA, to Peru, including Gulf of California and offshore islands. Found inshore over rocky bottoms and close to shelter. Small specimens in estuaries.

Colorado Snapper - *Lutjanus colorado* Jordan & Gilbert, 1882

FEATURES: Dark brownish red dorsally, fading to deep orange red on sides and below. Small specimens may have narrow pale bars on upper sides. Some with a blue streak under eyes. Teeth at tip of mouth enlarged and canine-like. Scales above lateral line on back in horizontal rows. Soft dorsal fin bluntly pointed posteriorly. HABITAT: Morro Bay, CA, to Panama, including Gulf of California. Possibly to Peru. Occur inshore and over offshore rocky reefs from surface to about 295 ft. Sometimes in estuaries.

Spotted Rose Snapper - *Lutjanus guttatus* (Steindachner, 1869)

FEATURES: Dusky pink to pale crimson or yellowish, with silvery sheen dorsally. Silvery white ventrally. Narrow yellowish to brown spots or stripes follow scales on back and sides. A large to relatively small blackish blotch or smudge present on back below middle portion of dorsal fin. Teeth at tip of mouth canine-like. Scales above lateral line in oblique rows. HABITAT: Southern CA to Peru, including Gulf of California. Rare in northern portion of range. Occur from inshore sandy bays, reefs, and hard bottoms from surface to about 350 ft. Also in brackish waters.

Pacific Dog Snapper - *Lutjanus novemfasciatus* Gill, 1862

FEATURES: Dark brownish red dorsally, fading to reddish or reddish orange below. Some overall reddish. About eight distinct to indistinct pale bars usually present on upper sides; very large specimens may lack bars. Teeth at tip of mouth enlarged and canine-like. Scales above lateral line in gently rising oblique rows. Soft dorsal fin rounded posteriorly. HABITAT: Morro Bay, CA, to Peru, including Gulf of California and Galápagos Islands. Occur from near shore and over rocky reefs to about 200 ft. Also common in estuaries; may travel far upstream.

Pacific Red Snapper - *Lutjanus peru* (Nichols & Murphy, 1922)

FEATURES: Pinkish red with iridescent sheen dorsally. Silvery white below. Caudal fin red, other fins pinkish red. Small specimens usually with several narrow, pale bars on back. Body otherwise unmarked. Teeth at tip of mouth enlarged and canine-like. Scales above lateral line on back in oblique rows. Dorsal and ventral profiles similar in shape. HABITAT: La Jolla, CA, and southern Baja California to Peru, including Gulf of California, where the species is abundant. Juveniles in shallow nearshore waters. Adults frequently found over reefs to about 295 ft. BIOLOGY: Feed on fishes, crustaceans, and squids. Adults sought commercially. Juveniles taken as shrimp trawl bycatch.

Amarillo Snapper
to 2.2 ft.

Colorado Snapper
to 3.5 ft.

Spotted Rose Snapper
to 2.6 ft.

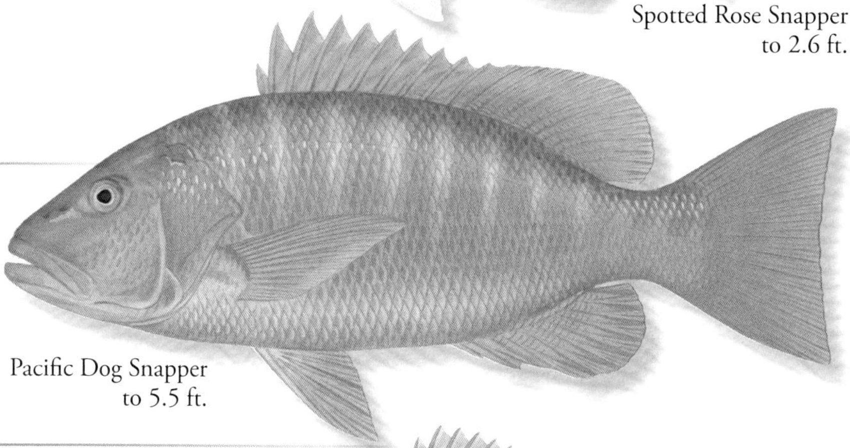

Pacific Dog Snapper
to 5.5 ft.

Pacific Red Snapper
to 3.1 ft.

Lobotidae - Tripletails

Pacific Tripletail - *Lobotes pacificus* Gilbert, 1898

FEATURES: Variably mottled, blotched, and spotted in shades of brown to silvery gray, or almost uniformly silvery gray to black. Juveniles mottled in shades of yellow to brown. Dark band through eyes. Snout short, jaws large. Eyes small. Head profile steep, somewhat concave over eyes. Soft dorsal, caudal, and anal fins broad and rounded. Body deep and laterally compressed. HABITAT: Possibly circumglobal. In eastern Pacific from San Pedro Breakwater, CA, to Peru. Found from brackish estuaries and bays to well offshore. Often occur around floating seaweeds, flotsam, and jetsam. BIOLOGY: May swim or lie on its side near surface. Sought as game and food fish. Some authors recognize this species as a junior synonym of *Lobotes surinamensis*.

Gerreidae - Mojarras

Pacific Flagfin Mojarra - *Eucinostomus currani* Zahuranec, 1980

FEATURES: Iridescent dusky dorsally, fading to silvery on sides and below. Spiny dorsal fin dusky at base, white at middle, and black along upper margin. Body otherwise unmarked. Snout short and pointed. Mouth highly protrusible. Scales extend across top of snout and around scaleless groove. Dorsal and anal fins with a scaly sheath along base. Body comparatively deep. HABITAT: Anaheim Bay, CA, to Peru, including Gulf of California and Galápagos Islands. Occur near bottom over sandy and muddy bottoms from freshwater streams and intertidal zone to about 330 ft. BIOLOGY: Schooling. Feed on bottom-dwellling invertebrates, detritus, and vegetation.

Pacific Spotfin Mojarra - *Eucinostomus dowii* (Gill, 1863)

FEATURES: Iridescent dusky dorsally, fading to silvery on sides and below. Body and fins unmarked. Snout short and pointed. Mouth highly protrusible. Scales extend across top of snout and around scaleless groove. Dorsal and anal fins with scaly sheath along base. Body comparatively shallow. HABITAT: Camp Pendleton Harbor, CA, to Peru, including Gulf of California. Found near bottom over sandy and muddy bottoms in estuaries and bays from near shore to about 375 ft. Also in tidepools.

Haemulidae - Grunts

Sargo - *Anisotremus davidsonii* (Steindachner, 1876)

FEATURES: Dusky silver to dusky gold or brassy dorsally. Sides and abdomen dusky silver to silvery. Fins similarly colored to dorsal area. All with a distinct, narrow black bar extending under fifth to seventh dorsal-fin spines to below pectoral fins. Small specimens also with two to three blackish stripes on sides. Snout short. Mouth small, with small, moderately thick lips. Preopercle finely serrated. Dorsal fin continuous, notched. Second anal-fin spine enlarged. HABITAT: Santa Cruz, CA, to southern Baja California, with an isolated population in Gulf of California. Found over seagrasses, kelp, and rocky and mixed bottoms from backwaters and surf zone to about 200 ft. BIOLOGY: Forage during the day on a wide variety of invertebrates. Adults usually form schools.

Tripletails

Pacific Tripletail
to 3.6 ft.

juvenile

Mojarras

Pacific Flagfin Mojarra
to 8.3 in.

Pacific Spotfin Mojarra
to 7.9 in.

Grunts

juvenile

Sargo
to 23.6 in.

Haemulidae - Grunts, *cont.*

Armed Grunt - *Conodon serrifer* Jordan & Gilbert, 1882

FEATURES: Brownish on back, silvery on sides and below. Others described as bluish gray on back. All with six to eight blackish bars on upper sides that are more narrow than pale interspaces. Fins yellowish to greenish yellow. Mouth small, eyes large. Preopercle serrated, with an elongate spine at lower corner. Dorsal fin deeply notched. Second anal-fin spine long and strong. HABITAT: San Onofre, CA, to Peru, including Gulf of California. Occur over soft bottoms from near shore to about 235 ft.

Salema - *Haemulon californiensis* (Steindachner, 1876)

FEATURES: Bluish gray to greenish gray, with silvery highlights dorsally. Silvery white below. Back and sides with seven to eight brownish orange to dark yellowish stripes. Fins brownish orange to yellowish. Mouth small, eyes very large. Spiny and soft dorsal fins separate. HABITAT: Monterey Bay, CA, to northern Peru, including Gulf of California. Rare north of Point Conception. Found around rocks and in upper kelp beds from shore to about 110 ft. BIOLOGY: Nocturnal planktivores in nearshore waters. Some authors classify this species as *Xenistius californiensis.*

Cortez Grunt - *Haemulon flaviguttatum* Gill, 1862

FEATURES: Silvery brown to dusky gray dorsally and on sides, silvery below. Pearly white spots form oblique rows on back and upper sides. Dorsal, anal, and pelvic fins yellowish. Caudal fin blackish at base, yellowish at margins. Mouth small, eyes moderately large. Preopercle finely serrated. Dorsal fin continuous and notched. HABITAT: Rarely found in Newport, Mission, and San Diego bays, CA. Commonly found from Gulf of California to Peru over rocky reefs from shore to about 350 ft.

Wavyline Grunt - *Microlepidotus inornatus* Gill, 1862

FEATURES: Bluish, with silvery highlights dorsally. Silvery below. Orange spots form wavy lines along scale rows above lateral line and narrow horizontal stripes on sides. Mouth small. Preopercle finely serrated and toothed. Dorsal fin continuous and notched; spiny portion with a straight margin and 14 spines. HABITAT: Newport Bay, CA, to central Gulf of California and to Manzanillo, Mexico. Occur over sandy bottoms and around reefs from near shore to about 73 ft. BIOLOGY: Schooling.

Sparidae - Porgies

Pacific Porgy - *Calamus brachysomus* (Lockington, 1880)

FEATURES: Overall silvery or with several vague to distinct bars on upper body. Some washed with gold dorsally. Fins may be banded. Underside of pectoral fins blackish. Head profile steep. Mouth small, with rows of canine- and molar-like teeth. Pectoral fins long and caudal fin forked. Body comparatively deep. HABITAT: Oceanside, CA, to Peru, including Galápagos Islands and Gulf of California. Found from surface to near bottom from intertidal zone to about 260 ft. Usually occur over sandy bottoms in clear waters. BIOLOGY: Form small groups that forage on bottom-dwelling invertebrates.

Armed Grunt
to 11.8 in.

Salema
to 11.8 in.

Cortez Grunt
to 16.5 in.

Wavyline Grunt
to 17.7 in.

Porgies

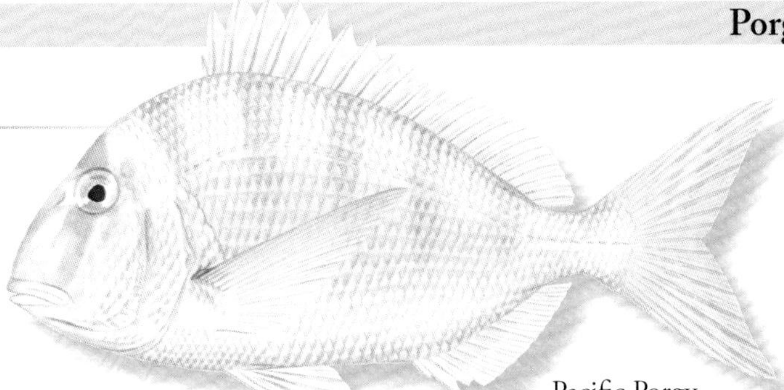

Pacific Porgy
to 24 in.

Polynemidae - Threadfins

Blue Bobo - *Polydactylus approximans* (Lay & Bennett, 1839)

FEATURES: Pale silvery blue on back, golden to silvery on sides. Ventral area silvery white. Pectoral fins dusky in juveniles, black in adults. Anal fin dusky to yellowish. Lower pectoral fins with five to six long, separate, and filamentous rays. Anal-fin origin below soft dorsal-fin origin. Lateral line with 55–59 scales. HABITAT: Monterey Bay, CA (very rare), to Chile, including Gulf of California and Galápagos Islands. Occur over soft bottoms from river mouths, along beaches, and in surf to about 350 ft.

Yellow Bobo - *Polydactylus opercularis* (Gill, 1863)

FEATURES: Pale silvery greenish, bronze to brownish on back, golden to silvery on sides. Ventral area silvery white. Pectoral fins yellowish. Anal fin dusky to yellowish. Lower pectoral fins with eight or nine long, separate, and filamentous rays. Anal-fin origin under middle of soft dorsal fin. Lateral line with 67–75 scales. HABITAT: Los Angeles Harbor, CA (very rare), to Chile, including Gulf of California. Occur over soft bottoms from estuaries, bays, and river mouths and along beaches to about 350 ft.

Sciaenidae - Drums and Croakers

White Seabass - *Atractoscion nobilis* (Ayres, 1860)

FEATURES: Shades of gray to bronze, with iridescent highlights dorsally. Silvery below. Upper sides usually with 7–8 dusky bars. Dorsal and caudal fins dusky yellow. Juveniles brownish green with darker bars. Roof of mouth with a row of small teeth. Pectoral fins comparatively long. Caudal-fin margin nearly straight. HABITAT: Juneau and Boca de Quadra, AK, to southern Baja California. Occur near bottom to surface over rocky bottoms and around kelp from bays and surf zone to about 400 ft. BIOLOGY: Juveniles feed on invertebrates and small fishes. Adults feed on fishes and squids. Sought as food fish.

Black Croaker - *Cheilotrema saturnum* (Girard, 1858)

FEATURES: Color and pattern vary. Uniformly brownish, blackish, coppery, to purplish, dorsally and on sides, with one to five pale bars on back. Silvery below. All with a black blotch on upper rear opercle. Juveniles whitish with black stripes. Snout overhangs mouth. Chin barbel absent. Dorsal fin deeply notched. Second anal-fin spine strong. HABITAT: Point Conception, CA, to S Baja California. Juveniles school over eelgrass and around reefs but may also live singly in caves and crevices. Adults occur near shore over rocky bottoms and in open, turbid bays from surface and surf zone to about 330 ft.

Shortfin Corvina - *Cynoscion parvipinnis* Ayres, 1861

FEATURES: Dusky gray to bluish gray, with iridescent highlights dorsally. Dorsal and caudal fins dusky. Inside of mouth yellowish orange. Underside of pectoral-fin base dusky. Body otherwise unmarked. Upper jaw with one or two canine-like teeth. Pectoral fins short, not reaching pelvic-fin margin. Caudal-fin margin nearly straight. HABITAT: Huntington Beach, CA, to Mazatlán, Mexico, including Gulf of California. Usually found off sandy beaches from intertidal zone to about 330 ft.

Threadfins

Blue Bobo
to 15.7 in.

Yellow Bobo
to 17.7 in.

Drums and Croakers

juvenile

White Seabass
to 5.5 ft.

variants

Black Croaker
to 17.7 in.

juvenile

Shortfin Corvina
to 23.6 in.

White Croaker - *Genyonemus lineatus* (Ayres, 1855)

FEATURES: Iridescent brassy to gray with indistinct wavy lines along scales dorsally. Silvery below. Dorsal, caudal, and pectoral fins dusky. Anal fin white to yellowish. Usually with a small black spot under upper pectoral-fin base. Snout rounded and overhanging upper jaw. Chin often with minute barbels along center fold. Spiny dorsal fin with 12–15 spines. Anal fin with 10–12 rays. HABITAT: Barkley Sound, British Columbia, to southern Baja California. Occur over soft bottoms from estuaries and surf zone to about 780 ft. BIOLOGY: Schooling. Sought as food fish and bait. Listed as Near Threatened.

California Corbina - *Menticirrhus undulatus* (Girard, 1854)

FEATURES: Shades of iridescent brassy to gray with indistinct wavy lines along scales dorsally. Ventral area silvery white. May display brownish saddles and blotches. Snout bluntly pointed and overhanging upper jaw. Chin with a small barbel. Dorsal fin with 10–11 spines. Anal fin with a single weak spine and 7–9 rays. Pectoral fins broad. Body elongate and flattened ventrally. HABITAT: Point Conception, CA, to Gulf of California. Also reported from Ecuador to northern Peru. Occur near bottom in bays and estuaries and in surf to about 65 ft. BIOLOGY: Solitary or in schools.

Spotfin Croaker - *Roncador stearnsii* (Steindachner, 1876)

FEATURES: Iridescent silvery to brassy with indistinct wavy lines along scales dorsally. Ventral area silvery white. Body may be unmarked or may display several vague to distinct brownish bars. During spawning males become golden and females have black streaks on abdomen. All with a large black blotch at pectoral-fin base. Snout steeply rounded. Chin barbel absent. Spiny dorsal fin with 9–10 spines. Anal fin with 7–9 rays. Body comparatively deep. HABITAT: Point Conception, CA, to tip of Baja California. Occur over sandy bottoms from surf zone to about 70 ft. Sometimes found over reefs. BIOLOGY: Usually solitary but may form schools.

Queenfish - *Seriphus politus* Ayres, 1860

FEATURES: Bluish dorsally, silvery on sides and below. Also described as tan or brownish dorsally. Fins yellowish. Small blackish spot present at upper pectoral-fin base. Body otherwise unmarked. Mouth large, with upper maxilla extending to rear margin of eyes. Dorsal fins separate. Anal fin long-based with 21–23 rays. HABITAT: Burrard Island, British Columbia, to southern Gulf of California. Occur over soft bottoms from surf zone to about 600 ft. Also in open water, bays, and harbors.

Yellowfin Croaker - *Umbrina roncador* Jordan & Gilbert, 1882

FEATURES: Iridescent dusky to pale grayish with distinct wavy lines along scale rows dorsally and on sides. May display faint to dark bars. Ventral area silvery white. Caudal, anal, pelvic, and pectoral fins yellowish. Snout bluntly rounded. Chin with a small barbel. Spiny dorsal fin with 10–11 spines. Anal fin with 6–7 rays. HABITAT: Point Conception, CA, to southern Baja California and in the Gulf of California. Occur over sandy bottoms in lagoons, quiet and open coastal waters from surf zone to about 150 ft.

White Croaker
to 16.3 in.

California Corbina
to 2.7 ft.

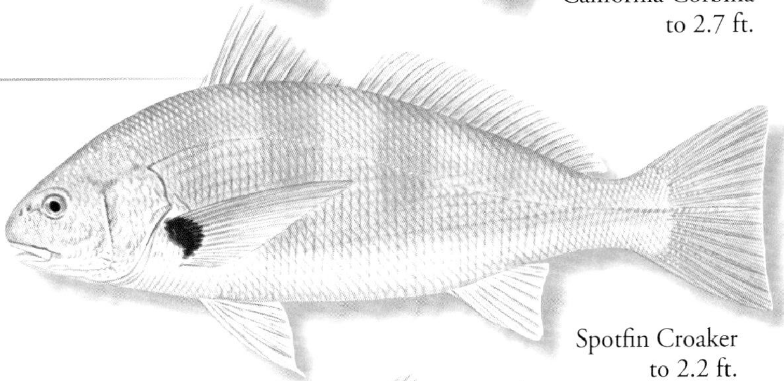

Spotfin Croaker
to 2.2 ft.

Queenfish
to 12 in.

Yellowfin Croaker
to 22 in.

Mullidae - Goatfishes

Bigscale Goatfish - *Pseudupeneus grandisquamis* (Gill, 1863)

FEATURES: Pale pinkish to brownish red, with irregular reddish bars, to nearly uniformly red dorsally. Blackish patch below spiny dorsal fin. Head with pearly white lines and spots and upper back with several rows of pearly white spots along scales. May quickly change color. Head steeply sloping. Mouth small, with a pair of long barbels under chin. HABITAT: San Onofre, CA (rare), to Chile, including Gulf of California and Galápagos Islands. Usually demersal over soft and sandy bottoms from near shore to about 220 ft. BIOLOGY: Use snout and barbels to rout prey.

Kyphosidae - Sea Chubs

Opaleye - *Girella nigricans* (Ayres, 1860)

FEATURES: Color highly variable. Shades of green or brown to grayish green dorsally and white below, to almost uniformly dark olive brown. One to three pale spots present under dorsal fin; very large specimens may lack spots. Often with a pale stripe in front of eyes. Eyes bright blue. Rear dorsal- and anal-fin margins rounded. Caudal-fin margin nearly straight. HABITAT: Otter Rock, OR, to southern Baja California. Occur over rocky reefs and seagrass beds, around kelp beds, and in lagoons, sloughs, and estuaries from intertidal zone to about 105 ft. BIOLOGY: Schooling. Feed primarily on algae.

Zebraperch - *Hermosilla azurea* Jenkins & Evermann, 1889

FEATURES: Silvery or pale green to golden with about 12 faint to distinct dark brown to blackish bars on sides. Head with dark, wavy bands. Opercle with a black to dark blue spot. Pectoral fins with a black blotch above and below base. Dorsal-fin base with a scaled sheath. Caudal-fin margin shallowly forked. HABITAT: Klamath River estuary, CA, to Gulf of California. Occur around rocks, seagrasses, and kelp beds from intertidal zone to about 50 ft in open water. BIOLOGY: Seek shelter in crevices at night.

Bluestriped Chub - *Kyphosus ocyurus* (Jordan & Gilbert, 1882)

FEATURES: Iridescent pale olive yellow dorsally, with a bright blue stripe under dorsal fin. Two blue stripes on head. Midbody with a blue stripe above a yellow stripe. Ventral area pearly white. Dorsal and anal fins low and long-based. Caudal fin deeply forked. Body relatively elongate. HABITAT: Redondo Beach, CA, to Ecuador, including Galápagos Islands. Found inshore to well offshore over reefs, along drop-offs, and under flotsam. Pelagic. BIOLOGY: Form small groups and large, fast-moving schools.

Blue-bronze Chub - *Kyphosus vaigiensis* (Quoy & Gaimard, 1824)

FEATURES: Grayish blue, with iridescent highlights and orange to brassy stripes on head and along scale rows. Small juveniles display silvery white spots. Soft dorsal and anal fins comparatively low and nearly uniform in height. Soft dorsal fin with 14 rays. Anal fin with 13–14 rays. Caudal fin shallowly forked. HABITAT: Redondo Beach, CA, to Peru, including Gulf of California and Galápagos Islands. Occur over hard bottoms and reefs and around kelp from surface to about 60 ft. BIOLOGY: Schooling.

Goatfishes

Bigscale Goatfish
to 12 in.

Sea Chubs

variants

Opaleye
to 2 ft.

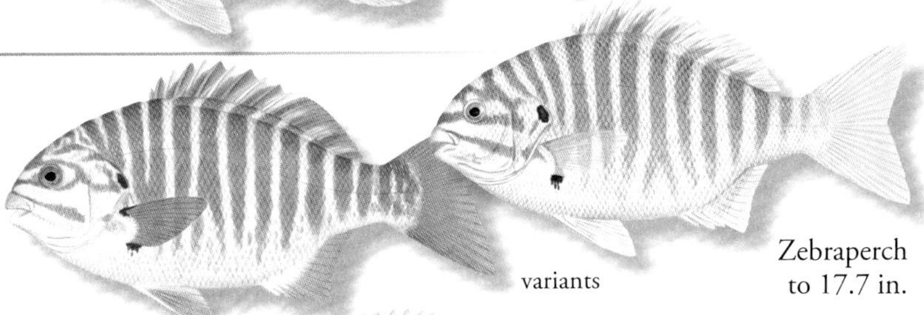

variants

Zebraperch
to 17.7 in.

Bluestriped Chub
to 2.3 ft.

Blue-bronze Chub
to 17.8 in.

Kyphosidae - Sea Chubs, *cont.*

Halfmoon - *Medialuna californiensis* (Steindachner, 1876)

FEATURES: Shades of blue to dark gray dorsally, fading to silvery below. Others overall gray blue. A large blackish blotch present above opercle. Sides with faint, oblique bars. Capable of color change. Soft dorsal, caudal, and anal fins scaled. Caudal-fin margin concave. HABITAT: Vancouver Island, British Columbia, to southern Baja California. Found over rocks and seagrass beds and around kelp, hard structures, and flotsam from surface and intertidal zone to about 145 ft. BIOLOGY: Form large schools.

Chaetodontidae - Butterflyfishes

Threebanded Butterflyfish - *Chaetodon humeralis* Günther, 1860

FEATURES: Silvery white to golden. Scales may have brownish margins. Dark brown to black bands run from nape through eyes, from nape through pectoral fins, and in front of caudal fin. Dorsal and anal fins with a broad brown to black margin. Caudal fin with three brown to black bars. Juveniles with a dark ocellus on soft dorsal fin. Snout short. Dorsal-fin spines increasing in height to fourth or fifth spines. Body deep, round in profile, and laterally compressed. HABITAT: King Harbor, CA, to Chile, including Gulf of California and Galápagos Islands. Found in tidepools and over inshore rocky reefs to about 350 ft. Also reported in estuaries and around mangroves.

Scythe Butterflyfish - *Prognathodes falcifer* (Hubbs & Rechnitzer, 1958)

FEATURES: Silvery to golden with a dark brown to black band from nape through eyes and a broad, inverted U shape from lower gill to anal-fin base. Dorsal and anal fins with dark brown to black inner margins. Pelvic fins mostly dark. Caudal fin pale, with dark upper and lower margins. Snout long and pointed. Dorsal-fin spines very tall and incised anteriorly. Body deep, rhomboid in profile, and laterally compressed. HABITAT: Santa Cruz Island, CA, to the Galápagos and other offshore islands and the southern tip of Baja California. Occur around reefs and steep slopes from about 10 to 900 ft.

Pomacanthidae - Angelfishes

Cortez Angelfish - *Pomacanthus zonipectus* (Gill, 1862)

FEATURES: Color somewhat variable. Dusky gray, with black-centered scales at midbody, dark bluish gray to gray on head, and dark brown to blackish posteriorly. A broad yellow bar present behind head, followed by a narrow yellow bar that is bordered by a pair of blackish bars. Dorsal, caudal, and anal fins variably marked with pale to dark blue wavy lines. Juveniles blackish, with yellow and blue semi-circular bands on body. All with a stout spine at lower preopercular corner. Adults with a hump at nape. Body very deep and laterally compressed, round in profile. HABITAT: Redondo Beach, CA (very rare), to Peru, including Gulf of California and offshore islands. Found over rocky reefs from about 5 to 165 ft. BIOLOGY: Juveniles usually solitary and territorial; pick parasites from other fishes. Adults solitary or in small aggregations. Feed primarily on sponges.

Sea Chubs

Halfmoon
to 19 in.

Butterflyfishes

Threebanded Butterflyfish
to 10.6 in.

Scythe Butterflyfish
to 6.8 in.

Angelfishes

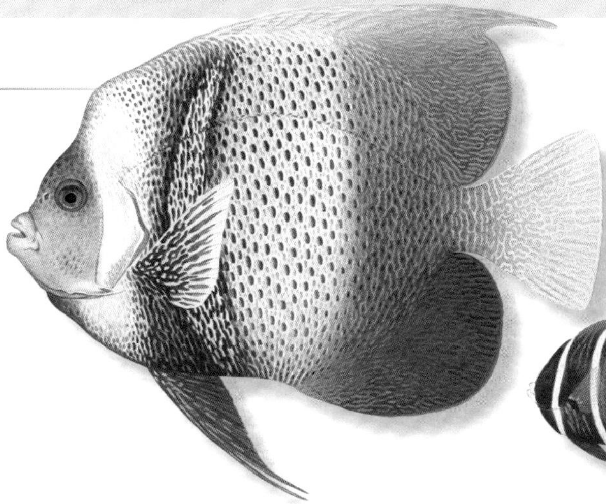

Cortez Angelfish
to 19.7 in.

juvenile

Barred Surfperch - *Amphistichus argenteus* Agassiz, 1854

FEATURES: Silvery to brassy and usually with 8–10 greenish or yellowish to rusty bars on sides. Often with irregular spots between bars. Some lack bars altogether. Fins grayish. Color darkens during breeding. Lower rim of eye above upper margin of upper jaw. Dorsal-fin spines shorter than longest dorsal-fin ray. Base of anal-fin with a row of scales. Body deep, oval in profile. HABITAT: Bodega Bay, CA, to north-central Baja California. Found in surf along coast to about 240 ft. Occasionally found in low-salinity estuaries. BIOLOGY: Usually found in schools. Sought recreationally.

Calico Surfperch - *Amphistichus koelzi* (Hubbs, 1933)

FEATURES: Silvery to brassy, with brownish to greenish or grayish spots that form broken bars or clusters on sides. Some lack spots. Body may be tinged with red. Fins grayish pink, with caudal fin pinkish to dark red. Mouth small. Lower rim of eye below upper margin of upper jaw. Dorsal fin notched, with longest spine about equal to longest rays. Base of anal fin with a row of scales. Body very deep and humped at top. HABITAT: Cape Flattery, WA, to northern Baja California. Occur along sandy beaches of open coast from surf zone to about 30 ft. BIOLOGY: A common sport fish.

Redtail Surfperch - *Amphistichus rhodoterus* (Agassiz, 1854)

FEATURES: Silvery with variable, scattered rusty to greenish spots or with 8–11 irregular bars with interspersed spots on sides. Dorsal and caudal fins pinkish to red. Anal and pelvic fins with some pink or red. Dorsal-fin spines increasingly tall to 5th spine, with all taller than dorsal-fin rays. Mouth small. Base of anal fin with a row of scales. Body deep, oval in profile. HABITAT: Vancouver Island to Avila Beach, CA. Usually found in surf zone to about 25 ft but also over low-relief rocks. Some females enter bays and estuaries to give birth. BIOLOGY: Feed primarily on amphipods but also clams, crabs, shrimps, and other invertebrates.

Kelp Perch - *Brachyistius frenatus* Gill, 1862

FEATURES: Reddish brown to golden brown, with a silvery stripe through eyes and several silvery blotches on midside. Snout short, pointed, and upturned. Mouth very small, with lower jaw protruding. Dorsal-fin spines become increasingly tall posteriorly. HABITAT: Near Sitka, AK, to central Baja California. Occur in kelp and algae beds, over seagrasses and boulders, and in sandy channels from intertidal zone and surface to about 250 ft. BIOLOGY: Solitary, in small groups, and in large schools.

Shiner Perch - *Cymatogaster aggregata* Gibbons, 1854

FEATURES: Silvery to brassy dorsally, with or without brownish oblique bars. Females with three bright yellow oblique bars on sides, separated by rows of black spots. Males with black-edged scales. During spawning season scales darken and sometimes become nearly completely black. Soft dorsal fin may have a blackish blotch near center. Mouth small. Dorsal-fin base with a row of scales. Body comparatively deep. HABITAT: Sitka, AK, to northern Baja California. Occur in a wide range of habitats from surf zone to about 480 ft. Also in estuaries. BIOLOGY: Usually in large schools.

Barred Surfperch
to 17 in.

Calico Surfperch
to 12 in.

Redtail Surfperch
to 16 in.

Kelp Perch
to 8.5 in.

Shiner Perch
to 8 in.

Black Perch - *Embiotoca jacksoni* Agassiz, 1853

FEATURES: Color highly variable. Shades of dark brown, green or reddish to gray with or without darker bars and iridescent highlights on scales. Can rapidly change color and may become silvery over sandy bottoms. All with a dark to pale blue band along anal-fin base and enlarged scales between pectoral and pelvic fins. Dorsal-fin spines shorter than dorsal-fin rays. Scale rows present along anal-fin base. HABITAT: Fort Bragg, CA, to central Baja California. Occur over a wide variety of bottoms from surface and intertidal zone to about 150 ft. BIOLOGY: Found singly and in schools.

Striped Seaperch - *Embiotoca lateralis* Agassiz, 1854

FEATURES: Background color varies from blackish brown dorsally, fading to golden or dusky silver below, to nearly uniform brown or reddish orange with golden highlights. All with about 18 narrow bright blue to whitish stripes along scale rows. Head with bluish blotches and wavy lines. Dorsal-fin spines shorter than dorsal-fin rays. Anal fin long-based. HABITAT: Klakas Inlet, AK, to northern Baja California. Found in a wide variety of habitats from brackish waters, intertidal zone, and near surface to bottom at about 310 ft. BIOLOGY: Feed on a wide variety of invertebrates. Inactive at night.

Spotfin Surfperch - *Hyperprosopon anale* Agassiz, 1861

FEATURES: Dusky dorsally, silvery below. Spiny dorsal fin with a small to large blackish blotch at upper margin. Anal fin with a blackish blotch near center. Caudal fin sometimes pinkish. Mouth large and oblique, with protruding lower jaw. Body comparatively shallow. HABITAT: Seal Rock, OR, to central Baja California. Usually over soft bottoms but also found around kelp beds and in surf zone from surface to about 330 ft. BIOLOGY: Feed on small invertebrates and fishes. Females move inshore during summer months to birth young.

Walleye Surfperch - *Hyperprosopon argenteum* Gibbons, 1854

FEATURES: Greenish to bluish dorsally, silvery below. During spawning, sides show five to six dark bars that are faint to absent otherwise. Pelvic fins pale in females, yellow in large males; both with a black pelvic-fin margin and pale anal fin. Caudal fin may have blackish margin. Mouth large and oblique, with protruding lower jaw. Body comparatively deep, round in profile. HABITAT: Vancouver Island, British Columbia, to central Baja California. Occur in estuaries, bays, along sandy beaches, in surf, and over reefs and seagrass beds from surface to about 600 ft. BIOLOGY: Form large schools.

Silver Surfperch - *Hyperprosopon ellipticum* (Gibbons, 1854)

FEATURES: Grayish green to grayish dorsally, silvery below. May or may not have several short, faint to dark bars on sides. Spiny dorsal fin dusky to blackish at margin. Caudal fin pale to pinkish, with a blackish margin. Pelvic fins uniformly pale. Females with a gray to black area on anal fin; male anal-fin lobe pale or with a black or orange blotch. Mouth moderately large and oblique, with protruding lower jaw. Body oval in profile. HABITAT: Brooks Peninsula, British Columbia, to northern Baja California. Occur in quiet bays, surf zone, and kelp beds from intertidal area to aout 360 ft.

Black Perch
to 15.4 in.

variants

Striped Seaperch
to 15.6 in.

Spotfin Surfperch
to 8 in.

Walleye Surfperch
to 12 in.

♀ ♂

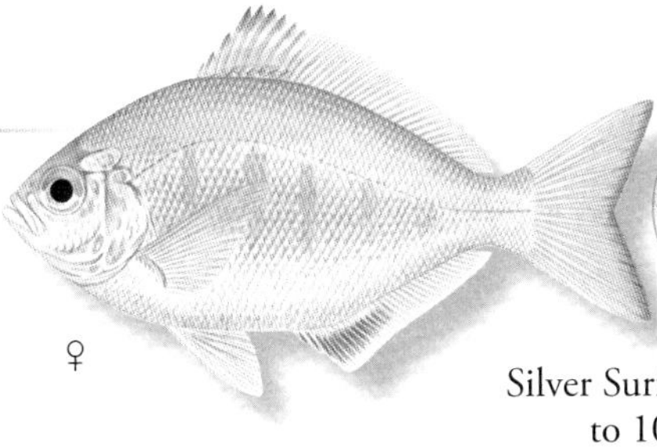

♀

Silver Surfperch
to 10.5 in.

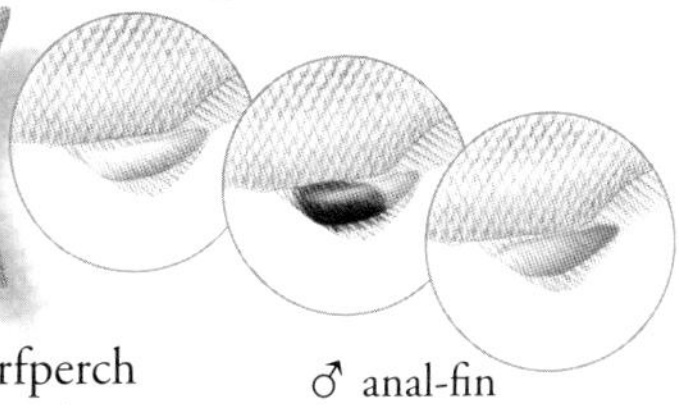

♂ anal-fin
variantions

Rainbow Seaperch - *Hypsurus caryi* (Agassiz, 1853)

FEATURES: Iridescent grayish to purplish with bright blue spots and numerous orange to dark reddish brown bars dorsally. Ventral area silvery, with yellow to orange spots and wavy lines on chest. Cheeks with bright blue lines and spots. A blackish blotch present at soft dorsal-fin base and on anal fin. Mouth small. Dorsal-fin spines somewhat shorter than longest dorsal-fin rays. Anal fin short-based. Caudal fin deeply forked. HABITAT: Cape Mendocino, CA, to northern Baja California. Occur over a variety of bottoms from surf zone and tidepools to about 165 ft. Usually near bottom but occasionally in kelp canopy. BIOLOGY: Inactive at night. Prefer waters below 52°F.

Reef Perch - *Micrometrus aurora* (Jordan & Gilbert, 1880)

FEATURES: Iridescent olivaceous to bluish or grayish dorsally. Sides with a golden to brassy stripe. Ventral area silvery to dusky silver. Scales behind and below pectoral fins with black, crescent-shaped margins. Pectoral-fin base with a triangular blackish mark. Mouth small. Anal fin short-based. Body slender and caudal peduncle long. HABITAT: Tomales Bay, CA, to central Baja California. Occur in channels, over seagrasses and algae, and over reefs from intertidal zone to about 20 ft.

Dwarf Perch - *Micrometrus minimus* (Gibbons, 1854)

FEATURES: Iridescent olivaceous to brassy dorsally, fading to silvery below. Sides with variable, irregular grayish to blackish blotches, sometimes forming a midlateral stripe posteriorly. Dorsal-fin base with a black blotch and pectoral-fin base with a black triangular mark. Anal fin variably marked with black and yellow or orange. Mouth small. Anal fin short-based. Body deep. Males with crescent-shaped depression on anal-fin base and large gland on anal fin. HABITAT: Bodega Bay, CA, to central Baja California. In bays, in estuaries, and along coast from surf zone to about 30 ft. BIOLOGY: Perform complex spawning rituals. Males are mature at birth; females mature soon after birth.

Sharpnose Seaperch - *Phanerodon atripes* (Jordan & Gilbert, 1880)

FEATURES: Silvery or with pinkish reflections. Dorsal area with reddish brown spots along scale rows. Pelvic fins with dusky to blackish tips. Anal fin may have a black blotch anteriorly. Mouth small. Dorsal-fin spines increasing in length posteriorly, but all shorter than tallest dorsal-fin rays. Caudal peduncle long and caudal fin deeply forked. HABITAT: Yaquina Bay, OR, to central Baja California. Occur over kelp and eelgrass beds, rocky reefs, and other structures from surface to about 750 ft. BIOLOGY: Form mixed schools with other Surfperches. Feed on a wide variety of invertebrates.

White Seaperch - *Phanerodon furcatus* Girard, 1854

FEATURES: Overall silvery, with a narrow dark band at dorsal-fin base. Fins pale and translucent to yellowish. Some with a dark blotch on anal fin and dark margin on caudal fin. Mouth small. Dorsal-fin spines increase in length posteriorly, but all shorter than tallest dorsal-fin rays. Caudal peduncle long and caudal fin deeply forked. HABITAT: Vancouver Island, British Columbia, to central Baja California. Occur in a wide range of habitats from bays, estuaries, and surf zone to about 230 ft. BIOLOGY: Schooling.

Rainbow Seaperch
to 12 in.

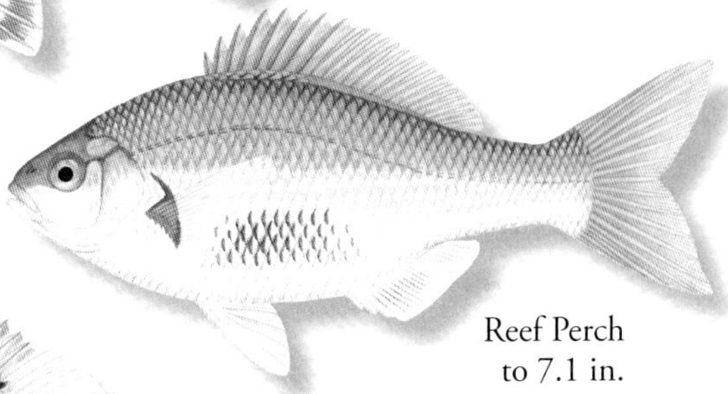

Reef Perch
to 7.1 in.

♀ Dwarf Perch
to 6.25 in.

Sharpnose Seaperch
to 11.5 in.

White Seaperch
to 12.4 in.

Rubberlip Seaperch - *Rhacochilus toxotes* Agassiz, 1854

FEATURES: Color varies. Iridescent brassy, brown, or olivaceous to golden, silvery, or black dorsally. Ventral color similar to dorsal or varies from pinkish, purplish, or golden to silver. Often with a darker midbody bar. Juveniles mostly pinkish. All with thick pink to whitish lips, with lower lip lobed underneath. Mouth highly protrusible. Anterior dorsal-fin rays slightly longer than dorsal-fin spines. HABITAT: Mendocino County, CA, to southern Baja California. Found over a wide variety of bottoms from surf zone to about 165 ft. Sometimes in bays and estuaries.

Pile Perch - *Rhacochilus vacca* (Girard, 1855)

FEATURES: Iridescent dusky to gray dorsally, with a brownish to blackish midbody bar that fades after capture. Scales along sides with dark centers. Pelvic and anal-fin lobes yellowish during breeding and may be tipped with black. Anterior dorsal-fin rays much taller than dorsal-fin spines. Caudal fin deeply forked. HABITAT: Southern British Columbia to central Baja California. Occur over a variety of hard bottoms and structures along open coast, in bays, and sometimes in brackish water from intertidal zone to about 295 ft. BIOLOGY: Feed on a wide range of invertebrates.

Pink Seaperch - *Zalembius rosaceus* (Jordan & Gilbert, 1880)

FEATURES: Silvery with pinkish to peachy luster dorsally. Silvery below. Back with two large brownish to blackish blotches between lateral line and upper profile. Fins translucent to pinkish. Mouth small. Caudal fin with a single trailing ray at upper tip. Anal fin with a trailing ray posteriorly. Body deep, oval in profile. HABITAT: Point Delgada, CA, to central Baja California. Occur near bottom over a variety of substrates from surf zone to about 750 ft. BIOLOGY: Found singly or in small groups.

Pomacentridae - Damselfishes

Panamic Sergeant Major - *Abudefduf troschelii* (Gill, 1862)

FEATURES: Yellow dorsally, fading to pale or dark blue below, with five bluish black bars on sides and possibly a broken bar on caudal peduncle. Mouth very small. Soft dorsal and anal fins bluntly pointed. Caudal fin deeply forked. Body deep, round in profile. HABITAT: Redondo Beach, CA, to Peru, including Gulf of California and offshore islands. Over reefs and rocky bottoms from intertidal zone to about 50 ft.

Swallow Damselfish - *Azurina hirundo* Jordan & McGregor, 1898

FEATURES: Steel blue to olive blue dorsally, fading to blue or silver below. Pectoral fins yellowish at tips. Dorsal, caudal, and anal fins with pale blue margins. Mouth small. Dorsal fin continuous, with 12 spines and 10–12 rays. Caudal fin deeply forked. Body slender and elongate. HABITAT: Anacapa, Santa Catalina, and San Clemente islands, CA. Also Isla Guadalupe, Rocas Alijos, and Islas Revillagigedo, Mexico. Occur over shallow, rocky inshore reefs and often near drop-offs. BIOLOGY: Listed as Near Threatened.

Surfperches

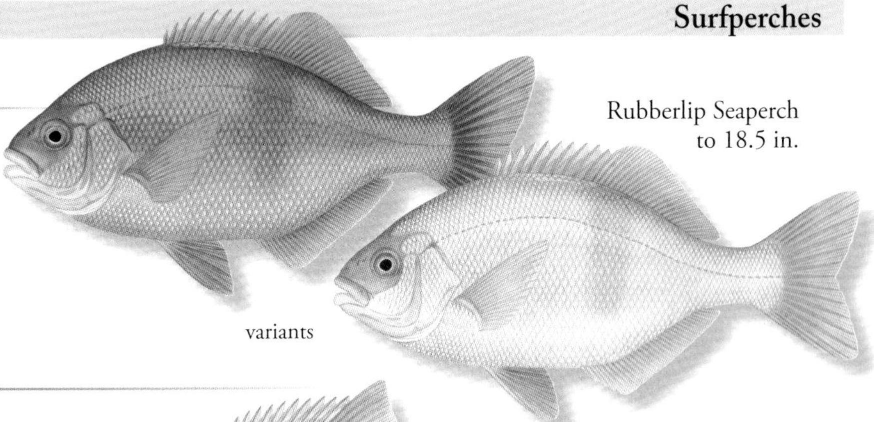

Rubberlip Seaperch
to 18.5 in.

variants

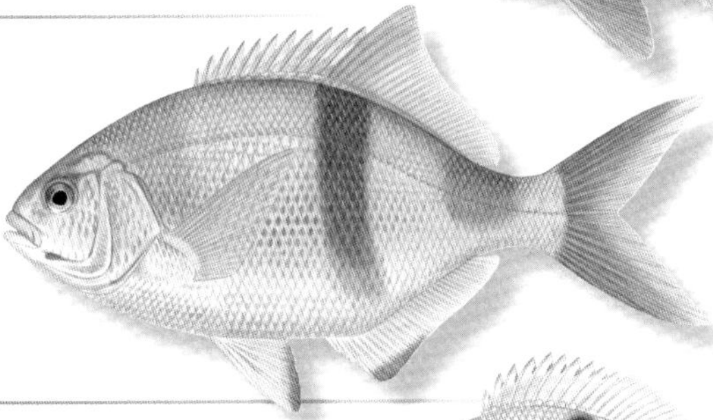

Pile Perch
to 17.4 in.

Pink Seaperch
to 8 in.

Damselfishes

Panamic Sergeant Major
to 9 in.

Swallow Damselfish
to 6.8 in.

Silverstripe Chromis - *Chromis alta* Greenfield & Woods, 1980

FEATURES: Adults pale to dark grayish brown, with darker, brown-edged scales. Snout usually with a yellowish V-shaped band in front of eyes. Upper rim of eyes may be blue. Body otherwise unmarked. Small juveniles dark blue, with bright blue spots on head and scales, a bright blue band above eyes, and a white band from first dorsal spine through caudal peduncle; bright blue spots are lost with age and ground color becomes grayish brown to bluish gray. Body very deep. HABITAT: Santa Catalina Island, CA, and central Baja California to Peru, including Gulf of California and many offshore islands. Found over coral and rocky reefs from shoreline to about 655 ft.

Blacksmith - *Chromis punctipinnis* (Cooper, 1863)

FEATURES: Shades of blue gray to violet gray or blue; often with shades of green along back and posteriorly. Scales with darker edges. Posterior portion of body, soft dorsal fin, and caudal fin with distinct black spots. Become blotched with white while resting. Breeding males with a dark bar between eyes. Nesting males become pale and mottled. Juveniles purple blue anteriorly, yellow orange posteriorly. Mouth small. Lateral line ends below soft dorsal fin. Body oblong in profile. HABITAT: Monterey Bay, CA, to central Baja California. Occur around a wide variety of hard structures and kelp and seagrass beds from near shore to about 200 ft. BIOLOGY: Feed on a wide variety of zooplankton. Males build nests and aerate and defend eggs.

Garibaldi - *Hypsypops rubicundus* (Girard, 1854)

FEATURES: Adults overall shades of bright orange. Dorsal, anal, and pelvic fins may have a narrow bright blue margin. Irises silvery. Body otherwise unmarked. Juveniles bright orange with numerous pale to dark blue spots and blue margins on fins. Mouth small. Soft dorsal- and anal-fin lobes may be elongate and pointed. Mature adults with a humped forehead. Body deep in profile. HABITAT: Monterey Bay, CA, to southwest corner of Baja California. Found around a wide range of hard structures and kelp and seagrass beds from intertidal zone to about 95 ft. BIOLOGY: Most are highly territorial. Males build nests and aerate and defend eggs. Protected in California.

Labridae - Wrasses and Parrotfishes

Rock Wrasse - *Halichoeres semicinctus* (Ayres, 1859)

FEATURES: Juveniles orange yellow to reddish brown, with two white stripes on body and an ocellated black spot on mid-dorsal fin. Initial phase shades of orange to pinkish or brown with rows of irregular black spots on dorsal scales; may display pale saddles. Mid-phase (not shown) described as similar to initial but with a dark bar behind pectoral fins. Terminal phase shades of green, gray, or orange to brown, with darker reddish to grayish vertical lines on scale rows and a large white-edged black area under bright yellow pectoral fins. Mouth with two pairs of forward-facing teeth in upper jaw. Dorsal and anal fins with 12 rays each. HABITAT: Diablo Cove, CA, to central Baja California. Also Gulf of California and Isla Guadalupe. Occur over sand, boulders, cobble, kelp, and seagrass beds from tidepools and surface to about 130 ft.

Damselfishes

Silverstripe Chromis
to 5.9 in.

juvenile

Blacksmith
to 12 in.

Garibaldi
to 14 in.

juvenile

Wrasses and Parrotfishes

Rock Wrasse
to 15 in.

initial

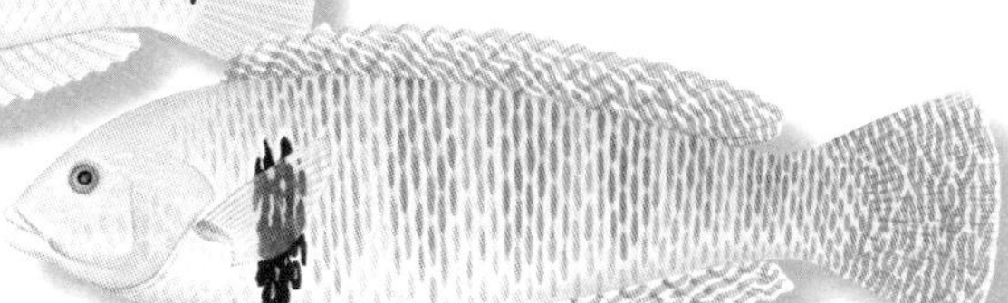

terminal

juvenile

Loosetooth Parrotfish - *Nicholsina denticulata* (Evermann & Radcliffe, 1917)

FEATURES: Juveniles and initial phase varying shades of reddish, greenish, or brown to grayish, with irregular darker and paler spots, blotches, and mottling on head, body, and fins; scales with pale margins. Terminal phase shades of green, with reddish chin, throat, and caudal fin; a dark blotch present on anterior dorsal fin; may become variably blotched and mottled. All with slightly protruding, unfused rows of flattened teeth and trailing filaments on spiny dorsal-fin tips. Caudal fin bluntly rounded. Body robust. HABITAT: Santa Catalina Island, CA. Also Islas San Benito, Isla Cedros, and Rocas Chester, Baja California, and Galápagos Islands to Peru. Occur near bottom over algae and grass beds and reefs from intertidal zone to about 33 ft. BIOLOGY: Feed by scraping corals or feed directly on algae.

Señorita - *Oxyjulis californica* (Günther, 1861)

FEATURES: Brownish to olivaceous dorsally, fading to yellow or pearly white below. Some may be pinkish to yellow. Scales with darker centers. Some with pale bars. May have several pale bands on cheeks. Base of caudal fin with a large, blackish blotch. Females and males similar in color and pattern. Upper jaw with small, forward-facing teeth at tip. Dorsal and anal fins with 13 rays each. Body elongate, comparatively slender. HABITAT: Salt Point, CA, to southern Baja California. Occur along outer coast and around seagrasses, rocks, and other structures from intertidal zone and near surface to about 320 ft. BIOLOGY: Feed on a wide variety of invertebrates and occasionally pick parasites. Bury in sand and pebbles at night.

California Sheephead - *Semicossyphus pulcher* (Ayres, 1854)

FEATURES: Juveniles shades of orange, with large black blotches on fins and a pearly white stripe on midine. Initial phase shades of pink, with a pale to white chin and a blackish spot on anterior dorsal fin. Terminal phase with varying shades of dark reddish black to black on head and posterior body with a broad red to orange area on midbody; chin white; head develops large hump over eyes with age. Lips fleshy. Mouth with pairs of pointed teeth. Spiny dorsal fin low, soft dorsal and anal fins pointed. HABITAT: Monterey Bay, CA, to southern Baja California, including Gulf of California and Isla Guadalupe. Occur around a variety of hard and soft bottoms and around seagrass beds from intertidal zone to about 295 ft. Other name: Goat. NOTE: Vulnerable.

Bathymasteridae - Ronquils

Alaskan Ronquil - *Bathymaster caeruleofasciatus* Gilbert & Burke, 1912

FEATURES: Dark blackish brown to reddish brown. May have faint to distinct, irregular pale grayish blue to blue blotches on upper and lower sides. May become reddish yellow with darker bars. Dorsal fin faintly banded. Anal fin may be blue at base. Juveniles reddish orange with grayish blue markings. Maxilla extends to or beyond rear of eyes. Inner pelvic-fin rays scaled. Caudal fin scaled for more than half its length. Body scales rough. HABITAT: Aleutian Islands to Queen Charlotte Islands, British Columbia. Demersal over rocky bottoms from about 15 to 740 ft.

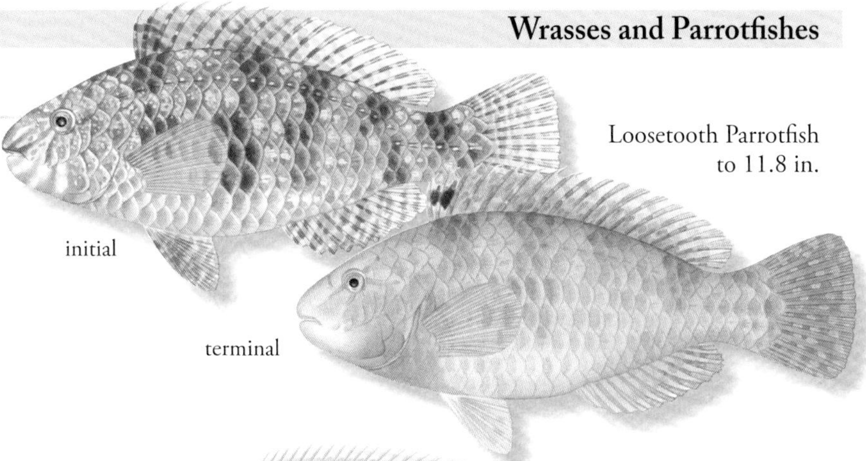

initial

terminal

Loosetooth Parrotfish
to 11.8 in.

Señorita
to 10 in.

initial

juvenile

terminal

California Sheephead
to 3 ft.

Ronquils

Alaskan Ronquil
to 12 in.

Smallmouth Ronquil - *Bathymaster leurolepis* McPhail, 1965

FEATURES: Overall dark brown to dark reddish brown. May have pale bluish bars on sides. Occasionally with small dark blue or blackish spots on opercle. Dorsal, caudal and anal fins irregularly banded. Small specimens with a dark blue or blackish blotch on anterior dorsal fin. Maxilla extends to below mid-eye or rear margin of pupil. Pelvic fins lack scales. Caudal fin scaled for less than half its length. Body scales nearly smooth. HABITAT: Pribilof and Aleutian Islands to eastern Gulf of Alaska. Also to western Hokkaido. Demersal in tidepools and shallow inshore waters to about 320 ft.

Searcher - *Bathymaster signatus* Cope, 1873

FEATURES: Pale tan or olive brown to dark green dorsally, with pale bars that merge with pale ventral area. Irises bright blue. Bright orange pores on preopecle. Black blotch on anterior dorsal fin. Lower pectoral rays, pelvic fins, and anal fin blackish. Maxilla reaching near to below rear of eye. Pectoral fins barely or not reaching anus. HABITAT: Navarin Canyon to Aleutian Islands and south to WA. Also east Siberian Sea to Hokkaido. Demersal on nearshore rock-and-algae bottoms to outer continental shelves over sandy and gravel bottoms to about 1,250 ft. Juveniles often at surface.

Stripefin Ronquil - *Rathbunella alleni* Gilbert, 1904

FEATURES: Color and pattern highly variable. Shades of gray to brown or pinkish, with irregular whitish blotches on sides and smaller dark blotches dorsally. Others very pale and nearly uniformly colored or with scattered dark speckles. All with a bright to dark blue inner stripe on anal fin and a blue blotch on anterior dorsal fin. Scales very rough and do not extend onto dorsal fin. Seven preopercular pores present. Canine teeth present in males only. See note below. HABITAT: Marin County, CA, to northern Baja California. Demersal from about 6 to 130 ft.

Bluebanded Ronquil - *Rathbunella hypoplecta* (Gilbert, 1890)

FEATURES: Color and pattern highly variable. Shades of gray to brown or pinkish, with regularly spaced whitish blotches ventrally and alternating whitish and dark brown to black blotches dorsally. Some pale, nearly unmarked. All with a blue band on each inner anal-fin membrane. Scales weakly rough and extend onto dorsal-fin base. 7+1 preopercular pores present. Canine teeth absent in both sexes. HABITAT: Point Conception, CA, to northern Baja California. Demersal from about 10 to 300 ft. NOTE: *R. alleni* and *R. hypoplecta* may be difficult to separate where ranges overlap.

Northern Ronquil - *Ronquilus jordani* (Gilbert, 1889)

FEATURES: Color and pattern variable. Uniformly grayish, pinkish gray, brownish, pale orange to olive, or variably blotched and barred in shades of brown. Fins variably tipped with orange. All with faint to dark orange bands below and behind eyes. Pectoral fins reach past anal-fin origin. Dorsal fin comparatively tall, with 41–48 rays. Scales nearly smooth. Body comparatively slender. HABITAT: Southeastern Bering Sea and Aleutian Islands to La Jolla, CA. Demersal over a variety of soft, hard, and mixed bottoms and in crevices from about 10 to 900 ft; usually above 490 ft.

Smallmouth Ronquil
to 8.25 in.

Searcher
to 15 in.

Stripefin Ronquil
to 6.3 in.

variants

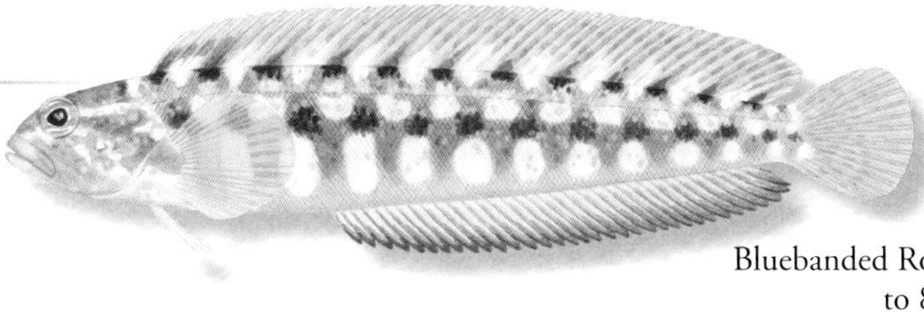

Bluebanded Ronquil
to 8.5 in.

Northern Ronquil
to 8 in.

variants

Twoline Eelpout - *Bothrocara brunneum* (Bean, 1890)

FEATURES: Shades of brown to gray. Inner margins of dorsal and anal fins bluish to blackish. Peritoneum black. Snout long, depressed. Mouth large, eyes moderately small. Head with large pits. Pectoral fins moderately long, but not reaching anus. Pelvic fins absent. Lateral line in two segments; lower segment extends to tail. Small scales cover body, abdomen, and tail; nape unscaled. Body robust anteriorly. HABITAT: Navarin Canyon, Bering Sea, and Aleutian Islands to northern Baja California. Demersal over continental slopes from about 420 to 6,000 ft. More common below about 1,300 ft.

Soft Eelpout - *Bothrocara molle* Bean, 1890

FEATURES: Shades of pale brown. Outer margins of posterior dorsal and anal fins blackish. Snout long, depressed. Mouth, eyes large. Head with large pits. Pectoral fins long and reaching to or past anus. Pelvic fins absent. Lateral line in two segments; lower segment extends to tail. Small scales cover nape, body, abdomen, and tail. Body comparatively slender anteriorly. HABITAT: Southern Bering Sea and Queen Charlotte Island, British Columbia, to northern Baja California. Also to Sea of Japan. Demersal from about 200 to 8,800 ft.

Alaska Eelpout - *Bothrocara pusillum* (Bean, 1890)

FEATURES: Head and body semi-translucent gray. Abdomen pale. Snout short and blunt, mouth and eyes large. Head with numerous pores. Pectoral fins moderately long. Pelvic fins and lateral line absent. Minute scales present on body but absent from nape. HABITAT: Eastern Bering Sea to OR. Found demersally and in water column just above bottom from about 260 to 7,200 ft.

Halfbarred Pout - *Gymnelus hemifasciatus* Andriashev, 1937

FEATURES: Color and pattern variable in all. Females brownish, with irregular pale bars, blotches, and spots, and dark vermiculations below dorsal fin. Males shades of orange, with pale bars, bands, and spots, and dark vermiculations below dorsal fin; often with two black ocellated spots on dorsal fin (some lack spots); anal fin with black margin. Pectoral fins usually with 10–11 rays. Pelvic fins absent. HABITAT: Beaufort Sea to northwest of Kodiak Island, AK. Also to Dease Strait, Canada, and to Sea of Okhotsk. Demersal over muddy and gravel bottoms from intertidal zone to about 655 ft. Usually between about 130 to 260 ft.

Fish Doctor - *Gymnelus viridis* (Fabricius, 1780)

FEATURES: Color and pattern highly variable. May be uniformly brown to green; yellowish orange with whitish bands; greenish brown with yellow bands; or greenish brown to reddish with whitish or yellowish mottling and reticulations. When present, bands always extend below lateral line. Mature males with blackish anal fin. One or more black ocellated spots may be present on dorsal fin. Dorsal and anal fins comparatively tall. Pectoral fins usually with 11–13 rays. Pelvic fins absent. HABITAT: Circumpolar. Beaufort Sea to Islands of Four Mountains, AK. Demersal over mud, sand, gravel, and rocky bottoms with cover from intertidal zone to about 1,040 ft.

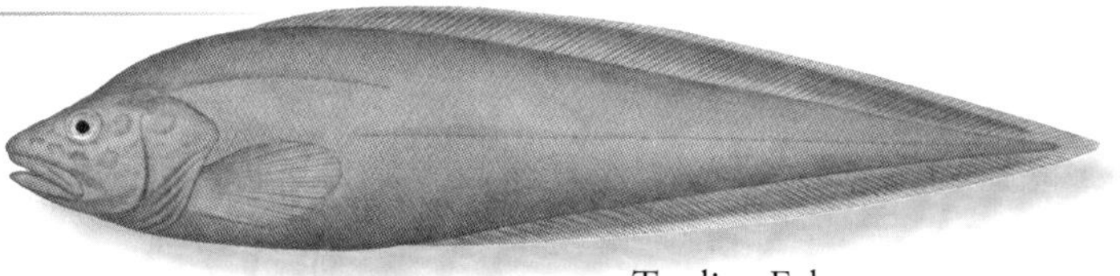

Twoline Eelpout
to 2.4 ft.

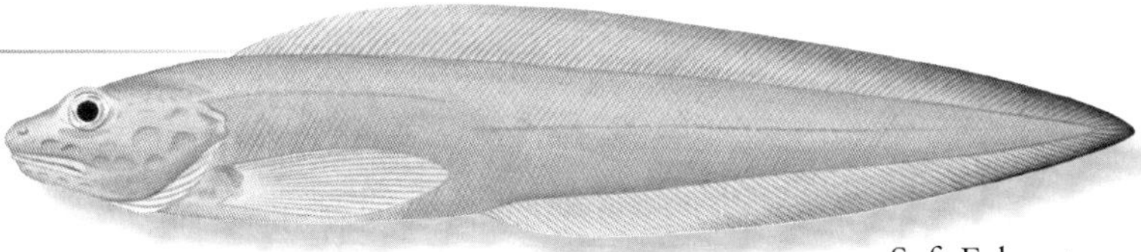

Soft Eelpout
to 23.2 in.

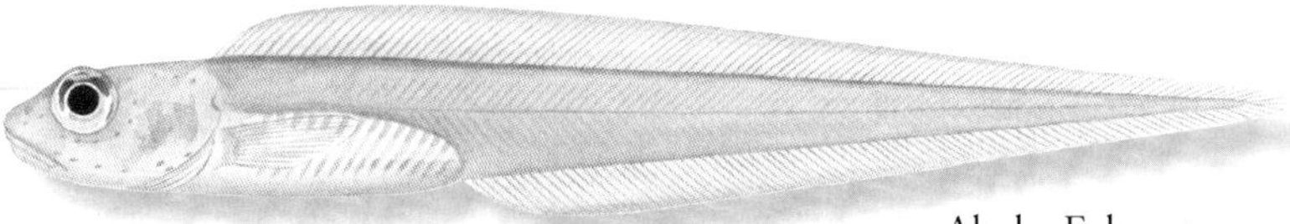

Alaska Eelpout
to 6.1 in.

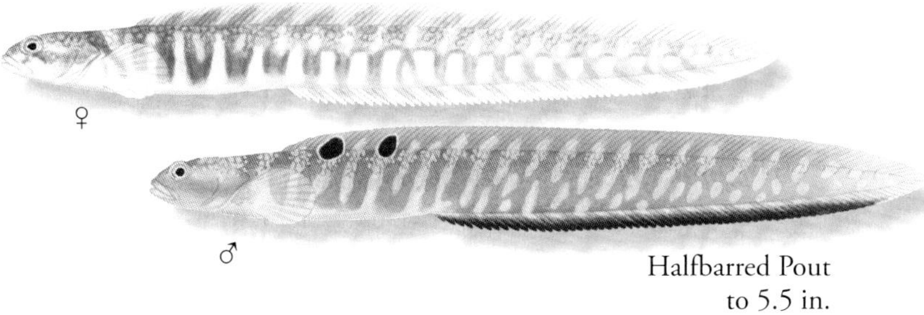

Halfbarred Pout
to 5.5 in.

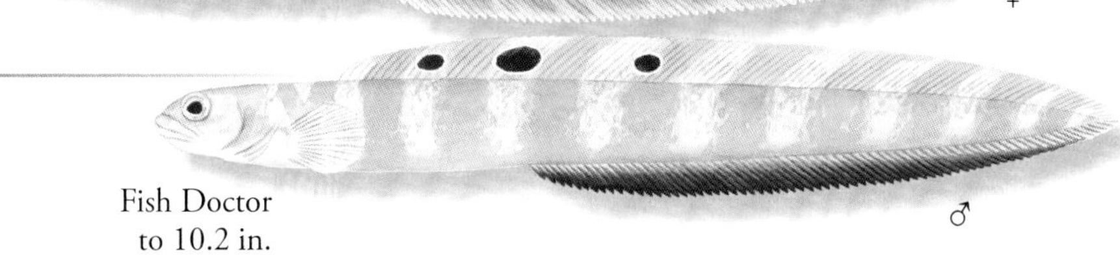

Fish Doctor
to 10.2 in.

Shortfin Eelpout - *Lycodes brevipes* Bean, 1890

FEATURES: Brown to gray brown with up to 13 narrow, pale bars that extend onto dorsal fin and a pale band on nape that connects opercles. Pale bars sometimes absent. Lower head and abdomen pale. Dorsal and anal fins with dark margins. Chin with low crests. Pectoral fins rounded or slightly notched. Pelvic fins tiny, less than half of eye diameter. Scales cover body and dorsal- and anal-fin bases. Lateral line runs ventrally. HABITAT: Eastern Chukchi Sea to eastern Aleutian Islands and south to Fort Bragg, CA. Demersal over sandy and muddy bottoms from about 40 to 3,200 ft.

Ebony Eelpout - *Lycodes concolor* Gill & Townsend, 1897

FEATURES: Uniformly dark brown, with blackish dorsal- and anal-fin margins to overall blackish. Head large, about 20% to 25% of total length. Chin with well developed crests. Pectoral fins large and rounded, middle rays longest. Pelvic fins small. Lateral line runs ventrally. Body covered with very small scales that extend onto dorsal and anal fins. Head scaleless. HABITAT: Eastern Bering Sea from Navarin Canyon along continental slope and Aleutian Islands to western Gulf of Alaska. Demersal from about 140 to 3,940 ft.

Bigfin Eelpout - *Lycodes cortezianus* (Gilbert, 1890)

FEATURES: Pale brown to bluish black dorsally, whitish below. Pectoral fins broad and dark with a pale margin or entirely pale in smaller specimens. Dorsal fin with a black blotch and margin. Anal fin with a posterior black margin. Head large, depressed. Chin with low crests. Pectoral fins large, rounded. Pelvic fins small. Lateral line runs along midline. Scales absent on head. HABITAT: Prince of Wales Island, AK, and Queen Charlotte Sound, British Columbia, to San Diego, CA. Demersal over mud, sand, and gravel from about 200 to 3,800 ft. Usually from about 330 to 1,500 ft.

Black Eelpout - *Lycodes diapterus* Gilbert, 1892

FEATURES: Bluish to brownish dorsally, with 8–9 whitish, inverted V-shaped marks or irregular pale bars on upper sides that merge onto dorsal fin. Markings fade with age. Dorsal fin with blackish margin. Pectoral, pelvic, and anal fins bluish black. Chin with low crests. Pectoral fins nearly straight-margined to notched. Pelvic fins small. Lateral line runs ventrally. HABITAT: Bering Sea from Navarin Canyon and Aleutian Islands to San Diego, CA. Also to Sea of Okhotsk and Sea of Japan. Demersal over muddy bottoms from about 40 to 4,300 ft.

Saddled Eelpout - *Lycodes mucosus* Richardson, 1855

FEATURES: Tannish, with variable brownish, dark-margined, V- or Y-shaped marks that extend onto fins and lower body. A pale band on nape connects opercles. Underside of head and abdomen pale. Markings fade with age. Head broad and flattened. Chin with well-developed crests. Pectoral fins large, fan-like. Pelvic fins small. Lateral line runs along midline. Scales weakly developed, absent anteriorly. HABITAT: Beaufort Sea to Pribilof Islands, AK. Also along Canadian and Russian Arctic. Demersal over soft bottoms from about 5 to 260 ft.

Shortfin Eelpout
to 12.9 in.

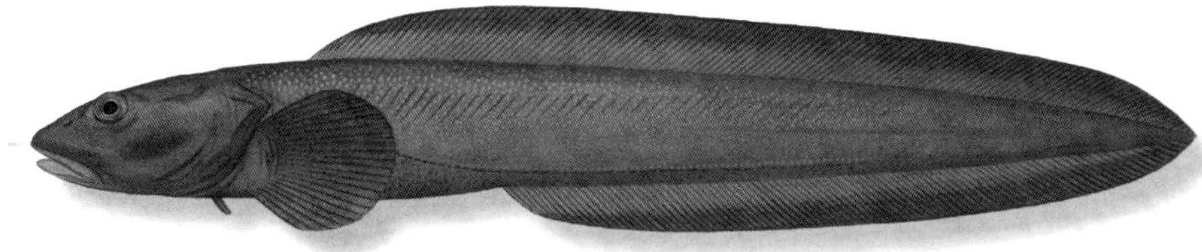

Ebony Eelpout
to 2.6 ft

Bigfin Eelpout
to 19.4 in.

Black Eelpout
to 14.6 in.

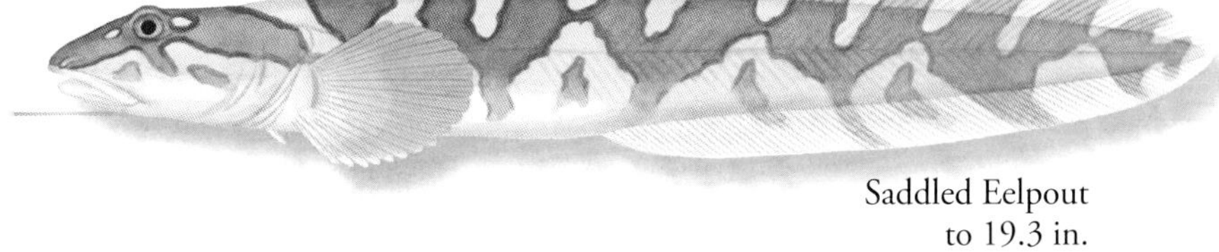

Saddled Eelpout
to 19.3 in.

Blackbelly Eelpout - *Lycodes pacificus* Collett, 1879

FEATURES: Shades of pale brown to pale gray dorsally. Abdomen pale. Peritoneum black and may be visible through skin. Young specimens with several narrow pale bars dorsally. Dorsal fin with variable, narrow black margin and a black blotch anteriorly. Pectoral fins comparatively pale. Males with larger head and mouth than females. Crests under chin inconspicuous. Lateral line runs along midline to above anterior anal fin. Body covered with small scales; head scaleless. HABITAT: Aleutian Islands to northern Baja California. Demersal over muddy bottoms from about 20 to 3,400 ft.

Wattled Eelpout - *Lycodes palearis* Gilbert, 1896

FEATURES: Shades of brown, with pectoral fins darker than body. Small specimens with pale bars in front of dorsal fin and pale body bars that extend onto fins. Bars fade with age. Anterior dorsal fin with a dark blotch that may fade with age. Chin with a pair of prominent, well-developed, projecting crests. Lateral line runs ventrally. Scales cover body and extend onto rear of head and onto dorsal and anal fins. HABITAT: Chukchi and Bering seas and Aleutian Islands to OR. Demersal over continental shelves over soft bottoms from about 7 to 3,000 ft; usually above 655 ft.

Canadian Eelpout - *Lycodes polaris* (Sabine, 1824)

FEATURES: Tannish to bright orange, with 9–11 dark bars with paler centers that merge onto fins. A pale band across nape may appear as three blotches. Bars may be poorly defined. Chin crests prominent. Lateral line runs along midline, with widely spaced sensory cells above. Scales usually absent. HABITAT: Nearly circumpolar along Arctic coasts. Beaufort and Chukchi seas to between St. Lawrence and Hall islands, AK. Demersal over mud and sandy mud bottoms from about 15 to 775 ft.

Marbled Eelpout - *Lycodes raridens* Taranetz & Andriashev, 1937

FEATURES: Yellowish tan, with irregular, wavy brown blotches that form loose bars on sides. Head with numerous wavy spots. Small specimens more evenly marked. Chin crests well developed. Scales cover body and extend onto dorsal and anal fins. Lateral line along midline. HABITAT: Chukchi and Bering seas to Bristol Bay, AK. Also to Sea of Okhotsk. Demersal over sand and clay bottoms from about 25 to 1,200 ft.

Midwater Eelpout - *Melanostigma pammelas* Gilbert, 1896

FEATURES: Color somewhat variable. Head and abdomen blackish. Body and fins shades of tan to brown or silvery blue. Snout blunt. Eyes large. Gill slit reduced to a small pore. Dorsal and anal fins embedded in skin. Pelvic fins absent. Skin loose and scaleless. Body very long, tapered and compressed posteriorly. HABITAT: Queen Charlotte Islands, British Columbia, to Gulf of Tehuantepec, Mexico. Found in midwater from about 320 to 6,600 ft.

ALSO IN THE AREA: Aleutian Pout, *Commandorella popovi;* Persimmon Eelpout, *Eucryphycus californicus*; Polar Eelpout, *Lycodes turneri*; Bearded Eelpout, *Lyconema barbatum;* pages 350–351.

Blackbelly Eelpout
to 18 in.

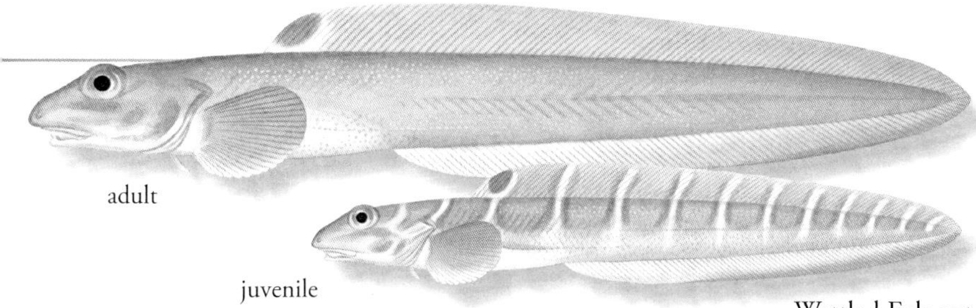

Wattled Eelpout
to 2 ft.

Canadian Eelpout
to 18 in.

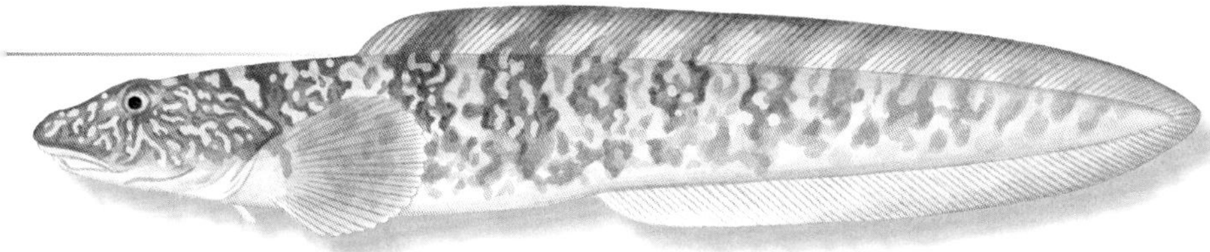

Marbled Eelpout
to 2.3 ft.

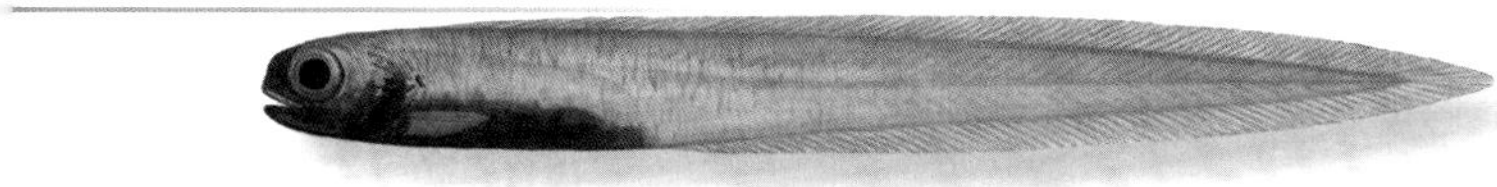

Midwater Eelpout
to 5 in.

Blackline Prickleback - *Acantholumpenus mackayi* (Gilbert, 1896)

FEATURES: Pale yellowish to pale brown dorsally, with a black stripe below dorsal fin and elongate blackish spots forming two broken stripes on sides. Whitish below. Caudal fin brownish, lacks bands. Eyes small. Dorsal fin very low anteriorly, with spines connected by membrane. Dorsal and anal fins not connected to caudal fin. Caudal-fin margin variable in shape. HABITAT: Circumpolar. Southeastern Chukchi Sea, eastern Bering Sea, and Aleutian Islands to northeastern Gulf of Alaska. Also to Sea of Japan. Demersal over soft bottoms from near shore and brackish waters to about 185 ft.

Stout Eelblenny - *Anisarchus medius* (Reinhardt, 1837)

FEATURES: Uniformly creamy, yellowish, to reddish, or with vague brownish spots on sides. Abdomen pale. Dorsal fin may have oblique bands. Caudal fin banded. Eyes relatively large. Dorsal fin moderately low anteriorly. Anal fin increasing in height posteriorly. Dorsal and anal fins connected to caudal fin. Caudal-fin margin rounded. HABITAT: Circumpolar. Beaufort Sea to Auke Bay, AK. Also to Sea of Okhotsk and Sea of Japan. Demersal over muddy bottoms from near shore to about 870 ft.

Slender Cockscomb - *Anoplarchus insignis* Gilbert & Burke, 1912

FEATURES: Color and pattern highly variable. Shades of gray, brown, purplish, or black to pinkish; usually with numerous irregular pale blotches along dorsal profile that merge onto dorsal fin. Irregular pale streak below eyes. Bluish to blackish blotch on anterior dorsal fin. Fins variably banded. Height of head crest variable but comparatively low on snout. Space between gill slits at throat 50% or less than eye diameter. Anal fin with 40–46 rays. HABITAT: SE Bering Sea and Aleutian Islands to Arena Cove, CA. Demersal over rocks and algae from intertidal zone to about 180 ft.

High Cockscomb - *Anoplarchus purpurescens* Gill, 1861

FEATURES: Color, pattern highly variable. Mottled in shades of brown, black, purple, or gray, with pale blotches along back that merge onto dorsal fin. Breeding males with reddish orange on back and fins. Irregular pale streak below eyes. Bluish blotch on anterior dorsal fin. Fins variably banded. Height of head crest variable but comparatively high on snout. Space between gill slits at throat usually more than 75% of eye diameter. Anal fin with 36–42 rays. HABITAT: Pribilof and Aleutian Islands to Santa Rosa Island, CA. Demersal over rocks and algae from intertidal zone to about 100 ft.

Nutcracker Prickleback - *Bryozoichthys lysimus* (Jordan & Snyder, 1902)

FEATURES: Brownish, with fine speckles and vermiculations forming vague dark bars and pale blotches on sides. Dark bar may be present below eyes. Pectoral fins often orange. Long cirrus on snout. Two pairs of multi-branched cirri above eyes and numerous long, branched or unbranched cirri on head and nape. Chin with a pair of cirri. Anterior dorsal fin with cirri. Anal fin with 47–50 rays. HABITAT: Scattered in eastern Bering Sea and around Aleutian Islands to western Gulf of Alaska. Demersal over soft bottoms from about 150 to 1,600 ft.

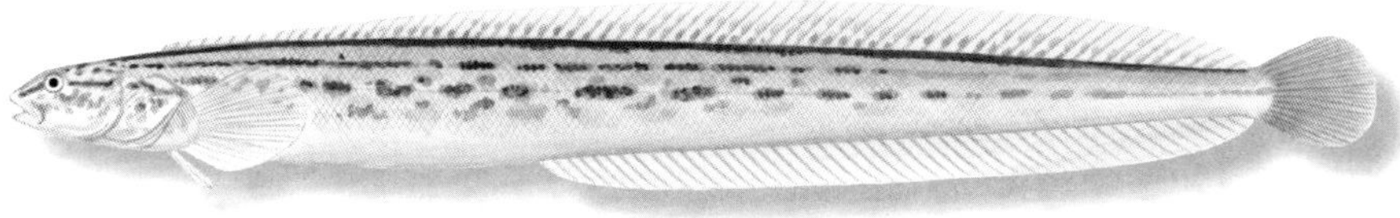

Blackline Prickleback
to 2.3 ft.

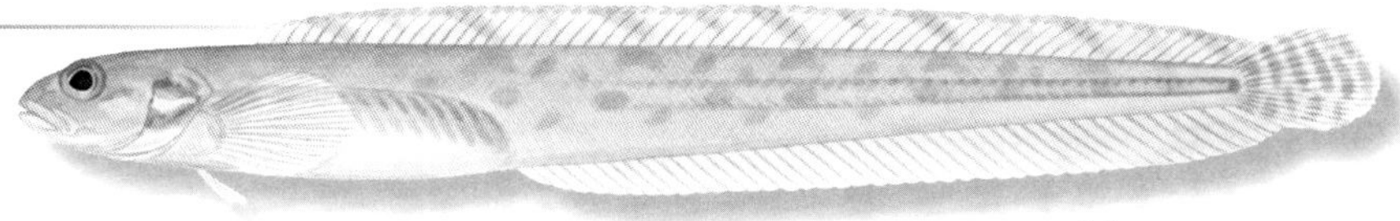

Stout Eelblenny
to 7.1 in.

Slender Cockscomb
to 4.6 in.

breeding ♂

High Cockscomb
to 7.8 in.

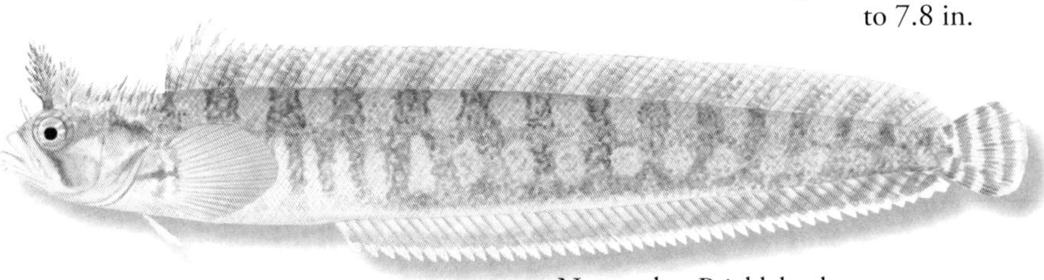

Nutcracker Prickleback
to 10.6 in.

Pearly Prickleback - *Bryozoichthys marjorius* McPhail, 1970

FEATURES: Pearly white, with fine brownish gray speckles forming about nine vague bars on upper sides. Dark bar may be present below eyes. Dorsal fin may have a dark anterior blotch. Pectoral fins pale, yellowish, or tan. Short cirrus on snout. A single pair of multi-branched cirri above eyes, followed by a row of three shorter branched cirri and 18–20 simple cirri on head and nape. Maxilla and lower jaw with a single cirrus. Anterior dorsal fin with cirri. Anal fin with 53–55 rays. HABITAT: Scattered along Aleutian Islands to S British Columbia. Demersal over soft bottoms from about 325 to 1,230 ft.

Monkeyface Prickleback - *Cebidichthys violaceus* (Girard, 1854)

FEATURES: Color and pattern highly variable. Nearly uniform shades of brown, gray, or olive to black, or mottled with variable irregular pale and dark blotches. Most with dark bars below eyes. May also have reddish blotches on side. Capable of rapid color change. Juveniles may have a whitish dorsal stripe. Adults with a lumpy head crest. Gill slits joined at throat. Pectoral fins moderately large. Dorsal fin with 22–25 spines and 40–43 rays. Dorsal and anal fins connected to caudal fin. Lateral line single, high on body. HABITAT: Southern OR to northern Baja California. Demersal over rocky bottoms, in caves and crevices, and around kelp beds from intertidal zone to about 80 ft.

Decorated Warbonnet - *Chirolophis decoratus* (Jordan & Snyder, 1902)

FEATURES: Color and pattern highly variable. Shades of tan to gray, with variable darker wavy bars and/or orange spots on body that merge onto dorsal fin. Body may also have scattered dark flecks. Pectoral, caudal, and anal fins banded. All with a single pair of very tall, complex cirri over eyes, followed by numerous simple or multi-branched cirri on head and nape. A few cirri on cheeks. Chin with a tiny pair of cirri. First 4–9 dorsal-fin spines with tall, multi-branched cirri. Anal fin with 44–47 rays. HABITAT: E Bering Sea and Aleutian Islands to Humboldt Bay, CA. Demersal over rocky bottoms, in crevices, and around weed beds from near shore to about 300 ft.

Mosshead Warbonnet - *Chirolophis nugator* (Jordan & Williams, 1895)

FEATURES: Color and pattern highly variable. Shades of red, brown, cream, or gray to orange, with alternating irregular pale and dark blotches, spots, and streaks on head and sides. All with 12–13 distinct to vague blackish ocellated spots or bars on dorsal fin. Branched cirrus on snout. Two pairs of moderately tall, multi-branched cirri over eyes and numerous mostly evenly tall simple or multi-branched cirri on head and nape. Anal fin with 37–42 rays. HABITAT: Aleutian Islands to San Miguel Island, CA. Demersal over a wide variety of bottoms from intertidal zone to about 265 ft.

Masked Prickleback - *Ernogrammus walkeri* Follett & Powell, 1988

FEATURES: Mottled tannish to grayish tan, with a dark brown stripe from tip of snout to tail that becomes wavy on body and broken posteriorly. Snout bluntly pointed. Cirri absent from head. Dorsal fin connected to caudal fin, anal fin barely connected. Four lateral line canals present. HABITAT: Monterey Bay to San Miguel Island, CA. Demersal from intertidal zone to about 70 ft.

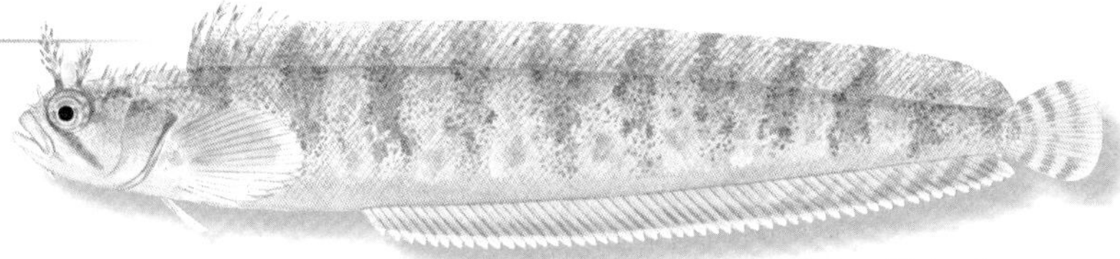

Pearly Prickleback
to 12 in.

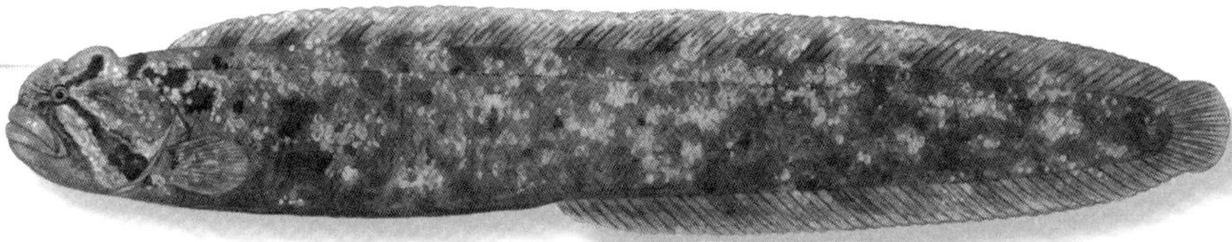

Monkeyface Prickleback
to 2.5 ft.

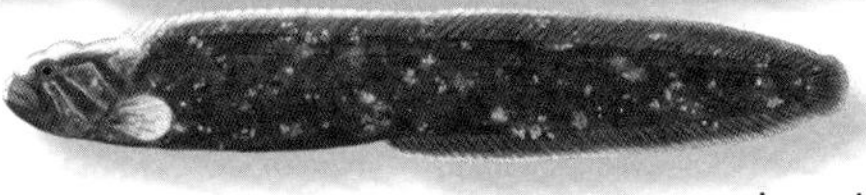

juvenile

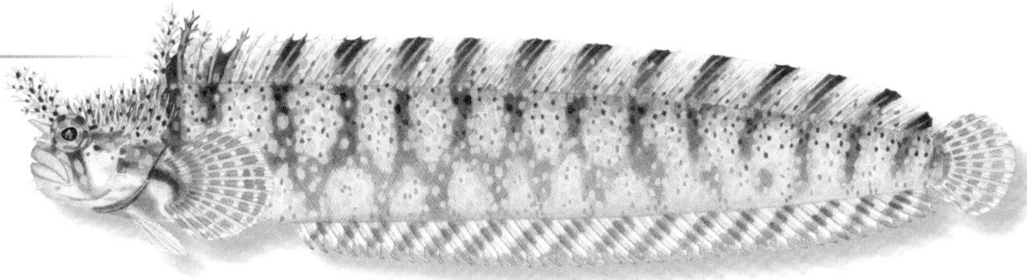

Decorated Warbonnet
to 16.5 in.

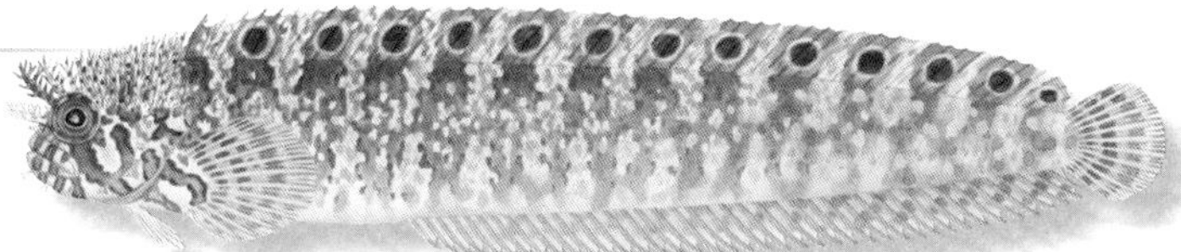

Mosshead Warbonnet
to 5.8 in.

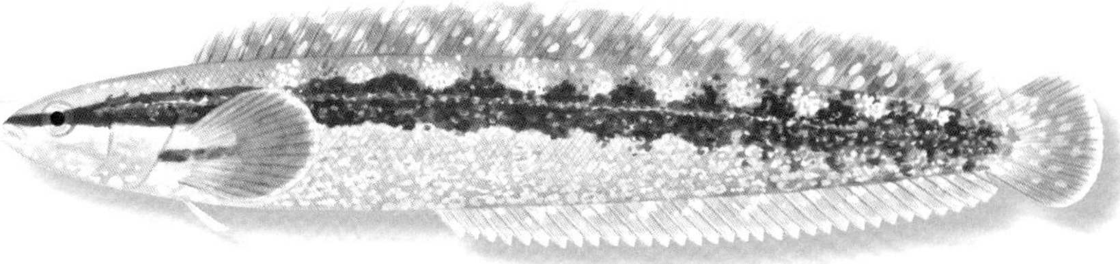

Masked Prickleback
to 11.3 in.

Fourline Snakeblenny - *Eumesogrammus praecisus* (Krøyer, 1836)

FEATURES: Shades of tan to dark brown or gray, with irregular, vague, dark bands on sides. Dorsal fin blackish along margin, with one to three ocellated spots anteriorly. Caudal-fin base yellowish, inner area black. Pectoral and anal fins black with pale margins. Snout pointed. Nape arched. Two lateral lines with two branches present but varying in length; some may be missing or partly missing. Body covered with small scales. HABITAT: Beaufort Sea to NE Bering Sea. Also to Sea of Okhotsk and Gulf of St. Lawrence. Demersal over soft and mixed bottoms from about 50 to 1,300 ft.

Sixspot Prickleback - *Kasatkia seigeli* Posner & Lavenberg, 1999

FEATURES: Shades of reddish orange, with a white stripe from snout to pectoral-fin base that may be bordered with brown to black above. Dorsal fin with five to six large ocellated spots; may have several smaller spots along base. Caudal fin with a clear margin. Lower jaw protrudes. Pelvic fins (when present) with a single, small spine. Pectoral fins large. Dorsal fin with 61–64 spines. Lateral line on midline with a dorsal branch. HABITAT: Mendocino County, CA, to Diablo Cove, CA. Demersal over hard bottoms and around vegetation from inshore to about 85 ft.

Daubed Shanny - *Leptoclinus maculatus* (Fries, 1838)

FEATURES: Tannish with darker spots and blotches on sides. Abdomen pale. Dorsal fin banded. Caudal fin barred. Mouth and eyes large. Snout blunt. Anterior dorsal-fin spines very short, gradually becoming taller at midbody and short toward tail. Lower pectoral-fin rays elongate, free at tips. Pelvic fins long, very slender. HABITAT: Beaufort Sea to Puget Sound, WA. Also to Sea of Japan and Cape Cod, MA. Demersal over soft or hard bottoms from about 7 to 1,600 ft.

Longsnout Prickleback - *Lumpenella longirostris* (Evermann & Goldsborough, 1907)

FEATURES: Pale to dark bluish brown or with irregular, vague, darker blotches and bars on upper sides. Pectoral and caudal fins blackish. Dorsal and anal fins with blackish margins. Snout long, overhangs lower jaw. Dorsal-fin spines stiff, with free tips. Lateral line indistinct. Head and body covered with minute scales. HABITAT: Bering Sea and Aleutian Islands to Burrard Inlet, British Columbia. Also to Sea of Okhotsk and Greenland. Demersal over outer shelves and upper slopes from about 80 to 3,920 ft.

Y-Prickleback - *Lumpenopsis hypochroma* (Hubbs & Schultz, 1932)

FEATURES: Shades of brown to grayish brown with variable darker blotches on lower sides that may resemble Y shapes. Upper sides with variable, small pale blue to gray spots and blotches. White streaks under eyes and a white slash on cheeks. Dorsal fin with about five dark blotches along base. Mouth, eyes large. Snout bluntly rounded. Dorsal fin low anteriorly, moderately high at middle. Pectoral fins moderately long. Scales cover body but not head. HABITAT: Southern British Columbia to CA. Demersal over hard and soft bottoms with mud, rock, and debris from about 100 to 330 ft.

Fourline Snakeblenny
to 9.1 in.

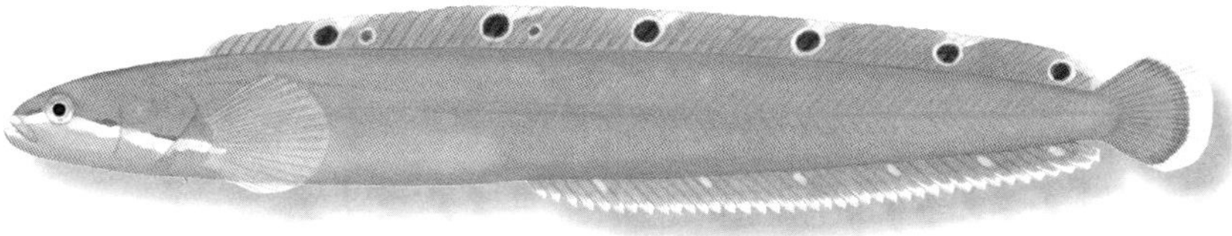

Sixspot Prickleback
to 5.5 in.

Daubed Shanny
to 7.9 in.

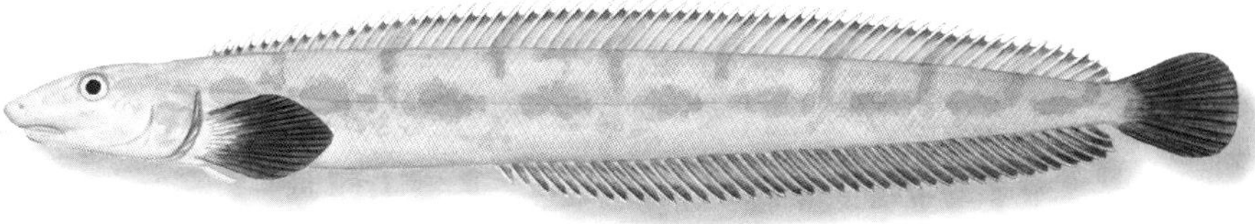

Longsnout Prickleback
to 17.2 in.

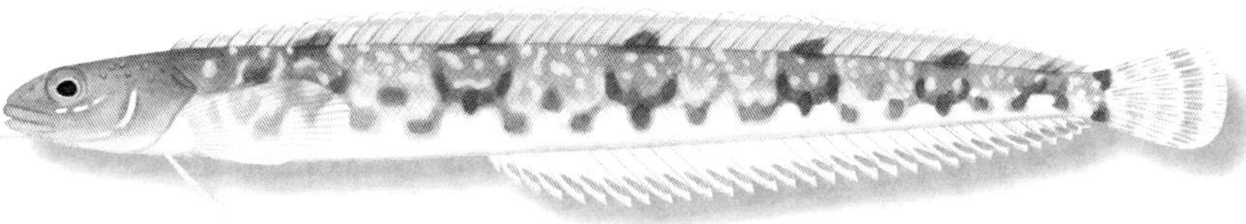

Y-Prickleback
to 2.9 in.

Slender Eelblenny - *Lumpenus fabricii* Reinhardt, 1836

FEATURES: Creamy to yellowish, with vague to distinct irregular pale to dark brownish blotches or broken diagonal bands on upper sides. Abdomen pale. Dorsal fin banded. Caudal fin unmarked or banded. Mouth horizontal. Dorsal and anal fins not connected to caudal fin. Posterior anal-fin rays comparatively short. Scales minute, present on cheeks. Body comparatively slender. HABITAT: Circumpolar. Beaufort Sea to Auke Bay, AK. Demersal over sandy to rocky bottoms from intertidal zone (rarely) to about 575 ft; usually shallower than 330 ft.

Snake Prickleback - *Lumpenus sagitta* Wilimovsky, 1956

FEATURES: Pale green to tannish or gray, with rows of irregular brown to gray or blackish blotches on upper sides. Uppermost row may merge onto dorsal fin. Caudal fin banded. Mouth nearly horizontal. First few dorsal-fin spines very short and free. Dorsal and anal fins not connected to caudal fin. Posterior anal-fin rays comparatively long. Scales minute, absent from cheeks. Body comparatively long and slender. HABITAT: Bering Sea and Aleutian Islands to Humboldt Bay, CA. Also to Sea of Japan. Demersal over sandy bottoms mixed with silt, stones, and pebbles from shore to about 655 ft.

Ribbon Prickleback - *Phytichthys chirus* (Jordan & Gilbert, 1880)

FEATURES: Color and pattern vary. Brownish, olivaceous, or bright green to dark red and with or without spots, lines, and marbling. Usually with a row of small whitish or dark spots on sides and dark streaks radiating from eyes. Anal fin sometimes with pale spots or dark bars. Snout short, mouth small and upturned. Pectoral fins about as large or slightly larger than eye diameter. Dorsal and anal fins confluent with caudal fin. Anal fin about half the length of body. Four branched lateral lines present. HABITAT: Aleutian Islands to southern CA. Demersal under rocks and in algae from intertidal zone to about 40 ft.

Bluebarred Prickleback - *Plectobranchus evides* Gilbert, 1890

FEATURES: Shades of tan to brown, with about 25 narrow blue bars and several dark blotches on body. Dorsal fin with oblique dark bands and two or three black blotches posteriorly. Pectoral, caudal, and anal fins with narrow white margins. Mouth moderately large. Snout short. Lower pectoral-fin tips free. Dorsal and anal fins not connected to caudal fin. Scales minute. HABITAT: Central British Columbia to San Diego, CA. Demersal over soft and mixed bottoms from about 240 to 1,200 ft.

Whitebarred Prickleback - *Poroclinus rothrocki* Bean, 1890

FEATURES: Dorsal area with alternating narrow white bars and tan to brown rectangular blocks or blotches that may have dark borders. Ventral area whitish. May have a dark stripe through eyes. Snout short and blunt. Dorsal and anal fins very low, not connected to caudal fin. Caudal fin long and pointed to rounded. Body covered with very small scales. HABITAT: Southern Bering Sea and Aleutian Islands to San Diego, CA. Demersal over rocky and soft bottoms from about 115 to 525 ft.

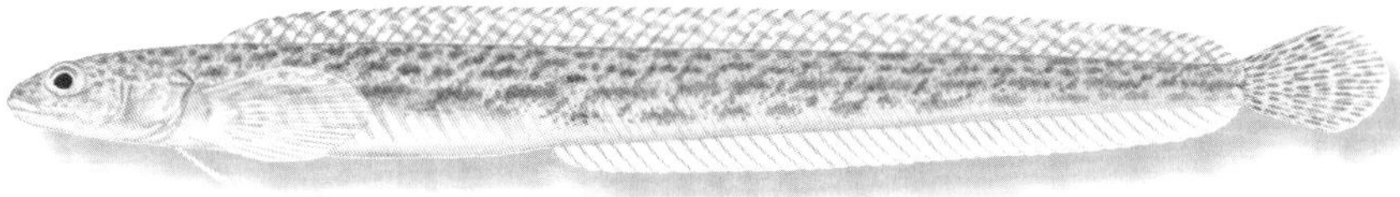

Slender Eelblenny
to 14.4 in.

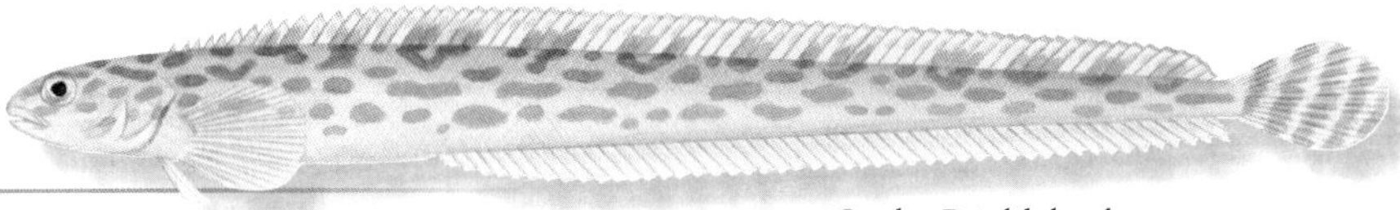

Snake Prickleback
to 20.1 in.

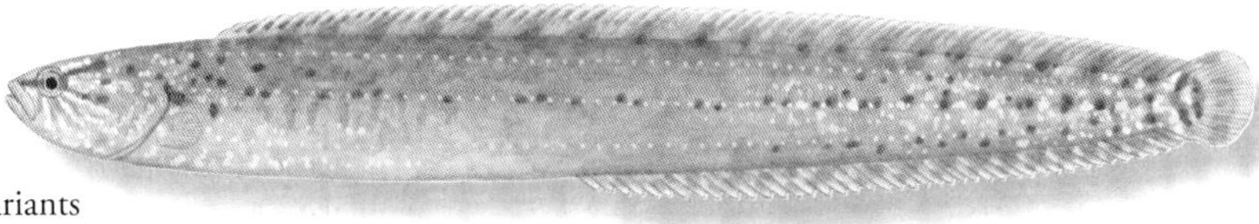

variants

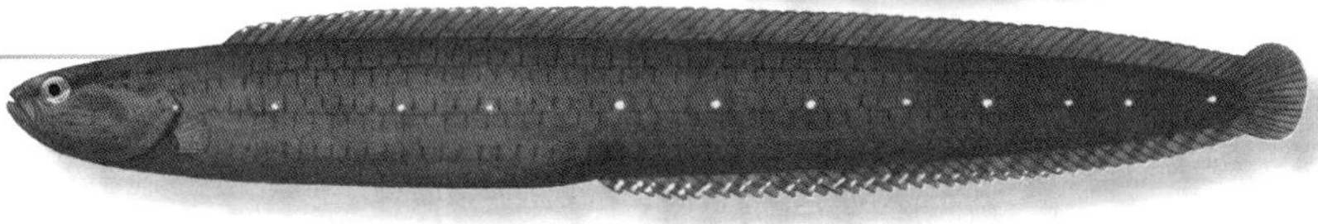

Ribbon Prickleback
to 8.3 in.

Bluebarred Prickleback
to 5.4 in.

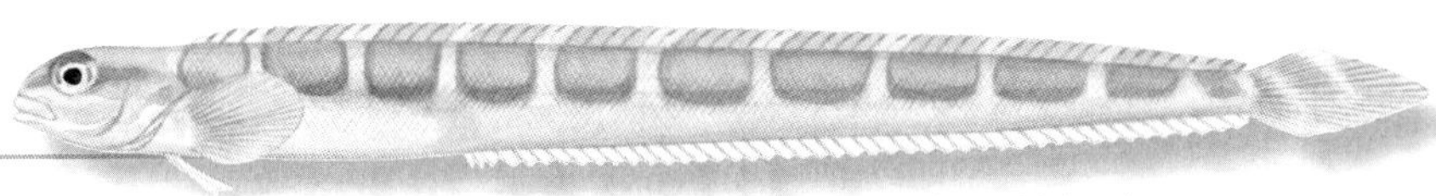

Whitebarred Prickleback
to 10 in.

Arctic Shanny - *Stichaeus punctatus* (Fabricius, 1780)

FEATURES: Color and pattern highly variable. Shades of tan, brown, olivaceous, pink, dark gray, or red to purplish, with variable, irregular pale and dark blotches and mottling on sides. All with dark bands behind eyes and on lower jaw. Dorsal fin with four to seven blackish spots with a posterior yellowish blotch. Snout bluntly pointed. Lateral line single, high on back, ending near midbody. HABITAT: Beaufort Sea to Skidgate Inlet, British Columbia, including Aleutian Islands. Also to Sea of Japan and Gulf of Maine. Demersal from shallow subtidal rocky and sandy bottoms to about 330 ft.

Black Prickleback - *Xiphister atropurpureus* (Kittlitz, 1858)

FEATURES: Almost uniformly colored or partially to completely mottled in shades of reddish purple to dark brown or black. Two dark brown bands with pale borders radiate below eyes. Usually with a pale bar at caudal-fin base. Pectoral fins minute. Dorsal fin originates well behind pectoral fins. Dorsal and anal fins confluent with caudal fin. Four branched lateral lines present. HABITAT: Gulf of Alaska to northern Baja California. Demersal around rocks and algae from intertidal zone to about 40 ft.

Rock Prickleback - *Xiphister mucosus* (Girard, 1858)

FEATURES: Nearly uniformly colored or mottled in shades of brown or dark olive to greenish black. Two dark-margined bands radiate below eyes. Usually with a pale bar at caudal-fin base. Pectoral fins minute. Dorsal fin originates above pectoral fins. Dorsal and anal fins confluent with caudal fin. Four branched lateral lines present. HABITAT: Gulf of Alaska to Santa Cruz Island, CA. Demersal around rocks and algae from intertidal zone to about 60 ft. Other name: Rock Eel.

NOTE: There are nine other rare or poorly recorded, shallow-water Pricklebacks in the area.

Cryptacanthodidae - Wrymouths

Dwarf Wrymouth - *Cryptacanthodes aleutensis* (Gilbert, 1896)

FEATURES: Uniformly red to pink or with irregular darker spots and stipples. Mouth large and steeply oblique. Nostril tube long. Dorsal and anal fins extend to caudal fin; sometimes extend onto caudal fin. Pelvic fins and scales absent. Lateral line indistinct. HABITAT: Southeastern Bering Sea and eastern Aleutian Islands to Eureka, CA. Demersal and often buried in soft bottoms from about 70 to 2,460 ft.

Giant Wrymouth - *Cryptacanthodes giganteus* (Kittlitz, 1858)

FEATURES: Shades of pinkish brown to pale brown or grayish, with darker spots forming rows. Sometimes tannish, with a dark band along lateral line. Mouth large, oblique, and with maxilla extending past eyes. Dorsal and anal fins extend to or onto caudal fin. Pelvic fins absent. Small, scattered scales on lateral line and posteriorly. HABITAT: Southeastern Bering Sea and eastern Aleutian Islands to Humboldt Bay, CA. Demersal and often buried in soft bottoms from about 20 to 1,090 ft.

Pricklebacks

Arctic Shanny
to 8.7 in.

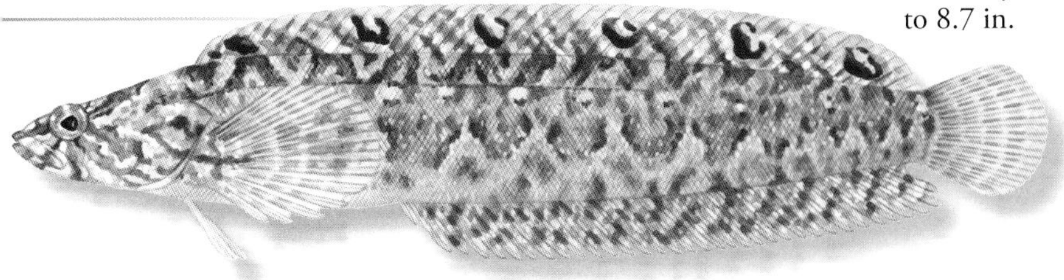

Black Prickleback
to 12.7 in.

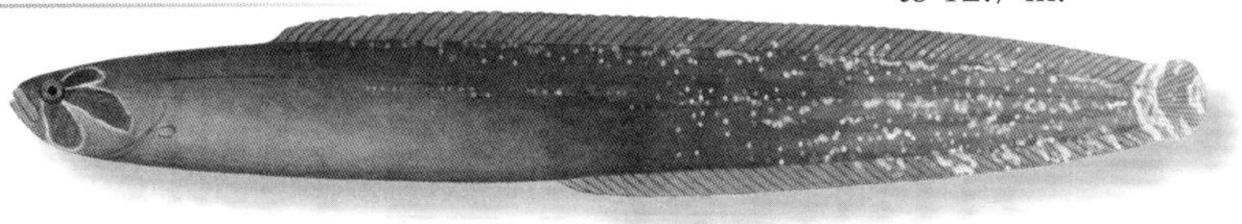

Rock Prickleback
to 23 in.

variants

Wrymouths

Dwarf Wrymouth
to 12.2 in.

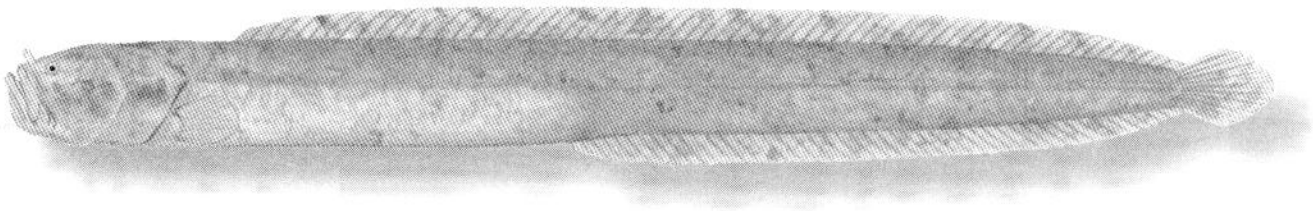

Giant Wrymouth
to 4.4 ft.

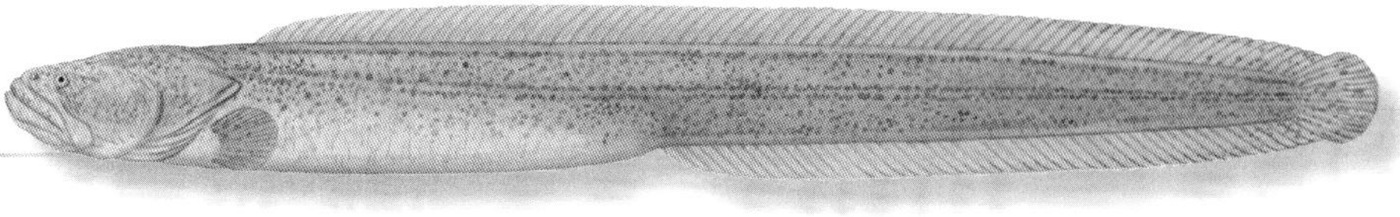

Penpoint Gunnel - *Apodichthys flavidus* Girard, 1854

FEATURES: Color highly variable. Usually uniformly colored in shades of green, orange, red, yellow, or brown to bronze, but may be mottled or have rows of dark and/or pale spots on sides. Usually with a dark streak below eyes and often another streak above eyes that is often broken. Sometimes with a dark or silvery white streak through eyes. Gill cover with a flap at corner. Pectoral fins moderately large, about two times eye diameter. Anal fin about one third of body length, with a strong, deeply grooved anterior spine that is recessed between fleshy folds. Pelvic fins absent. Body comparatively deep and compressed. HABITAT: Kodiak Island, AK, to Santa Barbara and Gaviota, CA. Demersal over a wide variety of habitats, including rockweed beds, from intertidal zone and tidepools to about 25 ft. Primarily in marine waters but can tolerate estuarine conditions.

Longfin Gunnel - *Pholis clemensi* Rosenblatt, 1964

FEATURES: Color varies from pale pink, bright red, or brownish to orangish or magenta, with about 15 irregular pale spots or saddles with dark interior spots along upper back. Ventral area with irregular pale streaks, blotches, or inverted U shapes, with or without dark centers. Some with a pale streak behind eyes. Pectoral fins large. Pelvic fins minute. Anal fin about half of body length. Body slender. HABITAT: Fillmore Island, AK, to Arena Cove, CA. Demersal over sandy, rocky, and mixed bottoms and with vegetation, hydroids, and bryozoans from about 25 to 210 ft.

Banded Gunnel - *Pholis fasciata* (Bloch & Schneider, 1801)

FEATURES: Dark orange or bright red to maroon with irregular, pale Y-shaped to hourglass-shaped marks with dark interior spots along upper back. Ventral area with vague to distinct irregular pale spots and darker bars. Single to double dark bar below eyes followed by a pale band. Head pores conspicuous. Pectoral fins moderately large. Pelvic fins occasionally absent. Anal fin about half of body length. Body moderately deep. HABITAT: Scattered in Bering Sea to Kodiak Island, AK. Also to Sea of Okhotsk and Sea of Japan and to Greenland. Demersal over rocky bottoms and around clumps of red algae from shallow subtidal zone to about 150 ft; usually shallower than 65 ft.

Crescent Gunnel - *Pholis laeta* (Cope, 1873)

FEATURES: Color highly variable. Nearly uniformly colored or mottled in shades of green, tan, or red to yellowish or brown. All with 13–15 whitish to pale yellow crescent-shaped to rounded marks bordered by dark parentheses-shaped marks along dorsal-fin base. Dark streak below eyes. Pectoral fins moderately large. Pelvic fins minute. Body moderately deep. HABITAT: Aleutian Islands and Alaskan Peninsula to Crescent City, CA. Also to Kamchatka. Demersal over a wide variety of bottoms from estuaries and intertidal zone to about 240 ft. Sometimes rise in the water column at night. BIOLOGY: Females lay eggs in nests under rocks that are usually guarded by both parents. Can live to about six years old.

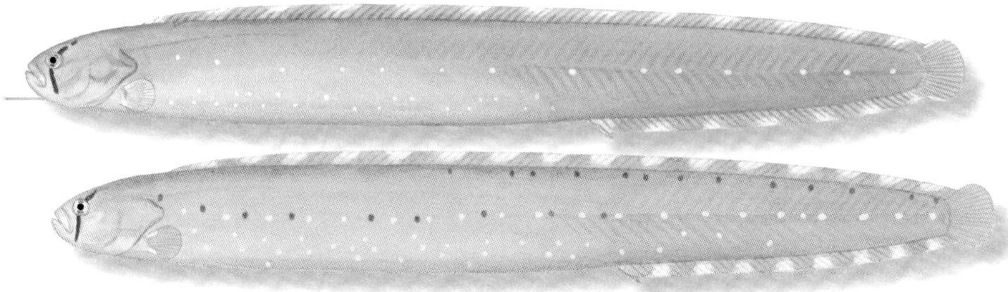

variants

Penpoint Gunnel
to 18 in.

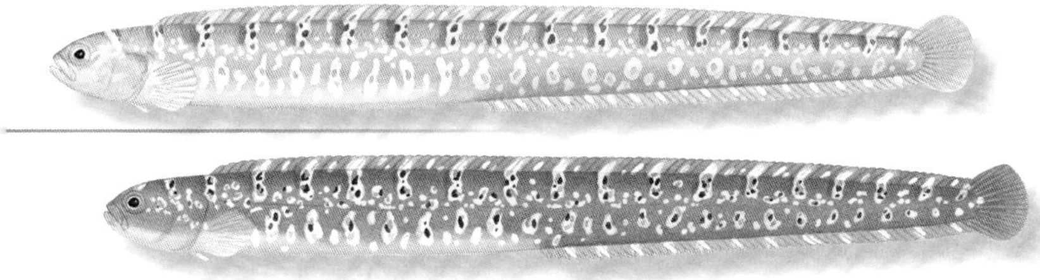

variants

Longfin Gunnel
to 5.3 in.

Banded Gunnel
to 11.8 in.

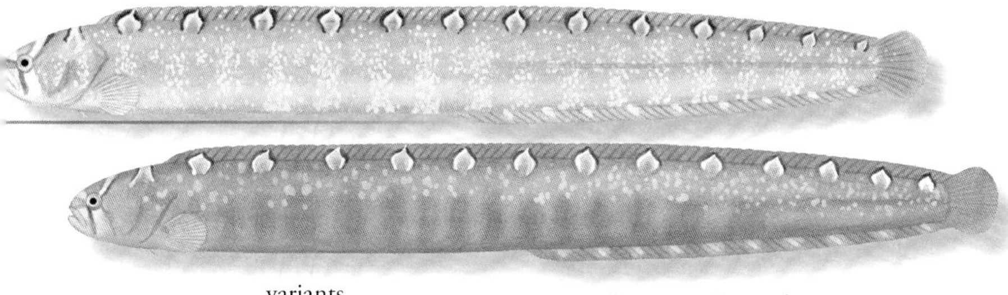

variants

Crescent Gunnel
to 10 in.

Saddleback Gunnel - *Pholis ornata* (Girard, 1854)

FEATURES: Color and pattern variable. Shades of red, green, or orange to brown or yellow, with about 10 darkly bordered, distinct to faint, U- or V-shaped marks along dorsal profile. Upper sides with distinct to faint, marbled, mottled, or vermiculated blotches that extend onto lower sides. Dark bar below eyes. Pectoral fins large. Pelvic fins minute. Body moderately deep. HABITAT: Vancouver Island, British Columbia, to Carmel Beach, CA. Demersal over soft, hard, mixed, and vegetated bottoms from intertidal zone to about 120 ft.

Red Gunnel - *Pholis schultzi* Schultz, 1931

FEATURES: Bright red to purplish or brownish, with about 13–16 irregular, small white marks along dorsal fin. Lower sides may have few to many small white spots. Some may be mottled or with small dark spots ventrally. Lower head white to pale red or brown and with two central dark bars. Anal fin often with white bands. Pectoral fins moderately large. Pelvic fins minute. Body slender. HABITAT: Queen Charlotte Islands, British Columbia, to Diablo Cove, CA. Demersal around rocks, kelp, grass beds, and surge areas from intertidal zone to about 80 ft.

Stippled Gunnel - *Rhodymenichthys dolichogaster* (Pallas, 1814)

FEATURES: Color and pattern highly variable. Shades of brown, red, or olivaceous to pale purple and often with scattered, minute, dark dots. May be mottled ventrally. Dorsal and anal fins with variable pale bands. All with a bright to vague white streak through eyes to pectoral fins. Pectoral fins moderately large. Pelvic fins minute. Body moderately deep. HABITAT: Southeastern Bering Sea and Aleutian Islands. Also to Sea of Okhotsk and Sea of Japan. Demersal around algae and rocks from intertidal zone and tidepools to about 485 ft.

Kelp Gunnel - *Ulvicola sanctaerosae* (Gilbert & Starks 1897)

FEATURES: Usually uniformly yellowish brown, but may be tan to reddish brown. Color corresponds to habitat. Pectoral and pelvic fins absent. Body comparatively long and slender. HABITAT: Pacific Grove, CA, to northern Baja California and Isla Guadalupe. Found demersally and in kelp canopies from intertidal zone to about 40 ft.

Rockweed Gunnel - *Xererpes fucorum* (Jordan & Gilbert, 1880)

FEATURES: Color variable. Uniformly green or reddish to yellowish brown, with or without a row of minute dark spots along midline. May have a dark streak below eyes. Red specimens can slowly change color, most green specimens cannot. Pectoral fins minute, about equal to eye diameter. Pelvic fins absent. Anal fin about ⅓ body length, with a strong anterior spine that arises from a pouch. Body comparatively slender. HABITAT: Banks Island, British Columbia, to north-central Baja California. Demersal around algae and rockweed beds from tidepools and inshore rocky areas to about 30 ft. BIOLOGY: Feed primarily on crustaceans. Can respire aerially and live out of water for as long as 12 hours. Previously known as *Apodichthys fucorum*.

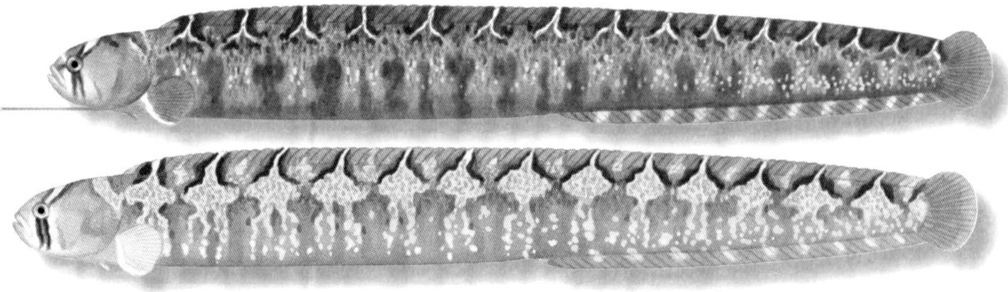

variants

Saddleback Gunnel
to 12 in.

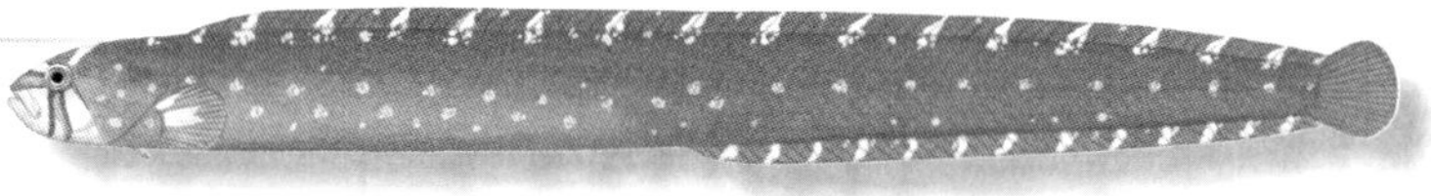

Red Gunnel
to 5 in.

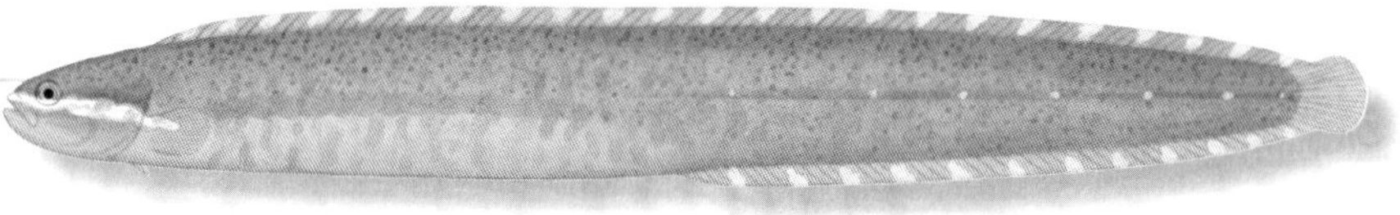

Stippled Gunnel
to 9.8 ft.

Kelp Gunnel
to 11.3 in.

variants

Rockweed Gunnel
to 9 in.

Anarhichadidae - Wolffishes

Bering Wolffish - *Anarhichas orientalis* Pallas, 1814

FEATURES: Adults shades of brown, reddish brown, black, or gray to yellowish or pinkish with irregular darker blotches and/or mottling. Juveniles whitish to golden, with brownish spots and broken wavy stripes. Mouth large with large canine-like teeth. Pectoral fins large and rounded. Dorsal and anal fins long-based. Head large, with body tapering to slender caudal peduncle. HABITAT: Beaufort and Chukchi seas to Prince William Sound, AK. Also to Sea of Japan and Bathurst Sound, Canada. Demersal over gravel and sand with stones and algae from intertidal zone to about 330 ft.

Wolf-eel - *Anarrhichthys ocellatus* Ayres, 1855

FEATURES: Color and pattern variable. Adults blue gray or gray to reddish brown or nearly white. Some with few to numerous, large, dark ocellated spots overlying dark bars; others covered with minute speckles. Females sometimes mottled brownish. Juveniles pale to dark orange, red, yellow, or purplish and covered with variable wavy spots on body and with rows of spots on dorsal fin. Mouth large, with many canine-like teeth. Large males with enlarged lower jaw. Pectoral fins large and rounded. Dorsal and anal fins long, tapering, and connected at pointed tail. Body very long and eel-like. HABITAT: Southeast Bering Sea and Aleutian Islands to northern Baja California. Juveniles may be pelagic; adults demersal around a variety of hard bottoms and structures and usually in caves and crevices from intertidal zone to about 740 ft.

Ptilichthyidae - Quillfish

Quillfish - *Ptilichthys goodei* Bean, 1881

FEATURES: Color and pattern vary. Uniformly yellowish, orange or green or with brownish red stripes along sides. Lower jaw projecting, with a fleshy tip. First dorsal fin consists of 81–90 isolated spines. Second dorsal and anal fins confluent with tail that tapers to a filament (when intact). Pelvic fins absent. Body very long and eel-like. HABITAT: Pribilof and Aleutian Islands to central OR. Also from Litke Strait and Commander Islands to Sea of Okhotsk and Sea of Japan. Found from surface to about 1,180 ft. Possibly also buried in bottom sediment during the day.

Zaproridae - Prowfish

Prowfish - *Zaprora silenus* Jordan, 1896

FEATURES: Color and pattern highly variable. Nearly uniformly colored to irregularly mottled, blotched, and/or spotted in shades of gray, green, brown, or blue to blackish. May have irregular yellowish spots and blotches or large areas of yellow. Mouth small, snout very short and blunt. Head rounded in profile. Rows of large white pores around eyes and on chin and cheeks. Pectoral fins large and broad. Dorsal and anal fins long-based. Pelvic fins absent. May have up to three lateral lines of widely spaced pores. HABITAT: Bering Sea and Aleutian Islands to San Miguel Island, CA. Also to Sea of Japan. Adults near bottom from near shore to about 2,630 ft. Juveniles pelagic.

Wolffishes

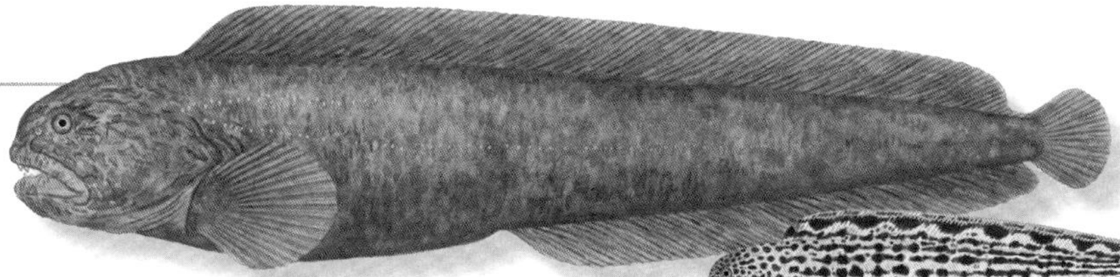

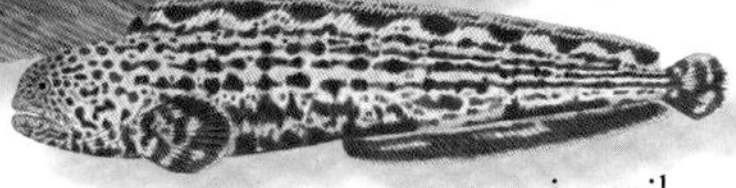

Bering Wolffish
to 4.1 ft.

juvenile

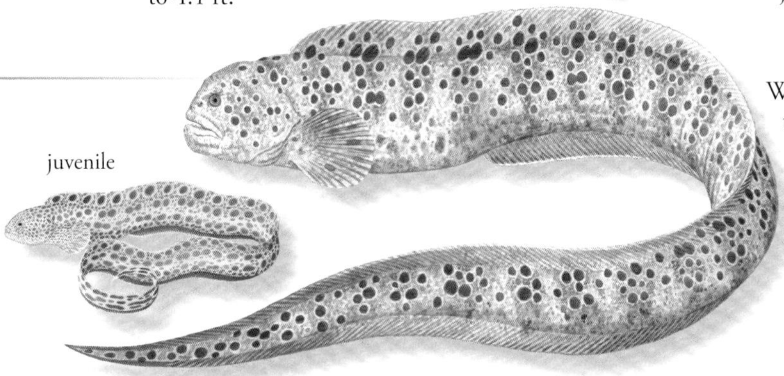

Wolf-eel
to 8 ft.

juvenile

Quillfish

Quillfish
to 15.3 in.

Prowfish

Prowfish
to 3.3 ft.

Scytalinidae - Graveldiver

Graveldiver - *Scytalina cerdale* Jordan & Gilbert, 1880

FEATURES: Tan, yellow, or greenish with darker mottling and speckling. Mouth large, upturned. Eyes very small and set high on the head. Pectoral fins very small. Dorsal and anal fins begin near midbody and are confluent with the caudal fin. Pelvic fins, lateral line, and scales absent. Body elongate, round in cross-section anteriorly, and laterally compressed posteriorly. HABITAT: Aleutian Islands to San Luis Obispo County, CA. Demersal along shallow open coasts. Burrow in soft and mixed bottoms.

Trichodontidae - Sandfishes

Pacific Sandfish - *Trichodon trichodon* (Tilesius, 1813)

FEATURES: Greenish brown dorsally, with blackish spots and wavy lines along dorsal profile. Silvery below. First and sometimes second dorsal fins with dark streaks. Mouth large and nearly vertical, with fringed lips. Pectoral fins broad. Body deep, compressed. HABITAT: Bering Sea and Aleutian Islands to San Francisco Bay, CA. Demersal over soft bottoms from intertidal zone to about 1,230 ft. Burrow in sand or mud during the day.

Ammodytidae - Sand Lances

Pacific Sand Lance - *Ammodytes hexapterus* Pallas, 1814

FEATURES: Green or blue to brown dorsally. Silvery below. Tip of lower jaw and snout pointed. Dorsal fin long-based. Caudal fin forked. Sides with numerous oblique folds. Lateral line high on long, slender body. HABITAT: Beaufort Sea to off southern CA. Also to Sea of Japan. Occur from intertidal zone to about 330 ft when inshore and near surface over deep water when offshore. BIOLOGY: Bury in sand and form large schools.

Uranoscopidae - Stargazers

Pacific Stargazer - *Astroscopus zephyreus* Gilbert & Starks, 1897

FEATURES: Grayish brown dorsally, with numerous minute whitish spots. Ventral area whitish. First dorsal fin black. Caudal fin with variable alternating dark and pale stripes. Pectoral fins grayish brown. Mouth large, vertical. Head very large. Cleithral spine covered with skin present above bluntly pointed pectoral fins. Pelvic fins thoracic. Body covered with small scales. HABITAT: Huntington Beach, CA, to Peru, including Gulf of California. Demersal over sandy bottoms from about 20 to 1,260 ft.

Smooth Stargazer - *Kathetostoma averruncus* Jordan & Bollman, 1890

FEATURES: Dark brownish to greenish black or gray dorsally with numerous pale spots that are minute anteriorly, large posteriorly. Ventral area whitish. Dorsal and caudal fins variably banded. Pectoral fins dark. Mouth large and vertical. Head very large. Prominent cleitheral spine present above broad pectoral fins. Pelvic fins thoracic. Spiny dorsal fin absent. Body scaleless. HABITAT: Piedras Blanca Point, CA, to Peru, including Gulf of California. Demersal over soft bottoms from about 40 to 2,000 ft.

Graveldiver

Graveldiver
to 6 in.

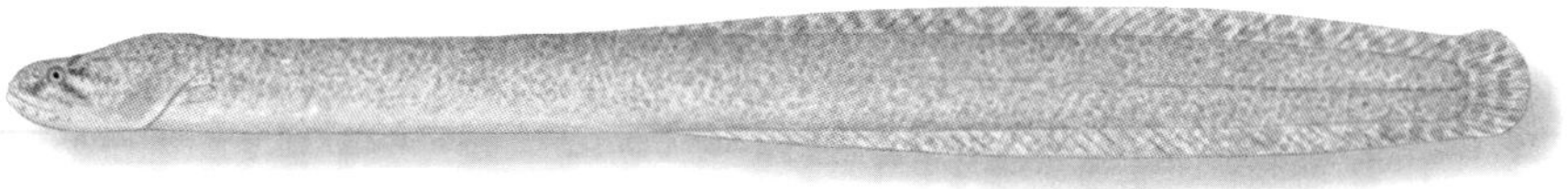

Sandfishes

Pacific Sandfish
to 12 in.

Sand Lances

Pacific Sand Lance
to 11 in.

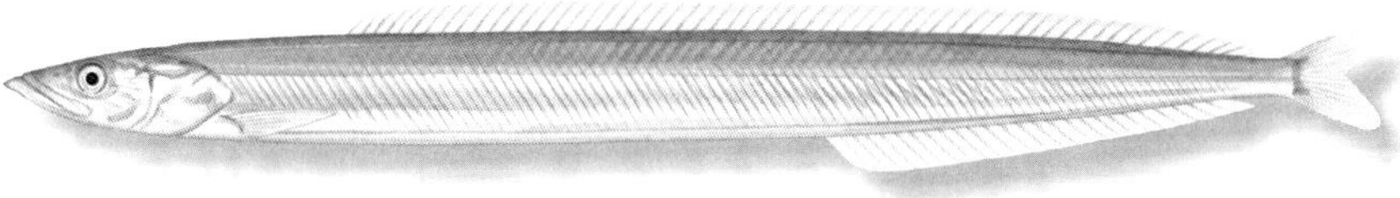

Stargazers

Pacific Stargazer
to 20.5 in.

Smooth Stargazer
to 12.6 in.

Bay Blenny - *Hypsoblennius gentilis* (Girard, 1854)

FEATURES: Color variable. Shades of green to reddish brown, with reddish brown to dark brown spots that form irregular clusters or saddles along back. Ventral area with pale spots and/or wavy lines that may form rows. Some with red or red streaks under chin, most evident during breeding. Dorsal fin with a blue to black or buff-colored spot anteriorly. Fins variably banded and spotted. Can quickly change color to match habitat. Snout steep and blunt. Nostrils tube-like. Cirri above eyes single, flap-like, and serrated to fringed along posterior edge. Cirri tall in males, shorter in females. Head evenly sloping. HABITAT: Monterey Bay to Gulf of California. Demersal over a variety of bottoms in bays, estuaries, and mangroves from intertidal zone to about 80 ft.

Rockpool Blenny - *Hypsoblennius gilberti* (Jordan, 1882)

FEATURES: Grayish to olive brown, with irregular dark brown to brownish green spots, blotches, and dark saddles along back. Often with dark streaks below eyes. Dorsal and anal fins dark with pale spots. May quickly change color with surroundings. During spawning, females turn golden, males turn blackish. Snout steep and blunt. Nostrils tube-like. Clusters of slender cirri above eyes that are closely set at bases. Head profile notched behind eyes. Lateral line ends at about midbody; some with lateral line extending to caudal fin. HABITAT: Morro Bay, CA, to southern Baja California. Demersal in crevices and tidepools, over rocks, along exposed coasts, and sometimes in sloughs and embayments, from intertidal zone to about 60 ft; usually found in very shallow water.

Mussel Blenny - *Hypsoblennius jenkinsi* (Jordan & Evermann, 1896)

FEATURES: Color and pattern highly variable. Pale to dark grayish or brownish, with irregular brown to greenish mottling and numerous to few reddish to orange brown spots on head and body. May have about eight dark saddles along back. Fins variably spotted or banded. All usually with a purple to blue or blackish crescent shape on upper cheeks. Cirri above eyes fleshy, stalked, and tree-like, or joined at base. Snout steep and blunt. Nostrils tube-like. Cirri reported to be longer in males. Head profile evenly sloping. Lateral line ends to above anterior anal fin. HABITAT: Morro Bay, CA, to Puerta Marquis, Mexico, including Gulf of California. Demersal over a wide variety of hard and soft substrates, usually in sheltered waters, from intertidal zone to about 70 ft.

Sabertooth Blenny - *Plagiotremus azaleus* (Jordan & Bollman, 1890)

FEATURES: Color and pattern variable. Dorsal area reddish brown to olivaceous or blackish with a narrow to broad yellow to white stripe along upper back. Dark areas may form stripes. White band from snout through eyes. Underside of head yellow to white. Dorsal fin with a black inner band. Snout short, blunt. Mouth small, directed ventrally, and with sharp, slender teeth. Dorsal fin low and long-based. Upper, middle, and lower caudal-fin rays elongate. Body long and slender. HABITAT: King Harbor, CA, Rocas Chester and Isla Cedros, Baja California, to Peru. Also Gulf of California. Found near bottom around rocky reefs and in crevices from about 7 to 85 ft. BIOLOGY: Feed by rapidly attacking other fishes and removing tissue, mucus, and scales.

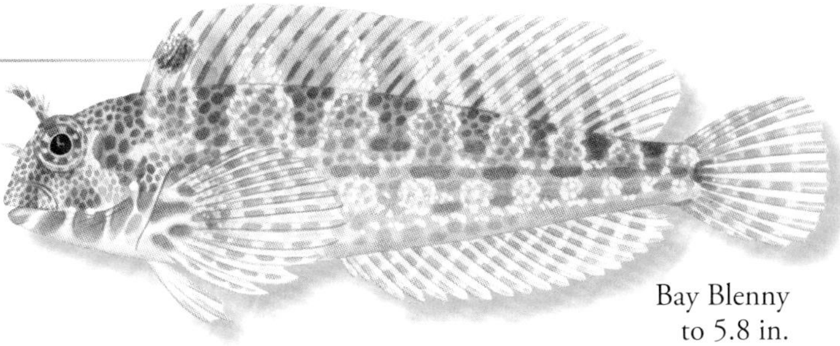

Bay Blenny
to 5.8 in.

Rockpool Blenny
to 6.8 in.

cirri variation

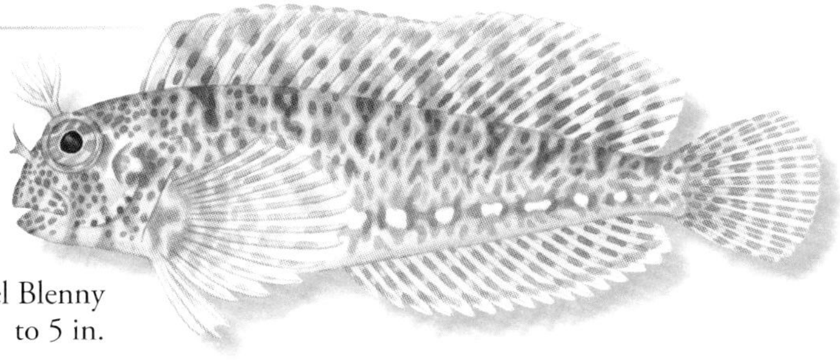

Mussel Blenny
to 5 in.

Sabertooth Blenny
to 4 in.

Spotted Kelpfish - *Gibbonsia elegans* (Cooper, 1864)

FEATURES: Color and pattern highly variable; can change color with habitat. May be nearly uniformly colored or highly mottled, spotted, blotched, and/or banded in shades of brown or red to green or gray. Sides with one to three (usually two) ocellated spots. Fins plain to banded. Mouth small. Snout short, bluntly pointed. A small, tufted cirri present above each eye. Pectoral fins comparatively long. Anterior dorsal-fin spines tall. Soft dorsal fin with 5–8 rays, with posterior rays more widely spaced than anterior rays. Scales extend well onto caudal fin. HABITAT: Piedras Blancas Point, CA, to southern Baja California. Demersal over rocks, vegetation, and mixed bottoms from intertidal zone to about 185 ft. Females found in shallower water than males.

Striped Kelpfish - *Gibbonsia metzi* Hubbs, 1927

FEATURES: Color and pattern variable. Brown or olive brown to reddish and usually with irregular pale stripes or darker mottling. Distinct, ocellated spots usually absent; muted when present. Fins variably banded. Mouth small. Snout bluntly pointed. A small, tufted cirri present above each eye. Pectoral fins comparatively short, not reaching anal-fin origin. Anterior dorsal-fin spines tall. Soft dorsal fin with 7–10 rays that are evenly spaced. Pectoral fins comparatively short. Scales do not extend onto caudal fin. HABITAT: Vancouver Island to central Baja California. Demersal around seaweeds and kelp from intertidal zone and in tidepools to about 30 ft.

Crevice Kelpfish - *Gibbonsia montereyensis* Hubbs, 1927

FEATURES: Color and pattern highly variable. Usually shades of scarlet red, dark red, lavender, pink, reddish brown. May be almost uniformly colored, distinctly blotched, or highly mottled, speckled, and spotted. Often with several dark spots loosely arranged in an irregular row on upper sides; spots may be ocellated. Mouth small, snout bluntly pointed. A small tufted cirrus present above each eye. Pectoral fins reach anal-fin origin. Anterior dorsal-fin spines tall. Soft dorsal fin with 5–8 rays, with posterior rays more widely spaced than anterior rays. Scales extend onto caudal-fin base. HABITAT: British Columbia to central Baja California. Demersal over rocks, coral, and algae and in crevices from surface and intertidal zone to about 120 ft.

Giant Kelpfish - *Heterostichus rostratus* Girard, 1854

FEATURES: Color and pattern highly variable. Shades of brown, yellow, green, or red. Almost uniformly colored or irregularly barred, striped, or mottled. Males usually brown, occasionally olivaceous, and often striped or mottled. Females any color typical of the species and usually plain or barred. Juveniles usually green or brown. All capable of changing color and pattern to match habitat. Snout long and pointed. Dorsal fin notched anteriorly, with 33–38 spines and 11–13 rays. Caudal fin deeply forked. Body most robust under anterior dorsal fin, becoming very slender at caudal peduncle. HABITAT: British Columbia to southern Baja California. More common south of Point Conception, CA. Occur from bays to outer coast from surface to bottom and usually around vegetation. Commonly found from subtidal zone to about 80 ft, but occasionally in tidepools.

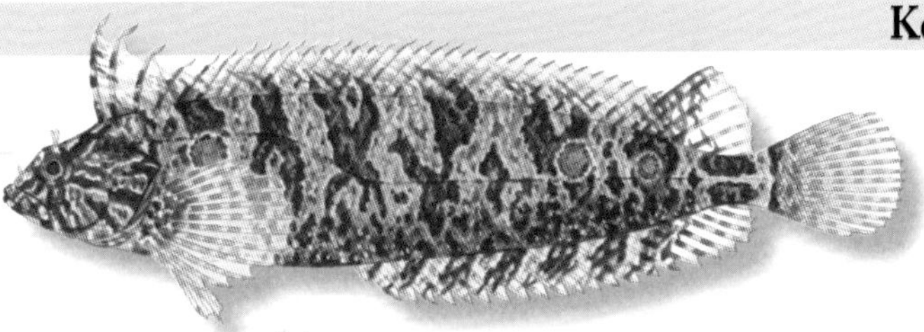

Spotted Kelpfish
to 6.2 in.

Striped Kelpfish
to 9.3 in.

Crevice Kelpfish
to 5.5 in.

variants

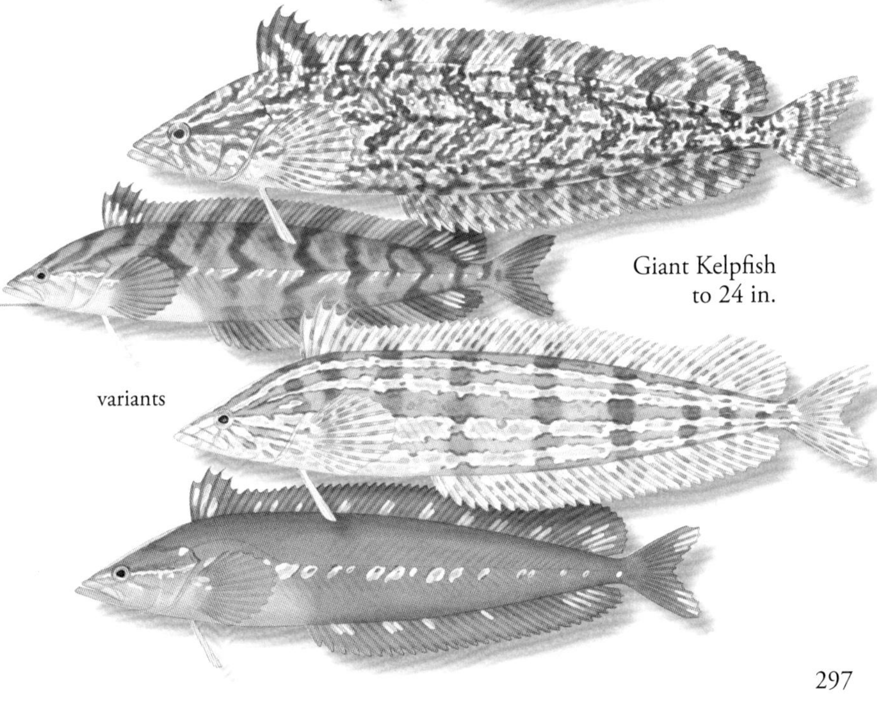

Giant Kelpfish
to 24 in.

variants

Labrisomidae - Labrisomid Blennies

Island Kelpfish - *Alloclinus holderi* (Lauderbach, 1907)

FEATURES: Color and pattern variable. Shades of orange, red, pink or purple to pale gray or tan with whitish flecks and irregular, large dark brown to black blotches on sides and back. Cheeks with a large pale blotch. Three-pronged cirri above eyes and on nape. Pectoral fins extend past anal-fin origin. Notch present between spiny and soft rays. HABITAT: San Miguel Island, CA, to southern Baja California. Demersal over hard and mixed bottoms and in crevices from intertidal zone to about 300 ft.

Deepwater Blenny - *Cryptotrema corallinum* Gilbert, 1890

FEATURES: Color and pattern variable. Pale greenish to grayish with irregular wavy stripes in shades of red, orange, pink, or yellowish. Usually with about six dark brown to blackish blotches on sides. Cheeks with wavy spots and lines. Simple, pronged cirri above eyes. Nape with single, simple cirri. Pectoral fins extend past anal-fin origin. Notch present between spiny and soft rays. HABITAT: Off Cook Point, San Miguel Island, CA, to northern Baja California. Demersal over rocky reefs and sandy and shelly interspaces from about 80 to 640 ft.

Reef Finspot - *Paraclinus integripinnis* (Smith, 1880)

FEATURES: Color highly variable. Uniformly colored, mottled, and/or banded in shades of whitish to brown, blackish, or purplish. All with a single, ocellated spot at about the 25th dorsal-fin spine. Fringed cirri above eyes and a pair of flap-like cirri on nape. Dorsal fin nearly evenly tall, with a slight notch behind third spine. Pectoral fins comparatively short. Scales large. HABITAT: Santa Cruz Island to southern Baja California. Demersal over rocky bottoms from intertidal zone to about 50 ft.

Chaenopsidae - Tube Blennies

Orangethroat Pikeblenny - *Chaenopsis alepidota* (Gilbert, 1890)

FEATURES: Females and juveniles tannish to greenish, with white spots and dark marbling. Males blackish to blackish blue or purplish, with blue to white spots and bright orange under head during breeding. Dorsal fin low in females, tall anteriorly in males, and confluent with caudal and anal fins. Snout long, depressed. Mouth gaping. Body long and slender. HABITAT: Santa Cruz Island and King Harbor, CA, to Gulf of California. Demersal over soft bottoms in burrows and tube worm holes from about 3 to 75 ft.

Sarcastic Fringehead - *Neoclinus blanchardi* Girard, 1858

FEATURES: Color and pattern highly variable. Shades of pink, brown, black, purple, gray, green, or red. May have paler and darker mottling, irregular bars, spots, and/or blotches. All with two distinct to vague spots on anterior dorsal fin that may be ocellated. Snout blunt and steep. Head large. Cirri over eyes simple, slender, and short to long. Fringed cirri present behind eyes. Jaws very large, with those in males extending to gill opening. HABITAT: San Francisco, CA, to central Baja California. Demersal in a wide range of dwellings along open coast and quiet backwaters from about 10 to 240 ft.

Labrisomid Blennies

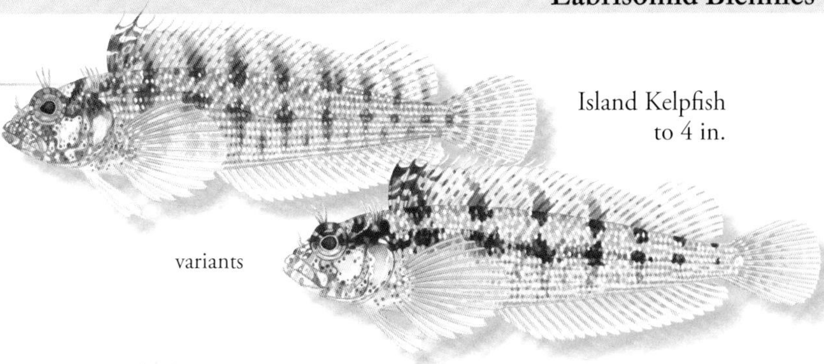

Island Kelpfish
to 4 in.

variants

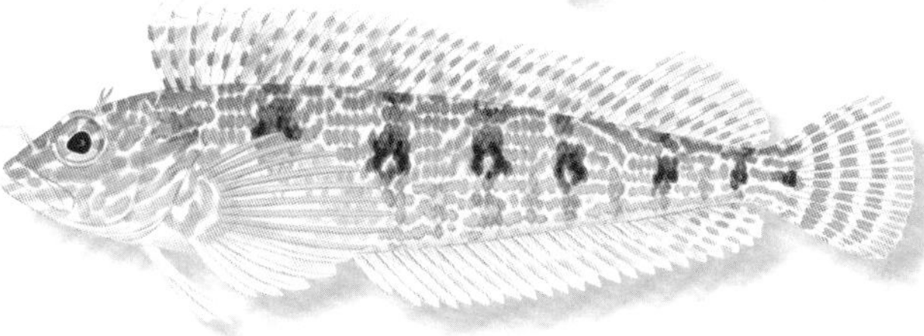

Deepwater
Blenny
to 5 in.

Reef Finspot
to 2.5 in.

Tube Blennies

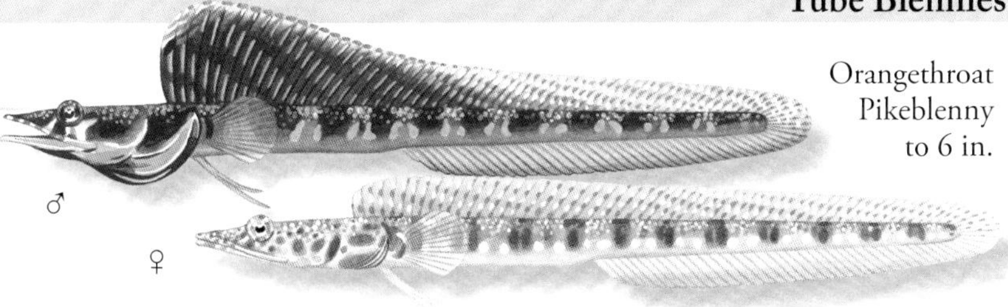

Orangethroat
Pikeblenny
to 6 in.

♂

♀

Sarcastic Fringehead
to 12 in.

♂

♀

Yellowfin Fringehead - *Neoclinus stephensae* Hubbs, 1953

FEATURES: Color and pattern highly variable. Shades of tan, brown, red, yellow, to orange or olive, with numerous white flecks and darker blotches and bars. Some nearly uniformly colored or very dark. Dorsal fin lacks distinct ocellated spots. Cirri over eyes in three pairs that are multi-branched to very bushy and may be short to long. Minute to relatively long simple cirri on nape. Upper jaw extending well past eyes. Dorsal fin nearly even in height. HABITAT: San Francisco, CA, to central Baja California. Demersal in and near empty shells and holes and crevices in mud and rocks. Also in bays near jetties. Occur from intertidal zone to about 90 ft.

Onespot Fringehead - *Neoclinus uninotatus* Hubbs, 1953

FEATURES: Color and pattern highly variable. Grayish to greenish, with blackish to reddish blotches and marbling. Some nearly entirely red with whitish blotches. Often with a dark stripe on sides. May be very dark to brightly colored. Anal fin with variable red banding. All with a single, vague to distinct ocellated spot between 1st and 2nd dorsal-fin spine. Anterior cirri over eyes tall and branched at tips. Several slender, branched cirri behind eyes. Upper jaw extends well past eyes. Dorsal fin nearly even in height. HABITAT: Bodega Bay, CA, to northern Baja California. Demersal inside a wide variety of empty objects and in holes from surf zone to about 180 ft.

Icosteidae - Ragfish

Ragfish - *Icosteus aenigmaticus* Lockington, 1880

FEATURES: Adults overall shades of reddish brown to purplish brown. Juveniles transparent to bluish or whitish with variable large and small pale to dark gray spots. Head and eyes small. Snout blunt. Dorsal and anal fins tall in juveniles, low in adults. Pelvic fins present in juveniles, absent in adults. Caudal fin rounded in juveniles, forked in adults. Juveniles deep bodied, adults elongate. Scales present in juveniles, absent in adults. All compressed and with limp flesh. HABITAT: Bering Sea and Pacific Ocean to Pt. Loma, CA. Young in shallow water or near surface offshore. Adults usually in deep water, with some near bottom to about 4,100 ft; sometimes found near shore.

Gobiesocidae - Clingfishes

Northern Clingfish - *Gobiesox maeandricus* (Girard, 1858)

FEATURES: Color, pattern highly variable. Nearly uniformly colored in shades of green, brown, gray, yellow, black, or red. May have numerous darker, web-like vermiculations, wavy lines, and/or mottling. Many with a pale bar between eyes. Head broad and rounded. Posterior nostrils just in front of eyes. Fleshy pectoral-fin pads present but not extending to edge of upper gill membrane. Dorsal fin with 14–16 rays (including 2–4 hidden) and connected to caudal fin. HABITAT: Noyes Island, AK, to central Baja California. Demersal over a wide range of hard and flexible substrates, in crevices, and under rocks from intertidal zone and tidepools to about 60 ft. Sometimes share dens with octopi.

Tube Blennies

Yellowfin Fringehead
to 4 in.

variants

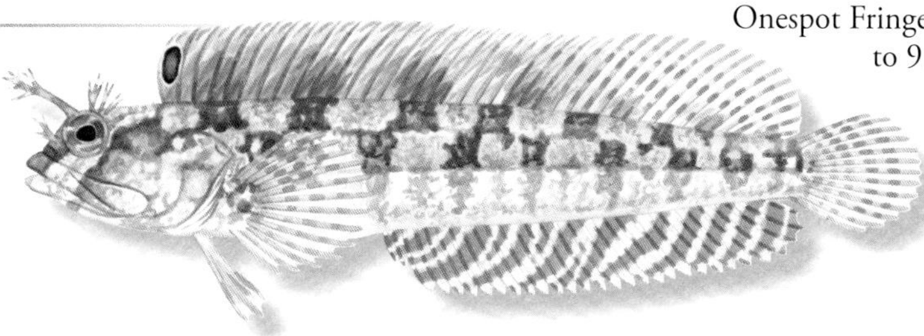

Onespot Fringehead
to 9.7 in.

Ragfish

Ragfish
to 7 ft.

juvenile

Clingfishes

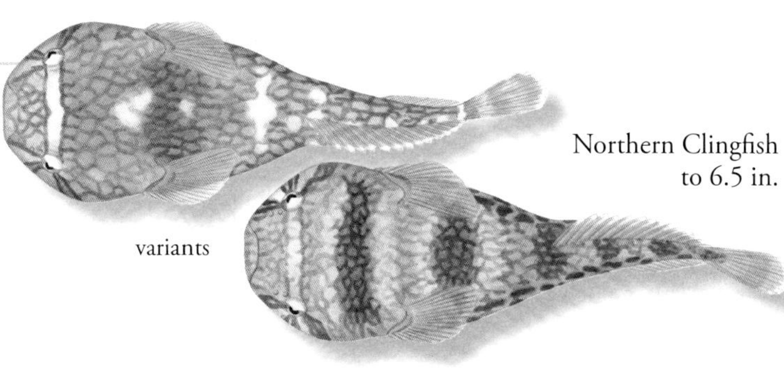

Northern Clingfish
to 6.5 in.

variants

Bearded Clingfish - *Gobiesox papillifer* Gilbert, 1890

FEATURES: Tannish to grayish with dense, irregular brown to blackish close-set spots. Dorsal and caudal fins banded. Underside of snout with rows of small, barbel-like papillae. Upper lip usually with three small papillae. Sucking disk with rows of flattened papillae. Dorsal fin with 14–15 rays (including 2–3 hidden) and connected to caudal fin. HABITAT: San Pedro, CA, to Panama, including Gulf of California. Demersal over rocky outcrops and in shallow, sandy areas from intertidal zone to about 20 ft.

California Clingfish - *Gobiesox rhessodon* Smith, 1881

FEATURES: Gray to brown, with darker spotting and paler marks. May also have a pale band between eyes and several pale bands on body. Head broad and rounded. Fleshy pectoral pad extends dorsally to a point equal to edge of upper gill membrane. Dorsal fin with 12–14 rays (including 2–4 hidden) and connected to caudal fin. HABITAT: Pismo Beach, CA, to southern Baja California. Demersal in intertidal and shallow subtidal waters to about 35 ft.

Slender Clingfish - *Rimicola eigenmanni* (Gilbert, 1890)

FEATURES: Color blends with habitat. Uniformly tan, green, olive, red, brown, or yellowish. May have a dark streak through eyes and/or a pale stripe on sides. Snout squared with two pairs of parallel pores at tip. Head and body slender. Dorsal fin with 5–8 rays and well separated from caudal fin. HABITAT: Palos Verdes, CA, to southern Baja California. Demersal in intertidal and shallow subtidal waters to about 50 ft.

Kelp Clingfish - *Rimicola muscarum* (Meek & Pierson, 1895)

FEATURES: Color blends with habitat. Uniformly yellowish, greenish, or brownish to reddish. May have a white streak through eyes and/or a dark stripe on sides. Snout angular, with two pairs of offset pores at tip. Head and body very slender. Dorsal fin with 6–8 rays and well separated from caudal fin. HABITAT: Glacier Bay, AK, to northern Baja California. Associated with kelp and eelgrass beds.

ALSO IN THE AREA: Lined Clingfish, *Gobiesox eugrammus*; Channel Islands Clingfish, *Rimicola cabrilloi*; Southern Clingfish, *Rimicola dimorpha*; page 351.

Eleotridae - Sleepers

Pacific Fat Sleeper - *Dormitator latifrons* (Richardson, 1844)

FEATURES: Color variable. Olive brown to purplish brown, with oblique pale yellow to reddish bands that follow scale rows. Dark bands radiate from eyes. Dorsal and anal fins usually spotted; may have dark to pale orange margins. All with a blackish blotch with iridescent highlights above pectoral fins. Snout short and blunt. Eyes small. Head broad and flattened above eyes. Pelvic fins not united. Scales large. HABITAT: Palos Verdes, CA, to Peru. Usually in fresh water but also in brackish water and shallow inshore areas to about 7 ft.

Clingfishes

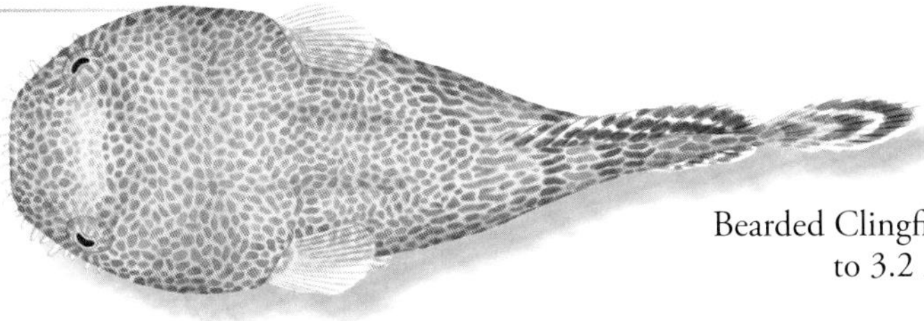

Bearded Clingfish
to 3.2 in.

California Clingfish
to 2.5 in.

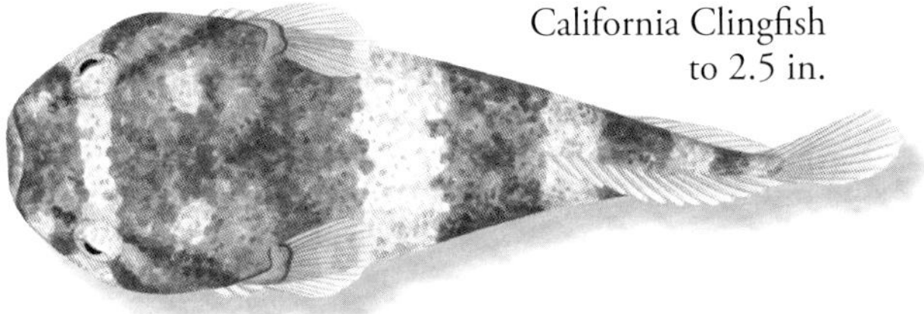

Slender Clingfish
to 2.3 in.

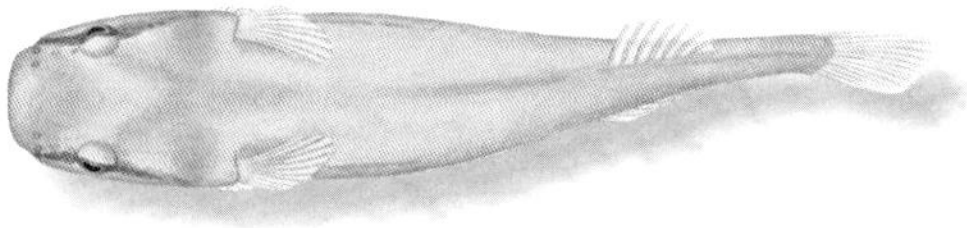

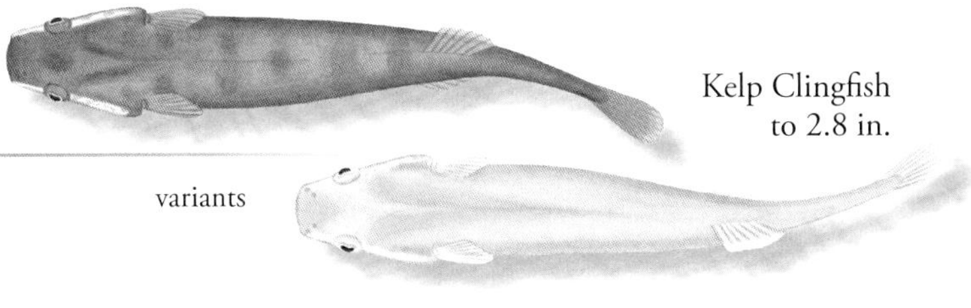

Kelp Clingfish
to 2.8 in.

variants

Sleepers

Pacific Fat Sleeper
to 2 ft.

Yellowfin Goby - *Acanthogobius flavimanus* (Temminck & Schlegel, 1845)

FEATURES: Tannish with irregular brownish spots and blotches that may cluster along midline. Usually with two dark streaks from eyes to upper jaw. Dorsal and caudal fins with spots that form bands. Anal fin yellowish. Mouth reaches to below anterior eye. Dorsal fins connected to barely separated. Anal fin with 11–12 rays. HABITAT: Introduced into the area from NW Pacific. San Francisco Bay to northern Baja California. Demersal in bays and estuaries from intertidal zone to about 20 ft. Also ascend rivers.

Arrow Goby - *Clevelandia ios* (Jordan & Gilbert, 1882)

FEATURES: Tan to pale gray or olivaceous with minute brown to black flecks that form clusters on back and sides. Some with white speckles. Dorsal and caudal fins variably banded. Anal fin with a blackish inner margin. Mouth large, extending well past eyes. Dorsal fins well separated. Anal fin moderately long-based wtih14–17 rays. HABITAT: Rivers Inlet, BC, to southern Baja California. Demersal over soft bottoms in estuaries, lagoons, and sloughs and in burrows from intertidal zone to about 35 ft.

Longtail Goby - *Ctenogobius sagittula* (Günther, 1862)

FEATURES: Tan with four oblong blackish spots along midline and another at caudal-fin base. Upper lip with a black spot, lower cheeks with a black stripe. Dorsal, caudal, and pectoral fins banded or with spots that form bands. Anal and pelvic fins plain. Dorsal fins barely separated. Caudal fin long and pointed. Anal fin long-based, with 13 rays. Body covered with small scales. HABITAT: Marina del Rey, CA, to Peru. Demersal in lagoons from intertidal zone to about 20 ft. Also in river mouths. NOTE: Previously known as *Gobionelluus sagittula* and *G. longicaudus*.

Tidewater Goby - *Eucyclogobius newberryi* (Girard, 1856)

FEATURES: Tannish to pale olive brown, with blackish speckles that form clusters or vague bars dorsally. Upper spiny dorsal-fin margin colorless. Soft dorsal and caudal fins banded. A pair of pores present between eyes. Top of gill opening below first pectoral-fin ray. Dorsal fins nearly attached to slightly separated. Anal fin short-based with 8–11 rays. Body with small scales. HABITAT: Mouth of Smith River to Agua Hedionda Lagoon, CA. Demersal over shallow soft bottoms in coastal lagoons, brackish bays, and river and stream mouths. May travel up to 5 miles upstream. NOTE: Near Threatened.

Longjaw Mudsucker - *Gillichthys mirabilis* Cooper, 1864

FEATURES: Shades of brown, olive, or black dorsally; may have paler and darker areas. Ventral area usually yellowish. Juveniles with about eight bars on sides and a dark blotch on posterior spiny dorsal fin. Males with very large jaws that extend to gill cover. Females with jaws that extend to about preopercle. Juveniles with short jaws. Dorsal fins well separated. Anal fin short-based, with 9–14 rays. Body covered with small scales. HABITAT: Tomales Bay, CA, to Gulf of California. More common south of San Francisco. Demersal in bays, salt marshes, estuaries, mangrove forests, and over mud flats from shore to about 15 ft. All ages and both sexes build or adopt burrows.

Yellowfin Goby
to 9.8 in.

Arrow Goby
to 2.3 in.

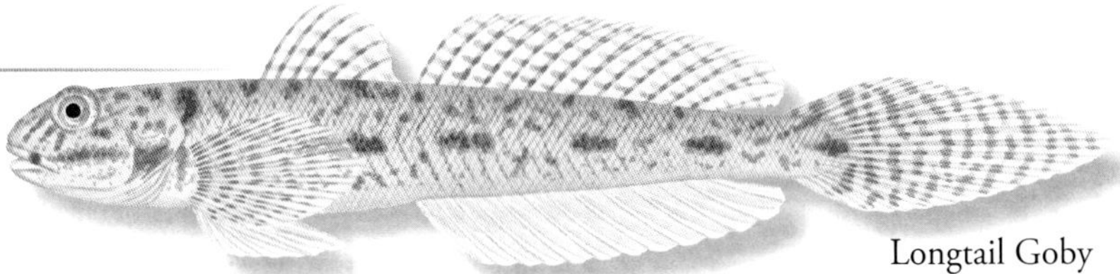

Longtail Goby
to 8 in.

Tidewater Goby
to 2.3 in.

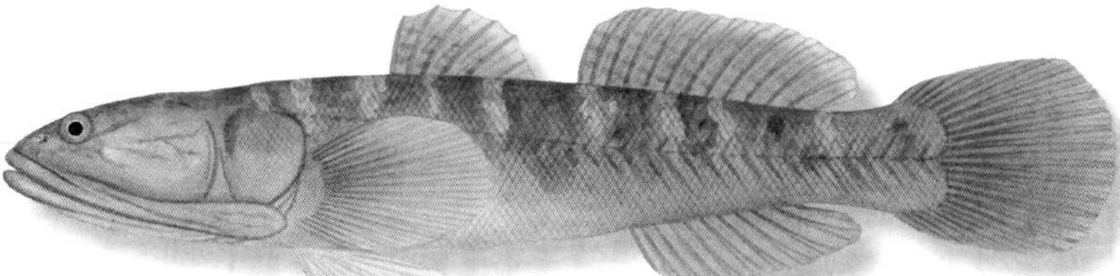

Longjaw Mudsucker
to 8.3 in.

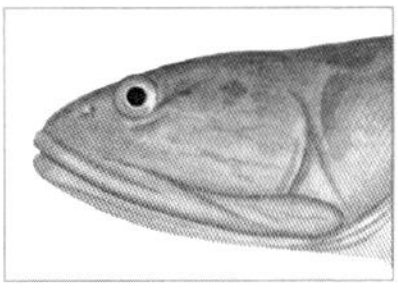

jaw of female

Cheekspot Goby - *Ilypnus gilberti* (Eigenmann & Eigenmann, 1889)

FEATURES: Pale gray to tan, with clusters of minute blackish speckles that form an irregular network. Anal fin with a blackish stripe. Opercle usually with a bluish black blotch. Maxilla extend to below middle of eyes. Pectoral fins usually shorter than pelvic fins. Anal fin with 12–16 rays. Scales small, embedded. HABITAT: Tomales Bay, CA, to southern Baja California. Demersal over sand and mudflats in shallow bays and estuaries.

Bay Goby - *Lepidogobius lepidus* (Girard, 1858)

FEATURES: Shades of pale gray, greenish, tan, or brownish, with dark gray to brown blotches on upper sides. Usually with a row of oblong blotches along midline. Spiny dorsal-fin margin grayish to blackish. Dorsal fins distinctly separate. Anal fin long-based, with 14–18 rays. Body covered with minute scales. HABITAT: Kegan Cove, AK, to central Baja California. Demersal over a wide variety of soft bottoms in quiet bays, in estuaries, on eelgrass beds, and along open coast from intertidal zone to about 1,000 ft.

Halfblind Goby - *Lethops connectens* Hubbs, 1926

FEATURES: Uniformly tan to pinkish. Mouth large, with upper maxilla extending well past eyes. Eyes very small and rudimentary. Spiny dorsal fin low, with six spines, and barely separate from soft dorsal fin. Body scaleless. HABITAT: Carmel, CA, to northern Baja California. Found among rocks and kelp holdfasts from intertidal zone to about 60 ft. Juveniles found in schools among kelp canopies.

Bluebanded Goby - *Lythrypnus dalli* (Gilbert, 1890)

FEATURES: Red to orange, with 2–9 (usually 3–5) narrow bright to dark blue bars on body that may be broken. Head with irregular blue bands and blotches. Spiny dorsal fin relatively low to very long and filamentous (especially in males). Anal fin with 12–14 rays. HABITAT: Morro Bay, CA, to Peru, including Gulf of California. Demersal over rocky slopes, coral reefs, in crevices, and around urchins from intertidal zone to about 250 ft.

Zebra Goby - *Lythrypnus zebra* (Gilbert, 1890)

FEATURES: Bright to dull red or orange red, with 12–18 bright blue, bright purple, to grayish blue bars that alternate with narrower bars on body. Spiny dorsal fin low to very tall and filamentous (especially in males). Anal fin with nine rays. HABITAT: Carmel Bay, CA, to southern Baja California. Demersal over reefs and rocks, in empty shells and crevices, under ledges, and among urchins and algae from intertidal zone to about 320 ft.

Shadow Goby - *Quietula y-cauda* (Jenkins & Evermann, 1889)

FEATURES: Body semi-transparent olivaceous to brownish, with clusters of dark speckles. Caudal peduncle with a sideways Y-shaped mark. Anal fin with a black stripe in mature males. Upper maxilla extends well beyond eyes. Dorsal fins distinctly separate. Anal fin long-based, with 12–15 rays. Scales small. HABITAT: Morro Bay, CA, to Gulf of California. Demersal over muddy and vegetated bottoms in lagoons, bays, estuaries, and river mouths from intertidal zone to about 20 ft. Adults primarily in burrows.

Cheekspot Goby
to 2.5 in.

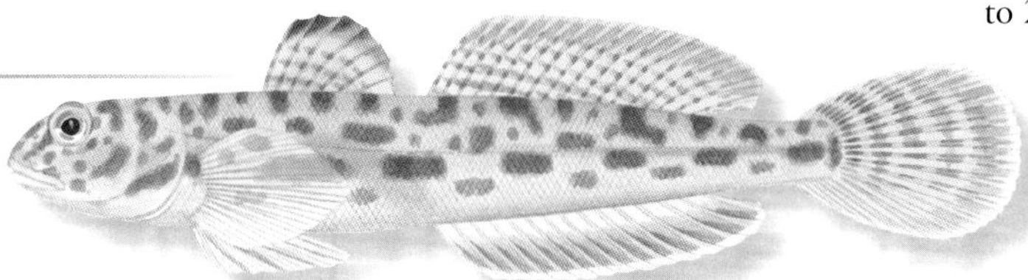

Bay Goby
to 4 in.

Halfblind Goby
to 2.5 in.

variants

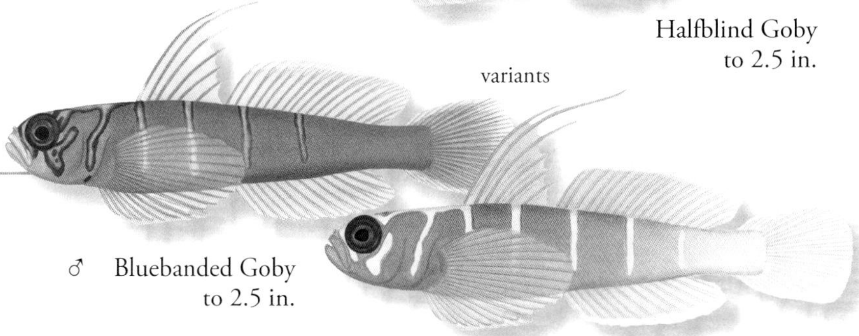

♂ Bluebanded Goby
to 2.5 in.

♂ Zebra Goby
to 2.3 in.

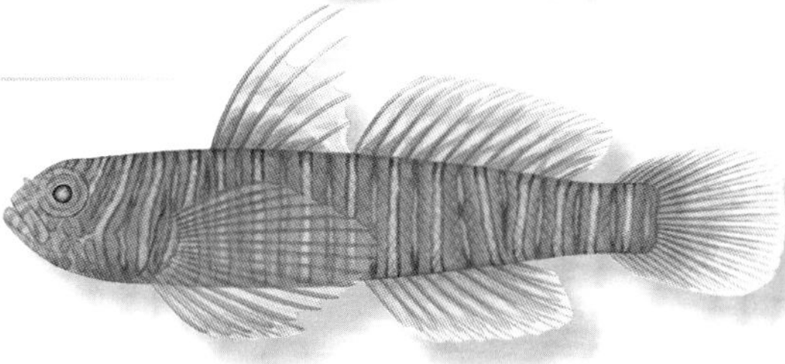

Shadow Goby
to 2.7 in.

Blackeye Goby - *Rhinogobiops nicholsii* (Bean, 1882)

FEATURES: Shades of white, gray, yellow, tan, orange, to red or pink. May be slightly to heavily mottled or blotched in shades of tan, gray, or brown, particularly at night. Eyes black or similarly colored to body. Pelvic fins black in breeding males. All with black on spiny dorsal-fin margin and a crest along top of head. Dorsal fins just touching at base. Pectoral fins reach past anal-fin origin. Anal fin with 11–12 rays. Scales comparatively large. HABITAT: Near Sitka, AK, to central Baja California. More common between S British Columbia and N Baja California. Demersal over a wide variety of soft, hard, and mixed bottoms from intertidal zone to about 2,100 ft. Usually to about 330 ft.

Shokihaze Goby - *Tridentiger barbatus* (Günther, 1861)

FEATURES: Shades of brown to gray, with about six irregular darker blotches and oblique bars. Spiny dorsal fin with an irregular blackish blotch. Dark spot at upper caudal-fin base. Upper and lower margins of mouth with a series of small barbels that extend onto cheeks and opercular margin. Head broad and flattened on top. Anal fin with 10 rays. HABITAT: Known from San Francisco Bay. Demersal in fresh, brackish, and marine waters to about 85 ft. Introduced into the area from NW Pacific.

Shimofuri Goby - *Tridentiger bifasciatus* Steindachner, 1881

FEATURES: Color and pattern highly variable. Shades of tan, green, or gray, with two vague to distinct dark stripes on sides; vaguely to distinctly blotched; or almost uniformly brown or blackish. All with small pale spots under chin and on throat. Usually with pale to dark orange margin on soft dorsal and anal fins and a black spot at upper caudal-fin base. Head wide and flattened on top. Dorsal fins nearly connected. Anal fin with about 10 rays. HABITAT: San Francisco Bay and adjacent freshwater drainages. Possibly drainages near Los Angeles. Demersal. Introduced into the area.

Chameleon Goby - *Tridentiger trigonocephalus* (Gill, 1859)

FEATURES: Can rapidly change color and pattern. Shades of whitish, yellow, tan, gray, or green with two dark stripes on sides; with vague to distinct blotches that may form a checker-board pattern; or almost uniformly brown to black. All usually with small pale spots on sides of head, a dark and pale bar at pectoral-fin base, and a black spot at upper caudal-fin base. Head wide, flattened on top. Dorsal fins nearly connected. Anal fin short-based, with 10–11 rays. HABITAT: San Francisco Bay, Los Angeles Harbor, and San Diego Bay. Demersal in marine and brackish waters from surf zone to about 50 ft. Also in fresh water. Introduced into the area from the NW Pacific.

Blind Goby - *Typhlogobius californiensis* Steindachner, 1879

FEATURES: Pale to bright pink, reddish, or with yellowish on head and tail. May have small, dark speckles dorsally. Eyes very small in juveniles, absent in adults. Head with enlarged cheeks. Spiny dorsal fin small, low, with two spines, and separate from soft dorsal fin. Skin scaleless, loose, and semi-transparent. HABITAT: San Simeon Point, CA, to southern Baja California. Demersal in ghost shrimp burrows, under and in holes in rocks, and around kelp beds from intertidal zone to about 50 ft.

Blackeye Goby
to 6 in.

variants

pelvic fins of
breeding male

Shokihaze Goby
to 4.4 in.

Shimofuri Goby
to 4.1 in.

variants

Chameleon Goby
to 4.5 in.

Blind Goby
to 3.3 in.

Ephippidae - Spadefishes

Pacific Spadefish - *Chaetodipterus zonatus* (Girard, 1858)

FEATURES: Adults silvery to dusky silver, with five to six brownish bars on head and body that may be faint to dark; become more uniformly silvery with age and while over sandy bottoms. Juveniles dark brown. Snout very short, mouth small. Preopercular margin finely serrated, lacks long spine at lower corner. Opercle with a blunt spine at rear margin. Spiny dorsal fin tall in juveniles, shortens with age. Second dorsal and anal fins with elongate anterior rays. Body disk-shaped in profile. HABITAT: San Diego, CA, to Peru, including Gulf of California. Occur over sand and rubble in bays and nearshore areas from shore to about 350 ft. Also found over reefs. BIOLOGY: Usually form small schools. Feed on bottom-dwelling invertebrates.

Luvaridae - Louvar

Louvar - *Luvarus imperialis* Rafinesque, 1810

FEATURES: Silvery bluish to pinkish dorsally. Silvery white below. All fins except pelvic fins reddish. Snout very short. A deep grove is present from snout to above eyes. The head is elevated; the forehead is compressed and keel-like. Dorsal fin originates behind eyes in juveniles. In adults, dorsal fin originates posterior to midline. Caudal peduncle with a slender keel. Pelvic fins reduced to small flaps. HABITAT: Circumglobal in tropical to temperate seas. In eastern Pacific from WA to Chile. Pelagic and oceanic in deep water over continental shelves from near surface to about 1,900 ft. Sometimes in shallower water near coast. BIOLOGY: Females may carry up to 47 million eggs. Solitary. Feed on plankton, medusae, and ctenophores.

Sphyraenidae - Barracudas

Pacific Barracuda - *Sphyraena argentea* Girard, 1854

FEATURES: Shades of gray to bluish gray or brownish dorsally, with a series of oblique bars. Silvery gray to silvery below. Caudal fin dusky yellow. Mouth large, with protruding lower jaw. Teeth long and sharp. Pelvic fins originate well behind pectoral fins and beneath or slightly anterior to spiny dorsal-fin origin. Body comparatively slender and elongate. HABITAT: Kodiak Island, AK, and Gulf of Alaska to southern Baja California. Pelagic along the coast from surf zone to about 125 ft. Young may enter bays. BIOLOGY: Often in small schools.

Mexican Barracuda - *Sphyraena ensis* Jordan & Gilbert, 1882

FEATURES: Shades of gray dorsally, with a series of oblique bars. Silvery below. Caudal fin grayish. Mouth large, with protruding lower jaw. Teeth long and sharp. Pelvic fins originate under tips of pectoral fins. Body comparatively deep. HABITAT: Oceanside, CA, to Chile, including lower Gulf of California. Pelagic along the coast to about 80 ft. Also in middle estuaries, mangroves, and lagoons. BIOLOGY: Usually in small schools.

Spadefishes

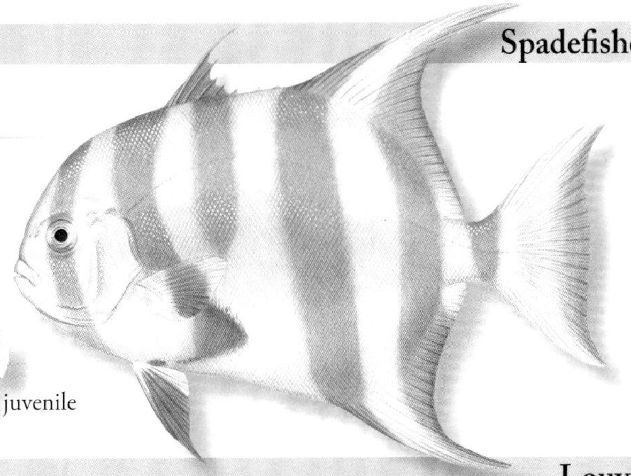

Pacific Spadefish
to 2.2 ft.

juvenile

Louvar

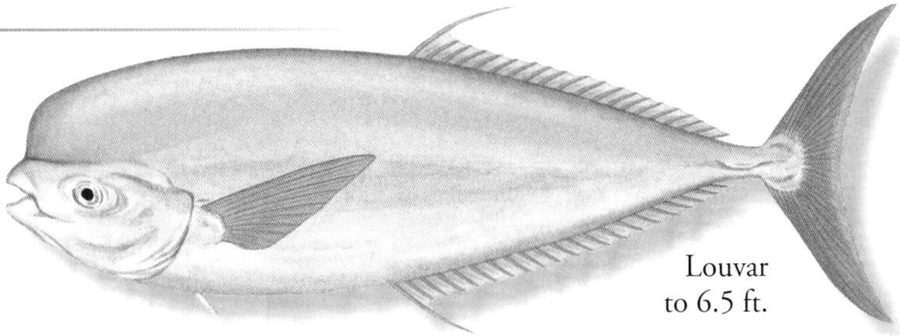

Louvar
to 6.5 ft.

Barracudas

Pacific Barracuda
to 5 ft.

Mexican Barracuda
to 4 ft.

Gempylidae - Snake Mackerels

Snake Mackerel - *Gempylus serpens* Cuvier, 1829

FEATURES: Steely gray to grayish brown, with a silvery luster. Fins with dark margins. Jaws large and toothy. Lower jaw with pointed tip. First dorsal fin long-based, with 26–32 spines. Second dorsal fin followed by 5–6 finlets. Anal fin followed by 6–7 finlets. Caudal fin forked. HABITAT: Circumglobal in tropical to temperate seas. In eastern Pacific from San Pedro, CA, to central Chile, including Gulf of California. Also Galápagos Islands. Offshore and oceanic from surface to about 655 ft.

Escolar - *Lepidocybium flavobrunneum* (Smith, 1843)

FEATURES: Almost uniformly dark brown, becoming black with age. First dorsal fin low, with 8–9 spines. Second dorsal fin followed by 4–6 finlets. Lateral line sinuous. Caudal peduncle with one strong keel and two smaller keels. HABITAT: Circumglobal in tropical to warm temperate seas. WA to Peru, including Gulf of California. Oceanic and pelagic over continental slopes from about 80 to 655 ft. BIOLOGY: Migrate to shallow water at night. Feed on squids and a variety of fishes.

Oilfish - *Ruvettus pretiosus* Cocco, 1833

FEATURES: Brown to dark brown. Tips of pectoral and pelvic fins blackish. First dorsal fin low, with 13–15 spines. Second dorsal and anal fins followed by two finlets. Lateral line straight posteriorly. Scales rough and spiny. HABITAT: Circumglobal in tropical to warm temperate seas. In eastern Pacific from Encinitas, CA, to southern Baja California and Guatemala to Chile. Found near bottom and in the water column over continental slopes and rises from about 170 to 3,800 ft.

ALSO IN THE AREA: Striped Escolar, *Diplospinus multistriatus*, page 351.

Trichiuridae - Cutlassfishes

Razorback Scabbardfish - *Assurger anzac* (Alexander, 1917)

FEATURES: Uniformly iridescent silver. Anterior dorsal fin blackish. Mouth large, with protruding lower jaw. Dorsal fin low, with 116–119 rays. Anal fin with 14–17 exposed rays. Caudal fin small and forked. Pelvic fins minute. Body extremely long, compressed, and tapering at tail. HABITAT: Scattered circumglobally in tropical to temperate seas. In eastern Pacific from Point Dome, CA, to Chile. Pelagic in deep water from about 490 to 1,300 ft.

Pacific Cutlassfish - *Trichiurus nitens* Garman, 1899

FEATURES: Iridescent silvery blue, steely blue, to silvery brown. Tips of pectoral fins dark. Mouth large, with protruding lower jaw. Dorsal fin tall anteriorly, with 118–127 rays. Pelvic and caudal fins absent. Anal fin consists of buried or nearly buried spinules. Body tapers to a pointed tail. HABITAT: Oxnard, CA, to Peru, including Gulf of California. Also to Sea of Japan. Primarily oceanic from surface to about 1,260 ft. Sometimes near shore. NOTE: Recently separated from *Trichiurus lepturus*.

Snake Mackerels

Snake Mackerel
to 3.5 ft.

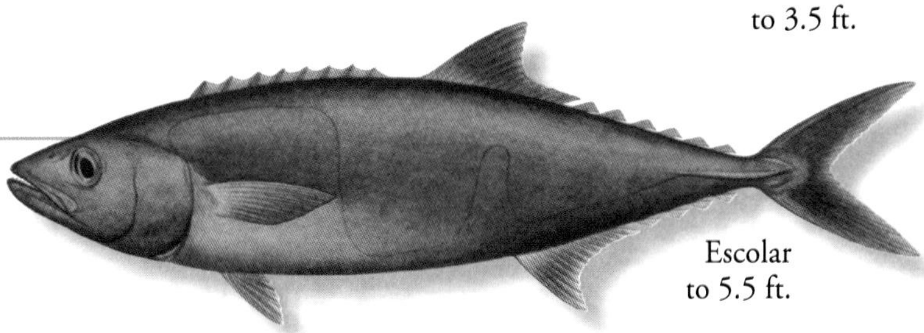

Escolar
to 5.5 ft.

Oilfish
to 9.8 ft.

Cutlassfishes

Razorback Scabbardfish
to 18.2 ft.

Pacific Cutlassfish
to 4.9 ft.

Wahoo - *Acanthocybium solandri* (Cuvier, 1832)

FEATURES: Iridescent bluish green dorsally, silvery below. Numerous irregular bars on sides. Snout long, pointed. First dorsal fin with 23–27 spines; originates over pectoral-fin base. Second dorsal fin followed by nine finlets. Body elongate and slightly compressed. HABITAT: Circumglobal in tropical to subtropical seas. Recently recorded from Newport Beach and San Diego, CA. Historically from south of the US-Mexican border to Peru, including southern Gulf of California and Galápagos Islands. Offshore in open ocean from surface to mid-water. BIOLOGY: Feed on squids and fishes. Females may carry up to 6 million eggs. Sought commercially and as sport fish.

Bullet Mackerel - *Auxis rochei* (Risso, 1810)

FEATURES: Dark blue dorsally, deep purple to blackish over head. Upper scaleless area with about 15 irregular, moderately broad, oblique bars; some bars may merge. Dorsal fins widely separated. Pectoral fins short, not extending past corselet. Corselet well developed posteriorly. Body comparatively shallow. HABITAT: Circumglobal in tropical to warm temperate seas. Redondo Beach, CA, to northern Chile, including Gulf of California and Galápagos Is. Occur in the water column over continental shelves and slopes from surface to about 655 ft. BIOLOGY: Form large schools. Sought commercially.

Frigate Mackerel - *Auxis thazard* (Lacepède, 1800)

FEATURES: Dark blue dorsally, deep purple to blackish over head. Upper scaleless area with 15 or more irregular, oblique, wavy bars. Bars may be whole, broken, or converging. Dorsal fins widely separated. Pectoral fins short, reaching to or slightly beyond corselet. Corselet narrow posteriorly. Body comparatively deep. HABITAT: Circumglobal in tropical to subtropical seas. Santa Catalina Island, CA, to Chile, including southern Gulf of California. Occur in upper water column over continental shelves and slopes from surface to about 655 ft. BIOLOGY: Schooling.

Kawakawa - *Euthynnus affinis* (Cantor, 1849)

FEATURES: Bluish black dorsally, with numerous oblique, wavy, and broken stripes on posterior upper body. Silvery below. Several gray to blackish blotches usually present above pelvic fins. Corselet extends to midbody. First dorsal fin tall anteriorly. Second dorsal fin followed by 8–10 finlets. HABITAT: Found in warm waters of the Indian and central Pacific oceans. A few stray specimens from eastern Pacific, including Los Angeles Harbor, CA. Occur in upper water column over continental shelves from surface to about 165 ft.

Black Skipjack - *Euthynnus lineatus* Kishinouye, 1920

FEATURES: Bluish black dorsally with about 3–9 nearly straight to slightly wavy black stripes on posterior upper body. Silvery below. Abdomen with several gray to black spots or long streaks. First dorsal fin tall anteriorly. Second dorsal fin followed by 8–10 finlets. Corselet extends well past pectoral fins. HABITAT: San Simeon, CA, and lower Gulf of California to Peru. Pelagic and oceanic from near shore to about 240 miles offshore. Primarily found near surface.

Wahoo
to 7.3 ft.

Bullet Mackerel
to 20 in.

Frigate Mackerel
to 2 ft.

Kawakawa
to 3.6 ft.

Black Skipjack
to 3 ft.

variants

Skipjack Tuna - *Katsuwonus pelamis* (Linnaeus, 1758)

FEATURES: Bluish black dorsally, silvery below. Upper scaleless area may have several oblique, iridescent blue bands. Lower sides with four to six grayish stripes that may appear as discontinuous lines of dark blotches. First dorsal fin tall anteriorly. Second dorsal fin followed by 7–9 finlets. Corselet extends well past pectoral fins. HABITAT: Circumglobal in tropical to warm temperate seas. In eastern Pacific from Yakutat Bay, AK, to Chile, including Gulf of California and Galápagos Islands. Occur in the water column over continental shelves and slopes from surface to about 850 ft. BIOLOGY: Form large schools. Often mix with Blackfin Tuna. Sought commercially and as sport fish.

Pacific Bonito - *Sarda chiliensis* (Cuvier, 1832)

FEATURES: Metallic turquoise green dorsally, with about 8–9 oblique black stripes on back that become broken below. Some lack dorsal stripes. Often with wide, obscure to distinct grayish bars on sides. Silvery below. First dorsal fin blackish, tall anteriorly, and decreasing in height posteriorly. Second dorsal fin followed by eight finlets. Corselet not extending past pectoral fins. HABITAT: Northern population from Gulf of Alaska to off southern Baja California. Southern population from Colombia to Chile. Pelagic in nearshore waters from bays and surf zone to well offshore. Also over reefs and around kelp beds. BIOLOGY: Form large schools. Highly migratory. Sought commercially.

Pacific Chub Mackerel - *Scomber japonicus* Houttuyn, 1782

FEATURES: Bright green to turquoise blue dorsally, with numerous oblique, wavy, and zig-zagging black bands. Often with a row of spots along midline. Ventral area silvery white. Eyes with a large adipose eyelid. First and second dorsal fins widely spaced. Second dorsal fin followed by five finlets. Body covered with relatively small scales. HABITAT: Western Gulf of Alaska to Gulf of California. Also from Panama to Chile and in the western Pacific. Pelagic from near surface and shore to over continental shelves and slopes to about 980 ft. BIOLOGY: Schooling. Sought commercially.

Gulf Sierra - *Scomberomorus concolor* (Lockington, 1879)

FEATURES: Iridescent bluish dorsally, fading to silvery below. First dorsal fin black, with white along posterior base. Females with two rows of golden orange spots on sides. Spots absent in males. Second dorsal fin followed by 6–9 (usually 8) finlets. Lateral line gradually curving downward along midbody. HABITAT: Historically from Soquel, CA, to central Mexico, including Gulf of California. Believed to be confined to central and northern Gulf of California. Occur near surface to about 50 ft. NOTE: Vulnerable.

Pacific Sierra - *Scomberomorus sierra* Jordan & Starks, 1895

FEATURES: Iridescent greenish to turquoise green dorsally, fading to silvery below. Sides with numerous rows of orange spots. Spots decrease in size and increase in number with age. First dorsal fin black, with white along posterior base. Second dorsal fin followed by 7–10 finlets. Lateral line gradually curving downward along midbody. HABITAT: Santa Monica, CA, to central Chile, including Gulf of California. Occur over continental shelves from near shore and surface to about 50 ft.

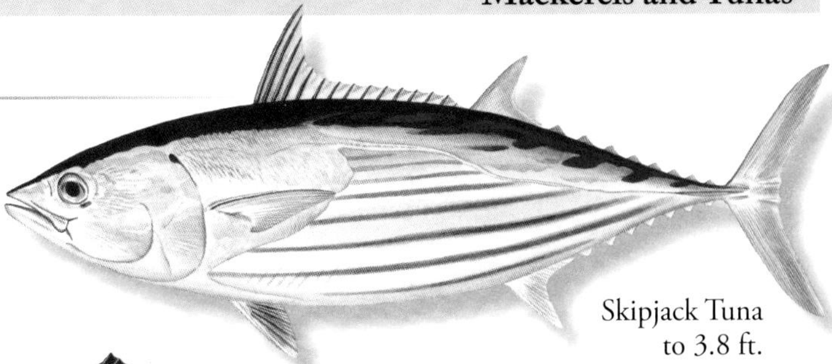

Skipjack Tuna
to 3.8 ft.

Pacific Bonito
to 4 ft.

Pacific
Chub Mackerel
to 25 in.

Gulf Sierra
to 2.8 ft.

Pacific Sierra
to 3.7 ft.

Albacore - *Thunnus alalunga* (Bonnaterre, 1788)

FEATURES: Metallic blackish blue dorsally, silvery white below. Second dorsal fin yellowish. Caudal-fin margin white. Pectoral fins extremely long, usually reaching to or beyond second dorsal finlet. HABITAT: Circumglobal in tropical to temperate seas. In eastern Pacific from northern and eastern Gulf of Alaska to Chile, including entrance to Gulf of California. Oceanic from surface to about 1,970 ft. Usually below thermocline. Other names: Longfin Tuna, Albie. NOTE: Currently listed as Near Threatened.

Yellowfin Tuna - *Thunnus albacares* (Bonnaterre, 1788)

FEATURES: Metallic blackish blue dorsally, yellow on upper sides, and silvery below. Lower sides often with about 20 rows of pale spots and nearly vertical lines. Second dorsal fin and anal fins yellow. Finlets yellow, with narrow black margin. Pectoral fins moderately long. Second dorsal and anal fins tall, becoming taller and more elongate with age. HABITAT: Circumglobal in tropical to subtropical seas. In eastern Pacific from Morro Bay, CA, to Peru, including lower Gulf of California. Oceanic, above and below thermocline in temperatures from about 64–87°F. NOTE: Near Threatened.

Bigeye Tuna - *Thunnus obesus* (Lowe, 1839)

FEATURES: Metallic blackish blue dorsally, silvery white below. Rows of pale spots may be present on lower sides. Second dorsal and anal fins silvery yellow and comparatively short. Finlets yellow, with broad black margin. Pectoral fins very long in small specimens, moderately long in large specimens. HABITAT: Circumglobal in tropical to subtropical seas. In eastern Pacific from Iron Springs, WA, to Chile. Oceanic and pelagic from surface to about 4,920 ft. NOTE: Listed as Vulnerable.

Pacific Bluefin Tuna - *Thunnus orientalis* (Temminck & Schlegel, 1844)

FEATURES: Metallic bluish black dorsally, silvery white below. Lower sides with rows of pale spots or lines. Anal fin and finlets dusky yellowish, with narrow black margins. Caudal keel blackish. Pectoral fins short. Corselet extends past pectoral fins. Body very large and deep. HABITAT: Found in tropical and temperate Pacific and Indo-Pacific oceans. In eastern Pacific from Shelikof Strait, Gulf of Alaska, to tip of Baja California. Oceanic and pelagic from surface to about 1,800 ft.

Xiphiidae - Swordfish

Swordfish - *Xiphias gladius* Linnaeus, 1758

FEATURES: Back and upper sides metallic bluish black to brownish black. Coppery below. Bars or spots absent. Long bill flattened in cross-section. Eyes very large. First dorsal fin long-based in juveniles, short-based in adults. Second dorsal and anal fins small. Single, broad lateral caudal keel present. Body robust. HABITAT: Circumglobal in tropical, temperate, and sometimes cold seas. In eastern Pacific from south of Vancouver Island, British Columbia, to Chile, including Gulf of California. Oceanic, but sometimes found in coastal waters, and generally above the thermocline.

Mackerels and Tunas

Albacore
to 4.4 ft.

Yellowfin Tuna
to 7.2 ft.

Bigeye Tuna
to 8.5 ft.

Pacific Bluefin Tuna
to 10 ft.

Swordfish

Swordfish
to 14.7 ft.

juvenile

Istiophoridae - Billfishes

Black Marlin - *Istiompax indica* (Cuvier, 1832)

FEATURES: Dark bluish black dorsally. Silvery white below. Distinct bars or spots on upper sides absent in adults. First dorsal fin blackish with blue to lavender highlights. Anterior lobe of first dorsal fin not taller than depth of body. Pectoral fins rigid, not able to fold against sides. Lateral line with series of loops or branches. Body comparatively robust. HABITAT: Found in the Pacific and Indo-Pacific oceans. Strays into SE Atlantic. In eastern Pacific from southern CA to northern Chile. Pelagic and oceanodromous from near shore to well offshore; usually in surface waters above the thermocline. NOTE: Previously classified as *Makaira indica*.

Sailfish - *Istiophorus platypterus* (Shaw, 1792)

FEATURES: Dark blue dorsally, coppery on sides, silvery white below. May be entirely dark and may display iridescent blue bars and spots on upper sides. Dorsal fin bluish black to dark brownish blue, with small dark spots; tall and sail-like. Pelvic fins very long. HABITAT: Circumglobal in tropical to temperate seas. In eastern Pacific from Dana Point, CA, to Chile, including Gulf of California. Oceanic and coastal; usually above thermocline and near surface. BIOLOGY: Very swift and migratory. Recorded hunting in coordinated groups. Sought commercially and for sport. NOTE: Previously classified as *Istiophorus albicans*.

Striped Marlin - *Kajikia audax* (Philippi, 1887)

FEATURES: Dark blue to bluish black dorsally, with about 15 cobalt blue bars or rows of spots. Silvery white below. Anterior lobe of first dorsal fin equal to or slightly taller than body depth. Pectoral fins comparatively long and pointed. Lateral line simple and single. HABITAT: Found in tropical to temperate waters of the Pacific and Indian oceans. Strays into SE Atlantic. Oceanic, pelagic, and usually above the thermocline. Rarely enter coastal waters. NOTE: Previously *Tetrapturus audax*. Near Threatened.

Blue Marlin - *Makaira nigricans* Lacepède, 1802

FEATURES: Dark blackish blue to blackish brown dorsally. Silvery white below. Upper sides with bright to faint blue bars. First dorsal fin dark blue to black. Anterior lobe of first dorsal fin not taller than depth of body. Pectoral fins able to fold against sides. Pelvic fins shorter than pectoral fins. Lateral line complex and reticulating. HABITAT: Circumglobal in tropical to warm seas. In eastern Pacific from southern CA to Chile. Oceanic. Usually in or above the thermocline from about 71–88°F. BIOLOGY: Highly migratory. Enter colder waters to feed. Sought commercially and as sport. Vulnerable.

Shortbill Spearfish - *Tetrapturus angustirostris* Tanaka, 1915

FEATURES: Cobalt blue to dark blue dorsally. Silvery below. Bars and spots absent. Bill very short. Pectoral fins short, pelvic fins long. Posterior portion of first dorsal fin nearly equally tall. Body comparatively long and slender. HABITAT: Found in the Pacific and Indian oceans. Strays into SE Atlantic. In eastern Pacific from Cape Mendocino, CA, to Chile. Oceanic, well offshore, and generally above the thermocline.

Black Marlin
to 16.3 ft.

Sailfish
to 11.7 ft.

Striped Marlin
to 13.8 ft.

Blue Marlin
to 16.4 ft.

Shortbill Spearfish
to 8 ft.

Centrolophidae - Medusafishes

Medusafish - *Icichthys lockingtoni* Jordan & Gilbert, 1880

FEATURES: Shades of bluish gray to brown, with dusky to black fins. Juveniles transparent to bluish or pinkish. Dorsal fin low, continuous, and originates well behind pectoral-fin base. Dorsal and anal fins fleshy at base. Caudal-fin margin rounded to notched. Adults elongate, juveniles deep bodied. Body limp. Scales absent in juveniles. HABITAT: South of Aleutian Islands and Gulf of Alaska to central Baja California. Also to Japan. Oceanic and pelagic from surface to about 3,300 ft. Juveniles with sea jellies.

Nomeidae - Driftfishes

Black Fathead - *Cubiceps baxteri* McCulloch, 1923

FEATURES: Overall dark brown. Large specimens black. Pectoral fins paler. Eyes dark blue. Snout short and blunt. Mouth small, eyes large. Spiny dorsal fin tall. Soft dorsal fin with 20–24 rays. Anal fin with 20–24 rays. Pectoral fins long, extending past anal-fin origin. Body covered with large scales. HABITAT: Occur in warm waters of Atlantic and Pacific oceans. In eastern Pacific from southern CA to Chile. Oceanic and pelagic.

Bluefin Driftfish - *Psenes pellucidus* Lütken, 1880

FEATURES: Adults uniformly bluish black, may be paler on abdomen. May have pale bars. Juveniles pale. Snout blunt, rounded. Mouth small. Eyes moderately large and with adipose tissue. Spiny dorsal fin low. Soft dorsal fin with 27–32 rays. Anal fin with 28–34 rays. Pectoral fins short. HABITAT: Circumglobal in warm seas. In eastern Pacific from Santa Catalina Island, CA, to northwest of Peru. Oceanic and pelagic.

Tetragonuridae - Squaretails

Smalleye Squaretail - *Tetragonurus cuvieri* Risso, 1810

FEATURES: Overall dark brown to black. Juveniles gray above, whitish below. Lower jaw with knife-like teeth; fits into upper jaw when mouth is closed. Eyes comparatively small. First dorsal fin low. Caudal peduncle square in cross-section, with two low keels. Scales ridged. HABITAT: In temperate Atlantic, Pacific, and Mediterranean. In eastern Pacific from south of Aleutians to Baja California. Oceanic to about 2,300 ft.

Stromateidae - Butterfishes

Pacific Pompano - *Peprilus simillimus* (Ayres, 1860)

FEATURES: Iridescent olivaceous to bluish green dorsally, silvery on sides and below. Juveniles may be golden. Snout very blunt and short. Mouth small. Dorsal fin single, long-based. Pectoral fins long. Caudal fin deeply forked. Pelvic fins absent. Body deep and compressed. Dorsal pores absent. HABITAT: Queen Charlotte Sound, British Columbia, to southern Baja California, including Gulf of California. Pelagic over the continental shelf from surf zone to about 300 ft. May enter brackish water. BIOLOGY: Sought as food fish. Juveniles associated with sea jellies.

Medusafishes

Medusafish
to 18.4 in.

Driftfishes

Black Fathead
to 3.8 ft.

Bluefin Driftfish
to 2.6 ft.

juvenile

Squaretails

Smalleye Squaretail
to 2.3 ft.

Butterfishes

Pacific Pompano
to 11 in.

Pacific Sanddab - *Citharichthys sordidus* (Girard, 1854)

FEATURES: Color, pattern highly variable. Eyed side shades of brown, gray, or tan; often with irregular orange, yellow, and/or whitish spots and/or blotches. Blind side whitish to tan. Maxilla extends to below anterior portion of lower eye. Interorbital area concave, with a bony ridge above lower eye. Pectoral fin on eyed side reaches to middle of eyes when folded forward. Caudal-fin margin fairly straight. Lateral line nearly straight. Blind side scales smooth. HABITAT: Kodiak Island, AK, to southern Baja California. Demersal over gravel, sand, shelly, and silty bottoms from intertidal zone to about 1,800 ft.

Speckled Sanddab - *Citharichthys stigmaeus* Jordan & Gilbert, 1882

FEATURES: Color and pattern highly variable. Eyed side shades of tan, gray, to brown; often densely speckled with black, brown, or white. Blind side white. Maxilla extends to anterior portion of lower eye. Interorbital area flat to concave; bony ridge absent. Pectoral fin on eyed side does not reach eyes when folded forward. Caudal-fin margin rounded. Lateral line nearly straight. Blind side scales mostly rough. HABITAT: Prince William Sound, AK, to southern Baja California. Demersal over a wide variety of bottoms from bays, estuaries, and open coast to about 1,200 ft. Prefer areas with some sand.

Longfin Sanddab - *Citharichthys xanthostigma* Gilbert, 1890

FEATURES: Eyed side shades of reddish brown, with irregular darker blotches that may form rosettes or vaguely ocellated spots with yellowish centers. Pectoral fin may be speckled or blackish. Blind side whitish. Maxilla extends to anterior portion of lower eye. Interorbital area with a sharp ridge above lower eye. Pectoral fin on eyed side extends past snout when folded forward. Caudal-fin margin rounded. Lateral line nearly straight. HABITAT: Monterey Bay, CA, to Costa Rica, including Gulf of California. More common in southern range. Demersal over soft bottoms from about 8 to 660 ft.

Bigmouth Sole - *Hippoglossina stomata* Eigenmann & Eigenmann, 1890

FEATURES: Eyed side shades of brown, with dark and pale speckles that form small eye spots. Sometimes with bluish speckles. Usually with large eye spots along upper and lower body margins. Blind side whitish. Snout humped dorsally. Mouth large, with maxilla extending below or beyond posterior lower eye. Upper and lower jaws lack teeth. Eyes large. Interorbital area with a narrow ridge. Lateral line arched over pectoral fin. Caudal fin angular to rounded. HABITAT: Monterey Bay, CA, to Gulf of California. Demersal over sandy and sandy mud bottoms from about 75 to 780 ft.

California Halibut - *Paralichthys californicus* (Ayres, 1859)

FEATURES: Color and pattern somewhat variable. Eyed side shades of brown, gray, tan, to olive with vague to distinct darker mottling and regularly spaced eye spots and/or crescent-shaped marks. Blind side usually white. Snout moderately long, bluntly pointed. Mouth large, with maxilla extending below or past rear margin of lower eye. Teeth large. Lateral line arched over pectoral fin. Caudal fin angular or straight to tri-lobed. Can be left- and right-eyed. HABITAT: Quillayute River, WA, to southern Baja California. Demersal over soft bottoms or near structures from surf zone to about 920 ft.

variants

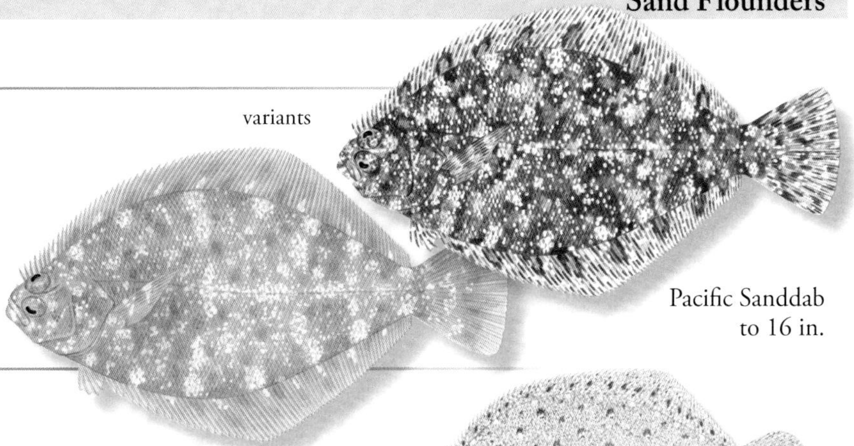

Pacific Sanddab
to 16 in.

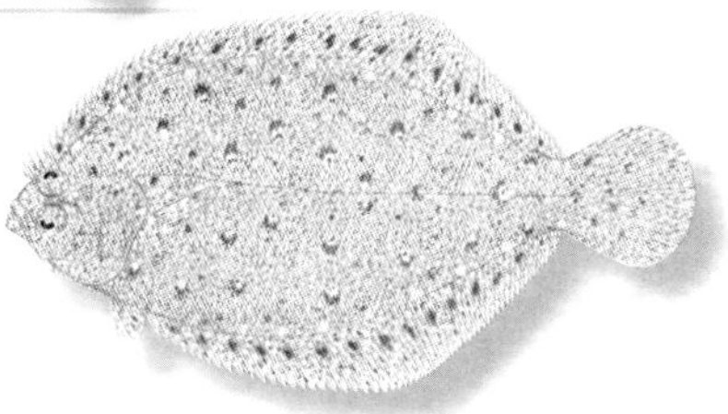

Speckled Sanddab
to 7 in.

Longfin Sanddab
to 10 in.

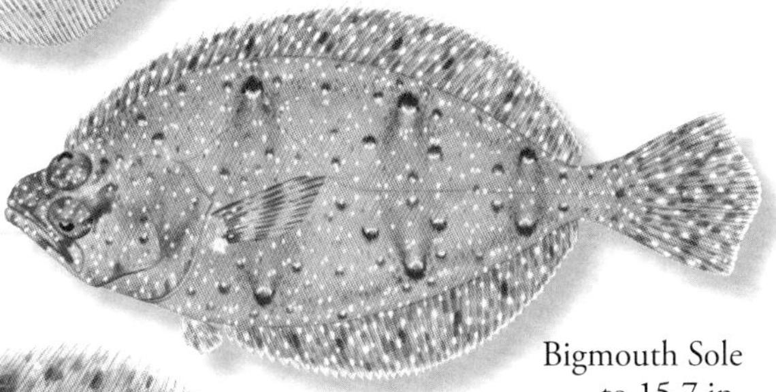

Bigmouth Sole
to 15.7 in.

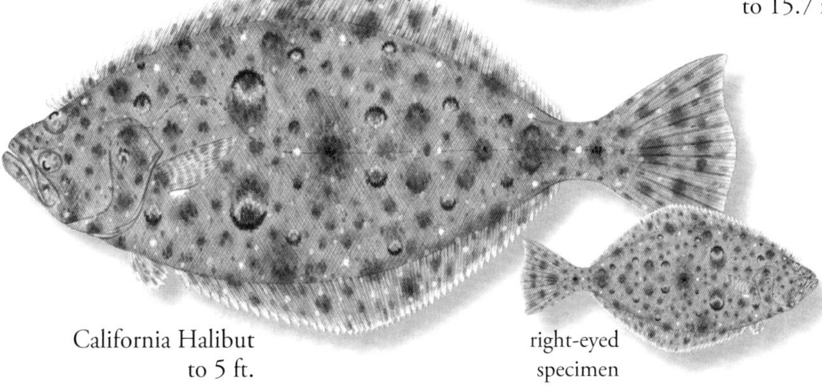

California Halibut
to 5 ft.

right-eyed
specimen

Paralichthyidae - Sand Flounders, *cont.*

Fantail Sole - *Xystreurys liolepis* Jordan & Gilbert, 1880

FEATURES: Eyed side shades of brown, gray, to olive with darker and paler mottling, scattered eye spots, and often with many bluish to pale gray spots. Usually with two large ocellated spots on midline. Blind side white. Mouth small, with maxilla extending to about middle of lower eye. Eyes moderately large. Pectoral fin about equal to length of head. Caudal-fin margin rounded to slightly angular. Lateral line highly arched over pectoral fin. Can be left- or right-eyed. HABITAT: Monterey Bay, CA, to Gulf of California. Demersal over sandy or muddy bottoms from surf zone to about 450 ft.

Pleuronectidae - Righteye Flounders

Kamchatka Flounder - *Atheresthes evermanni* Jordan & Starks, 1904

FEATURES: Eyed side uniformly brown or mottled and blotched. Blind side off-white to dirty white. Mouth large, with maxilla extending to or beyond rear margin of lower eye. Lower jaw with a single row of sharp teeth. Upper eye not visible from blind side. Caudal-fin margin concave. Lateral line slightly curved over pectoral fin. Scales slightly rough. HABITAT: Bering Sea from Gulf of Anadyr and Aleutian Islands. Also to Sea of Okhotsk and northern Sea of Japan. Demersal over soft bottoms from about 80 to 3,960 ft.

Arrowtooth Flounder - *Atheresthes stomias* (Jordan & Gilbert, 1880)

FEATURES: Eyed side uniformly shades of brownish or with darker and paler spots and blotches. Blind side shades of white to gray or finely dotted with black. Mouth large, with maxilla extending to or beyond rear margin of lower eye. Lower jaw with two rows of sharp teeth. Upper eye visible from blind side. Caudal-fin margin concave. Lateral line slightly curved over pectoral fin. Scales mostly smooth. HABITAT: Bering Sea from near Cape Navarin and Aleutian Islands to San Simeon, CA. Also to Commander Islands. Demersal over soft bottoms from about 30 to 3,760 ft.

Roughscale Sole - *Clidoderma asperrimum* (Temminck & Schlegel, 1846)

FEATURES: Eyed side shades of brown, with variable mottling. Blind side gray to grayish brown. Mouth small. Eyes large, with lower eye anterior to upper eye. Eyed side covered with irregular scaly plates and six or more rows of larger tubercles. Lateral line slightly arched over pectoral fin. Caudal-fin margin rounded. Body very deep. HABITAT: Scattered in Bering Sea, around Aleutian Islands, and to northern CA. Also to Korea. Demersal over muddy bottoms from about 50 to 6,230 ft.

Deepsea Sole - *Embassichthys bathybius* (Gilbert, 1890)

FEATURES: Eyed side shades of brown to maroon, with irregular bluish gray spots that cluster to form broad bars. Blind side dusky gray to brown. Mouth small. Eyes large. Interorbital area with a narrow ridge. Caudal fin small, with a rounded margin. Lateral line with a low arch over pectoral fin. HABITAT: Bering Sea and Aleutian Islands to northern Baja California. Also to Hokkaido, Japan. Demersal over muddy bottoms of continental slopes and sides of sea mounts from about 300 to 4,700 ft.

Sand Flounders

Fantail Sole
to 21 in.

right-eyed
specimen

Righteye Flounders

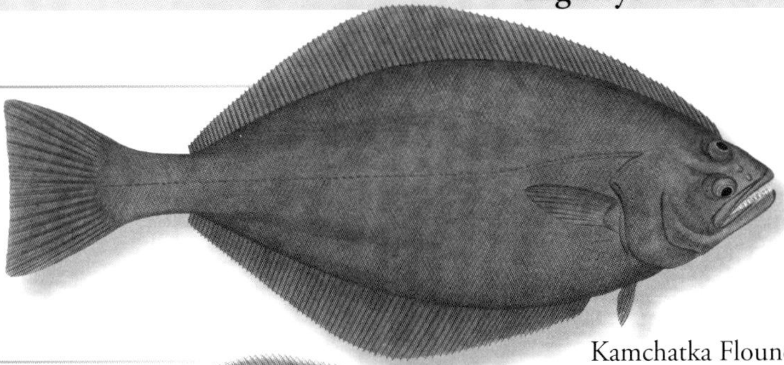

Kamchatka Flounder
to 3.6 ft.

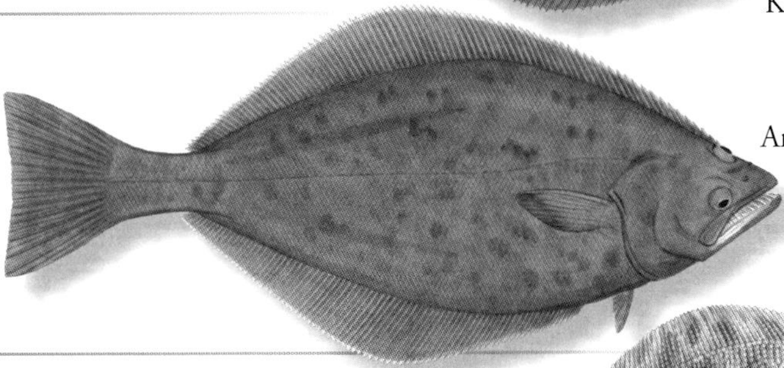

Arrowtooth Flounder
to 2.8 ft.

Roughscale Sole
to 24.4 in.

Deepsea Sole
to 20.5 in.

Petrale Sole - *Eopsetta jordani* (Lockington, 1879)

FEATURES: Eyed side variably blotched and spotted in shades of brown. Blind side white; sometimes with traces of pink. Mouth large, with maxilla extending to below middle of lower eye. Upper jaw with two rows of arrow-shaped teeth. Eyes large.Interorbital space wide and flat, with 4–5 scale rows. Caudal-fin margin slightly angular to rounded, with middle rays longest. Lateral line slightly arched over pectoral fin. HABITAT: Unalaska Island, AK, and Gulf of Alaska to northern Baja California. Demersal over sandy and muddy bottoms from surf zone to about 1,800 ft.

Rex Sole - *Glyptocephalus zachirus* Lockington, 1879

FEATURES: Eyed side shades of brown to gray with irregular and variable darker spots, speckles, and blotches. Lateral line may be pale. Pectoral fin mostly blackish. Blind side off-white to dusky. Mouth small with maxilla extending to below anterior lower eye or slightly beyond. Upper head profile arched. Pectoral fin on eyed side long. Caudal-fin margin rounded to slightly angular. Lateral line nearly straight. HABITAT: Bering Sea from Navarin Canyon and Aleutian Islands to central Baja California. Also to Kuril Islands. Demersal over sandy and muddy bottoms from surf zone to about 3,750 ft.

Flathead Sole - *Hippoglossoides elassodon* Jordan & Gilbert, 1880

FEATURES: Eyed side reddish brown to brownish with vague, diffuse mottling. Dorsal and anal fins blotched. Caudal-fin membranes pale. Blind side white with translucent areas. Mouth large, with maxilla extending to below middle of lower eye or just beyond. Eyes large. Interorbital area narrow, with a sharp ridge. Caudal-fin margin angular to rounded. Lateral line slightly arched over pectoral fin. HABITAT: Bering Sea and Aleutian Islands to Monterey, CA. Also to northern Sea of Japan. Demersal over muddy and silty bottoms from intertidal zone to about 3,445 ft.

Bering Flounder - *Hippoglossoides robustus* Gill & Townsend, 1897

FEATURES: Eyed side described as uniformly reddish brown to grayish but may have mottling. Blind side white. Fin membranes mostly pale to transparent. Mouth large, with maxilla extending to below or slightly beyond middle of lower eye. Eyes small. Interorbital space narrow and flat. Caudal-fin margin rounded to angular. Lateral line with a low arch over pectoral fin. HABITAT: Chukchi Sea, and possibly Beaufort Sea, to Bering Sea off Alaska Peninsula and west to Akutan Island. Demersal from intertidal zone to about 1,745 ft.

Pacific Halibut - *Hippoglossus stenolepis* Schmidt, 1904

FEATURES: Eyed side shades of brown to black with irregular paler and darker blotches and mottling. Blind side white to milky white. Mouth large, with maxilla extending to about middle of lower eye. Eyes small. Interorbital area flat to slightly concave. Caudal-fin margin crescent shaped to tri-lobed. Lateral line with a high arch over pectoral fin. Body thick and robust. HABITAT: Chukchi Sea to northern Baja California. Also to off Hokkaido, Japan. Demersal over a variety of bottoms from about 20 to 3,600 ft. Usually shallower than 980 ft.

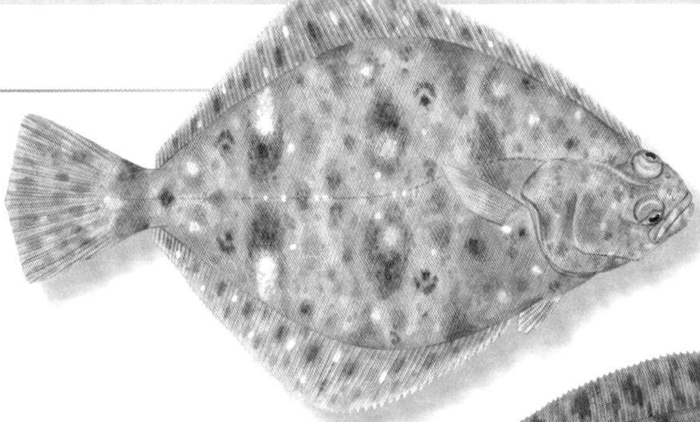

Petrale Sole
to 2.3 ft.

Rex Sole
to 24 in.

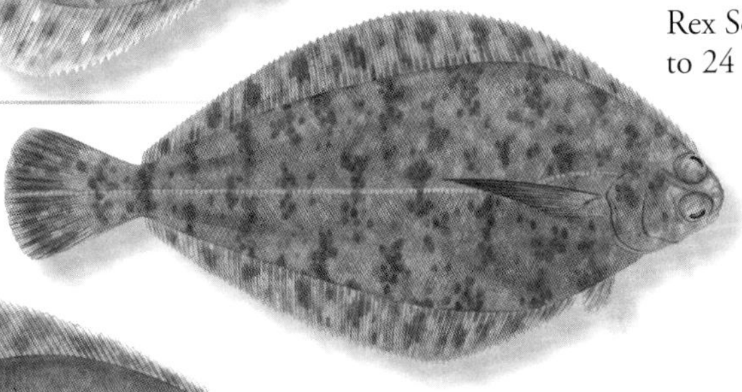

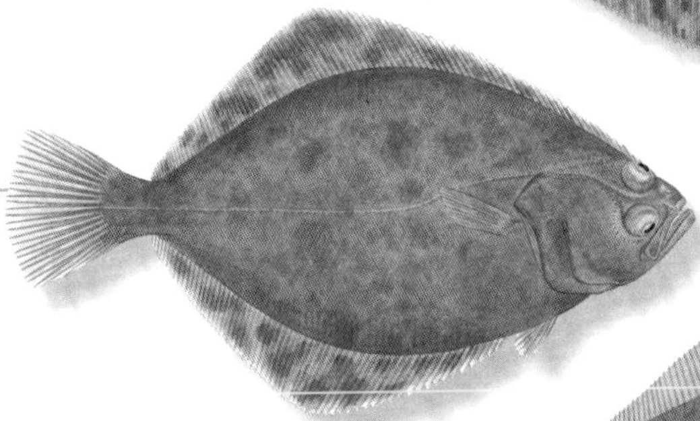

Flathead Sole
to 22 in.

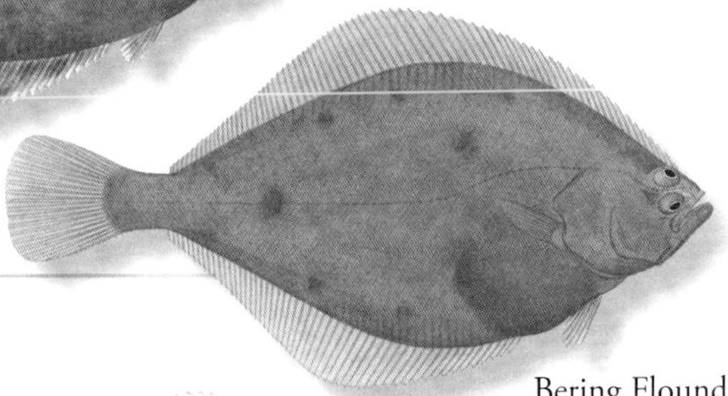

Bering Flounder
to 20.5 in.

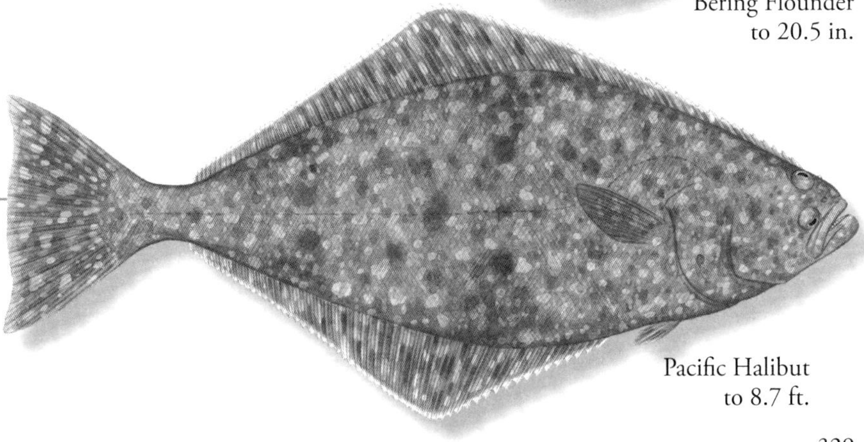

Pacific Halibut
to 8.7 ft.

Diamond Turbot - *Hypsopsetta guttulata* (Girard, 1856)

FEATURES: Eyed side shades of tan, brown, gray, to greenish gray with few to many irregular whitish, grayish, to pale bluish spots. Some with a few dark spots. Blind side white with yellow around mouth and head margin. Mouth small with maxilla extending to below anterior lower eye. Eyes large. Interorbital space with a low ridge. Lateral line with a low arch over pectoral fin and a long dorsal branch. Body very deep and diamond-shaped in profile. HABITAT: Cape Mendocino, CA, to southern Baja California. Demersal over sand and mud from surf zone to about 150 ft. Also in estuaries, lagoons, and mangroves. NOTE: Also classified by some authors as *Pleuronichthys guttulatus.*

Butter Sole - *Isopsetta isolepis* (Lockington, 1880)

FEATURES: Eyed side shades of brown to grayish brown with yellow to green blotches. Tips of dorsal and anal fin spines yellow. Blind side white. Mouth and eyes small. Interorbital area flat. Caudal-fin margin rounded to slightly angular. Lateral line slightly arched over pectoral fin and with a moderately long dorsal branch. HABITAT: SE Bering Sea and Aleutian Islands to Ventura, CA. More common from Umnak Island, AK, to northern CA. Demersal over muddy to silty bottoms from surf zone to about 1,400 ft. Sometimes in estuaries.

Southern Rock Sole - *Lepidopsetta bilineata* (Ayres, 1855)

FEATURES: Eyed side shades of gray, brown, olive, reddish, to blackish with irregular speckles, spots, blotches, and/or bands. Blind side white; glossy over muscle bands. Mouth small, with maxilla extending to below anterior lower eye. Eyes small. Interorbital space narrow, with up to three scales at narrowest part. Gill rakers broad, robust. Lateral line strongly arched over pectoral fin, with a moderately long dorsal branch. Caudal-fin margin rounded to angular. HABITAT: SE Bering Sea and Aleutian Islands to Cortes Bank, CA. Demersal over sandy and gravel bottoms from about 40 to 1,100 ft.

Northern Rock Sole - *Lepidopsetta polyxystra* Orr & Matarese, 2000

FEATURES: Eyed side shades of brown, reddish brown, to greenish brown with irregular speckles, spots, blotches and/or bands. Blind side white. Mouth small, with maxilla extending to below anterior lower eye. Eyes small. Interorbital space moderately wide, with up to five scales at narrowest part. Gill rakers slender, pointed. Lateral line strongly arched over pectoral fin, with a moderately long dorsal branch. Caudal-fin margin rounded. HABITAT: Bering Sea from St. Lawrence Is. and all Aleutian Islands to Puget Sound. Also to Sea of Okhotsk. Demersal over soft bottoms from about 10 to 1,700 ft.

Yellowfin Sole - *Limanda aspera* (Pallas, 1814)

FEATURES: Eyed side shades of brown to almost black, with scattered dark flecks that fade with age. Fins dusky yellow, yellow brown, to olivaceous. Dorsal and anal fins with a narrow black band at base. Blind side white. Mouth small. Eyes variable in size. Interorbital space convex. Lateral line highly arched over pectoral fin. Caudal-fin margin rounded. HABITAT: Beaufort Sea to Barkley Sound, British Columbia. Also to Korea. Demersal over soft and hard bottoms from about 30 to 1,970 ft.

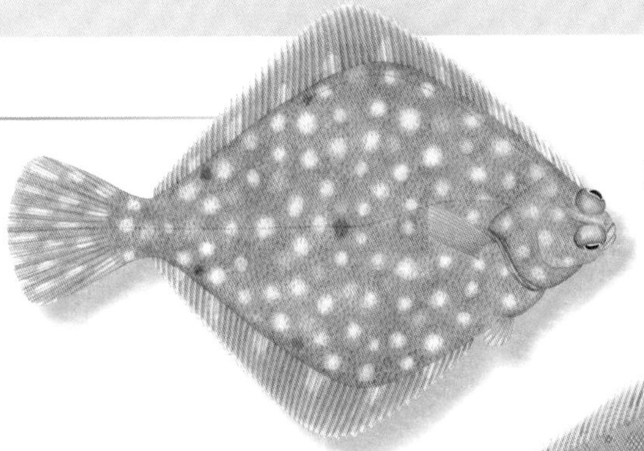

Diamond Turbot
to 18 in.

Butter Sole
to 22 in.

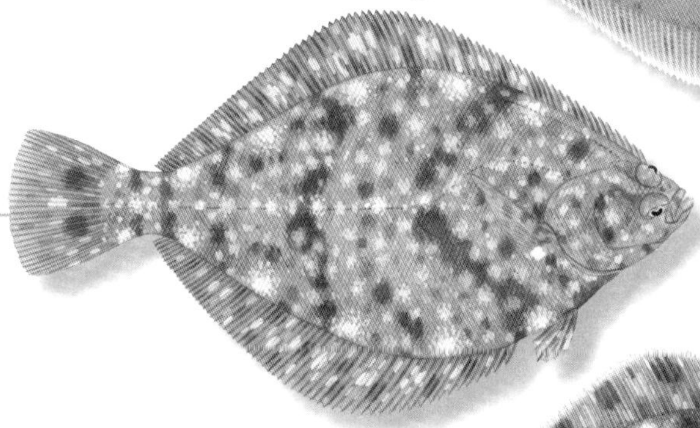

Southern Rock Sole
to 23 in.

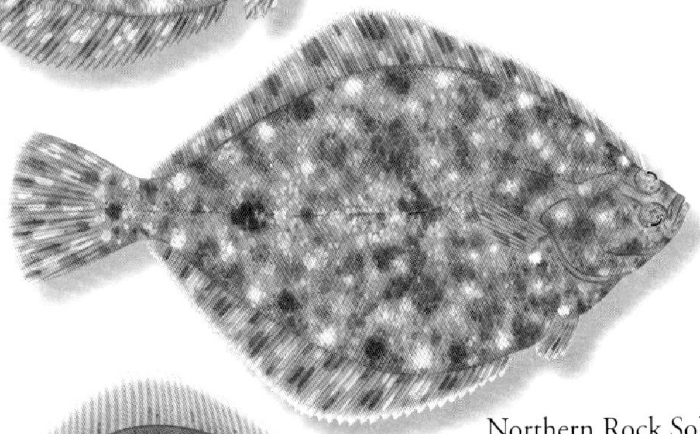

Northern Rock Sole
to 27 in.

Yellowfin Sole
to 19 in.

Longhead Dab - *Limanda proboscidea* Gilbert, 1896

FEATURES: Eyed side shades of greenish gray, brownish green, reddish brown, to blackish with irregular, indistinct spots and darker mottling. Blind side yellowish; yellow brighter along dorsal and ventral profiles. Snout long, upturned, and bluntly pointed. Mouth moderately large, with maxilla extending to front of lower eye. Head profile concave above eyes. Caudal-fin margin nearly straight to slightly rounded. Lateral line arched over pectoral fin. HABITAT: Beaufort Sea off Point Barrow to north of Unimak Island, AK. Also to Sea of Okhotsk. Demersal over soft bottoms from about 15 to 410 ft.

Arctic Flounder - *Liopsetta glacialis* (Pallas, 1776)

FEATURES: Eyed side shades of brown to greenish brown or black, with numerous darker speckles and small spots. Some uniformly colored. Fins paler and usually with dark spots. Blind side chalky white to lime green. Mouth small, with maxilla extending to below anterior lower eye. Interorbital area with a low ridge that becomes prominent posteriorly and ends in a large and small protuberance. Caudal fin slightly rounded. Lateral line nearly straight, rising slightly over pectoral fin. HABITAT: Beaufort Sea to SE Bering Sea and Aleutian Islands. Occur in shallow coastal waters from brackish bays and estuaries to about 60 ft. NOTE: Previously classified as *Pleuronectes glacialis*.

Slender Sole - *Lyopsetta exilis* (Jordan & Gilbert, 1880)

FEATURES: Eyed side shades of brown with variable darker and paler spots and blotches. Blind side whitish to pale orange. Mouth large, with maxilla extending to below middle of lower eye. Eyes large. Lateral line gently arched over pectoral fin. Caudal-fin margin angular to rounded. Body elongate. HABITAT: East of Kodiak Island, AK, to southern Baja California. Demersal over silty to muddy bottoms from about 30 to 3,760 ft.

Dover Sole - *Microstomus pacificus* (Lockington, 1879)

FEATURES: Eyed side highly variable in shades of brown to grayish brown, with diffuse to distinct dark or pale spots, blotches, and/or mottling. May be pale with dark markings, with ocellated spots, or almost uniformly colored. Blind side off-white to brownish gray. Dorsal and anal fins usually spotted. Mouth small, with maxilla reaching below anterior lower eye. Eyes large. Interorbital space convex. Lateral line nearly straight, with a low arch over pectoral fin and an indistinct branch on top of head. Caudal-fin margin rounded. HABITAT: Southeastern Bering Sea and Aleutian Islands to southern Baja California. Demersal over sandy and muddy bottoms from near shore to about 4,500 ft.

English Sole - *Parophrys vetulus* Girard, 1854

FEATURES: Eyed side highly variable in shades of tan, gray, or brown, with dark or pale spots, blotches, and speckles. May also be uniformly colored. Blind side white to pale yellow. Snout bluntly pointed. Mouth small, with maxilla reaching below anterior margin of lower eye. Eyes large, with upper eye visible from blind side. Caudal-fin margin nearly straight. Lateral line slightly arched over pectoral fin, with a long dorsal branch. HABITAT: Nunivak Island to Alaska Peninsula, and Aleutian Islands to central Baja California. Demersal over soft bottoms from intertidal zone to about 1,800 ft.

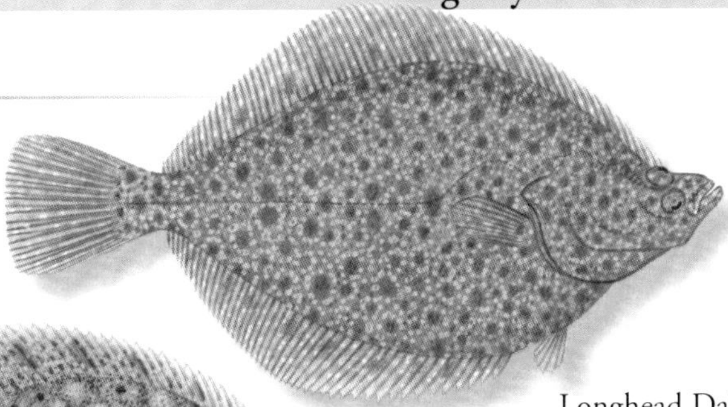

Longhead Dab
to 16 in.

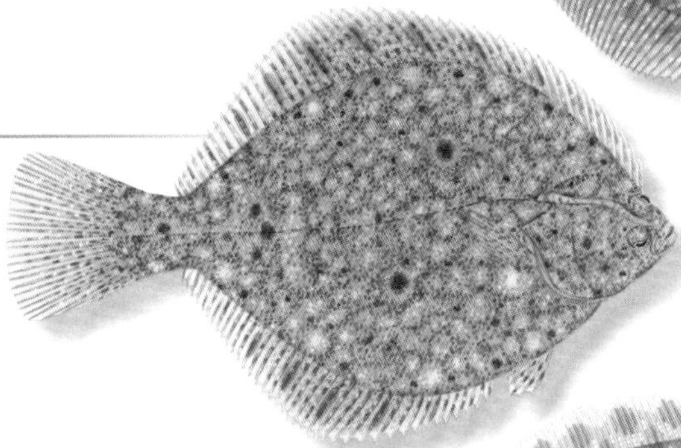

Arctic Flounder
to 13.8 in.

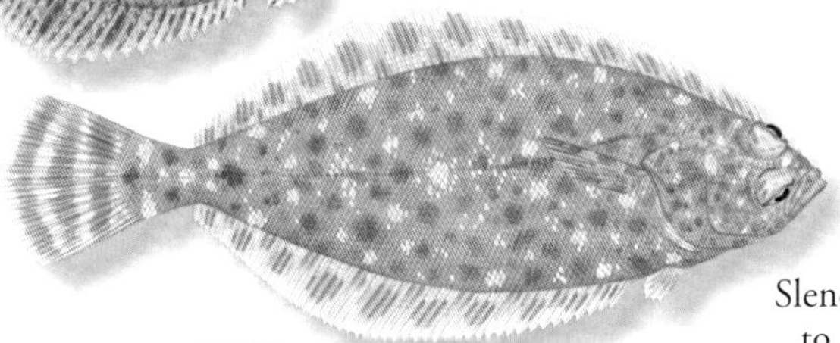

Slender Sole
to 13.7 in.

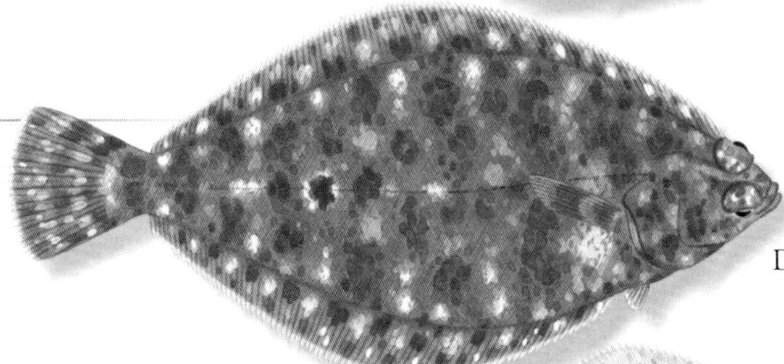

Dover Sole
to 2.5 ft.

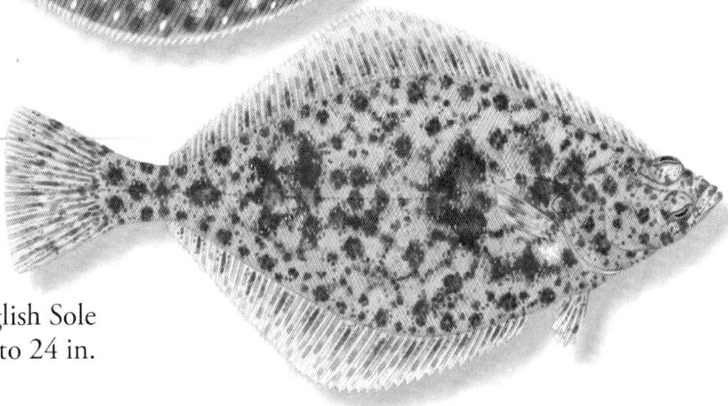

English Sole
to 24 in.

Starry Flounder - *Platichthys stellatus* (Pallas, 1787)

FEATURES: Eyed side highly variable in shades of tan, brown, olive, or black with irregular spots and blotches. May be uniformly colored. Blind side white or with some dark blotches. All with dorsal and anal fins that are white, cream, yellowish to greenish with blackish bars. Mouth small, with maxilla extending to below anterior lower eye. Eyes small. Caudal-fin margin nearly straight to slightly rounded. Body very deep and covered with large, irregularly spaced scales. Can be right- or left-eyed. HABITAT: Beaufort Sea to Los Angeles Harbor, CA. Also to Bathurst Inlet, Canada, and to off Korea. Demersal over soft and mixed bottoms from intertidal zone to about 1,970 ft. BIOLOGY: Known to hybridize with English Sole, *Parophrys vetulus*, to produce a similar-looking flatfish that has faint bars on dorsal and anal fins.

Alaska Plaice - *Pleuronectes quadrituberculatus* Pallas, 1814

FEATURES: Eyed side shades of dark brown or olive brown to black. May be nearly uniformly colored or with numerous small, pale speckles, especially in smaller specimens. Blind side yellow. Mouth small, with maxilla extending to below anterior lower eye. Eyes moderately large. Interorbital area with a distinct ridge that ends posteriorly with three to seven bony knobs. Snout bluntly pointed. Anterior head margin slightly arched. Lateral line nearly straight, with a low arch over pectoral fin and a short branch on top of head. Caudal-fin margin rounded. Body very deep and round in profile. HABITAT: Chukchi Sea to Unalaska Island and Gulf of Alaska to near Ketchikan, AK. Possibly in Beaufort Sea. Also to Sea of Japan. Demersal over soft bottoms from about 15 to 1,640 ft. Reported to about 2,790 ft in winter.

C-O Sole - *Pleuronichthys coenosus* Girard, 1854

FEATURES: Eyed side highly variable in shades of brown, tan, gray, to black, with heavy mottling, marbling, blotches, and spots. May have large areas of pink. Usually with a large vague to distinct spot on midline that may appear ocellated. All with small to large spot on center of caudal fin and an inverted "C" near caudal-fin base. Blind side creamy white, sometimes with dark blotches. Snout very short. Mouth very small. Eyes large. Interorbital area with a prominent ridge. First 5–6 dorsal-fin spines on blind side, originate at a point level with upper lip. Body very deep and round in profile. HABITAT: Southeastern AK to southern Baja California. Demersal over a wide variety of bottoms in estuaries to open coast from surf zone to about 1,150 ft.

Curlfin Sole - *Pleuronichthys decurrens* Jordan & Gilbert, 1881

FEATURES: Eyed side highly variable in shades of brown, yellow, to gray with irregular mottling, marbling, blotches, and spots. Blind side usually white, sometimes brown. Snout very short. Mouth very small. Eyes large. Interorbital area with ridge that bears a tublercle or spine at each end. Several tubercles present behind upper eye. First 9–12 dorsal-fin rays on blind side, originate at a point below mouth midline. Body very deep and round in profile. HABITAT: Northwest coast of Unimak Island and Gulf of Alaska to southern Baja California. Demersal over soft bottoms, along rock and sand interfaces, in estuaries, protected areas, and open coast to about 1,150 ft.

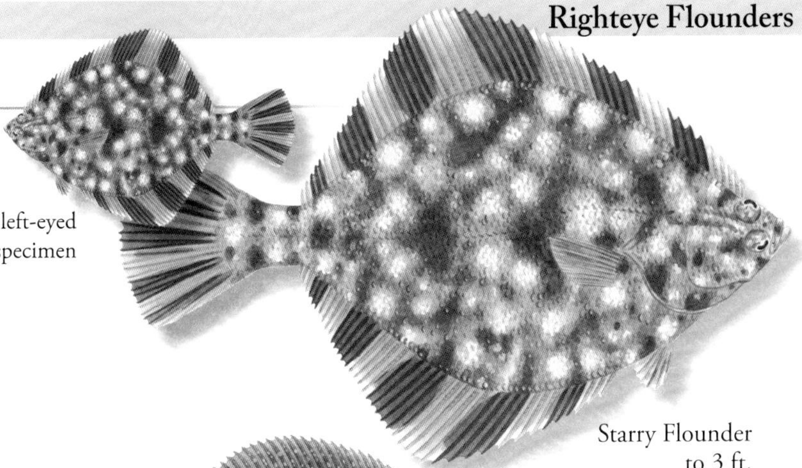

left-eyed specimen

Starry Flounder
to 3 ft.

Alaska Plaice
to 24.4 in.

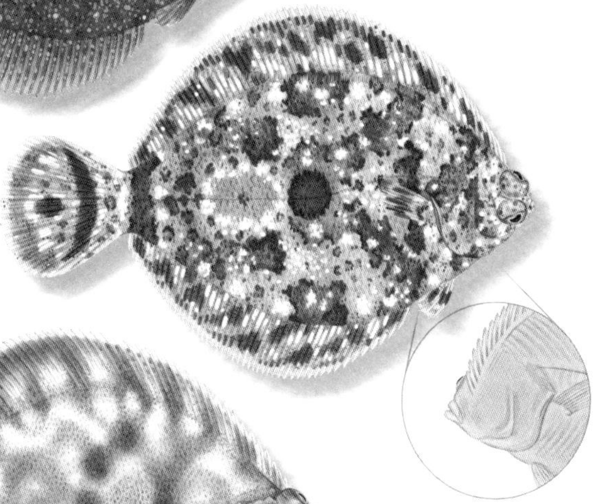

C-O Sole
to 14 in.

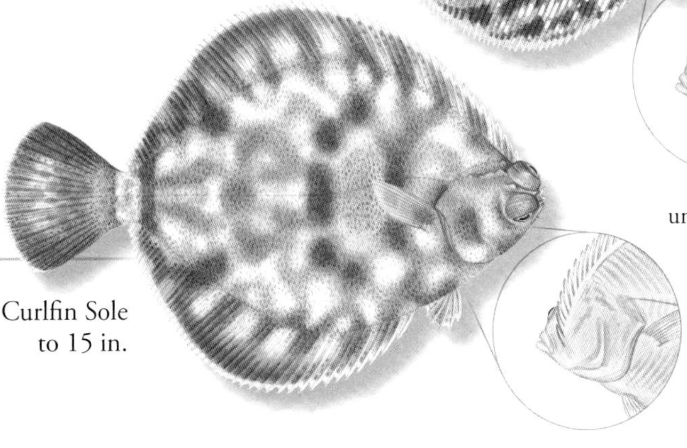

underside of head

Curlfin Sole
to 15 in.

Spotted Turbot - *Pleuronichthys ritteri* Starks & Morris, 1907

FEATURES: Eyed side highly variable in shades of brown to gray, with darker and paler blotches and irregular, scattered spots. Usually with one or two large, dark spots on midline and a dark spot at posterior dorsal- and anal-fin bases. Blind side white. Snout very short. Mouth very small. Eyes large. Interorbital area with a low ridge. First 5–6 dorsal-fin rays on blind side, originate above mouth midline. Body very deep, round in profile. HABITAT: Northern CA to southern Baja California. Demersal over soft bottoms from estuaries and open coast to about 650 ft. More common in shallow water. Some on grass beds.

Hornyhead Turbot - *Pleuronichthys verticalis* Jordan & Gilbert, 1880

FEATURES: Eyed side shades of tan to brown, with irregular darker blotches and mottling. Often with one or two dark blotches on midline that may be ocellated. Some also with pale spots that are usually along upper and lower profiles. Blind side white. Snout very short. Mouth very small. Eyes large. Interorbital area with a high, bony ridge that bears a blunt spine above the mouth and a sharp spine behind upper eye. First 4–6 dorsal-fin spines on blind side, originate above mouth. Body very deep, round in profile. HABITAT: OR to southern Baja California. Isolated population in northern Gulf of California. Demersal over soft bottoms in bays and sloughs from about 15 to 780 ft.

Sand Sole - *Psettichthys melanostictus* Girard, 1854

FEATURES: Eyed side shades of tan, brown, greenish, to gray, with minute darker speckles that may cover the body and fins or may cluster to form spots and vermiculations. May appear almost uniformly colored to blotchy. Blind side white. Mouth large, oblique, and with maxilla extending to below about middle of lower eye. Interorbital space flat. First dorsal-fin rays elongate and mostly free of membrane. Lateral line slightly arched over pectoral fin, with a short dorsal branch. Caudal-fin margin rounded. Scales rough. HABITAT: In Bering Sea from Unalaska Island to Port Heiden and Gulf of Alaska to Balboa Pier, CA. Demersal over sandy bottoms from intertidal zone to about 1,070 ft.

Greenland Halibut - *Reinhardtius hippoglossoides* (Walbaum, 1792)

FEATURES: Eyed side dark brown, blackish, grayish, to yellowish brown. Blind side paler. Mouth large, with maxilla extending to posterior margin of lower eye. Upper and lower jaws with a single row of large teeth. Eyes widely separated, with upper eye on dorsal profile and visible from blind side. Lateral line nearly straight over pectoral fin. Caudal-fin margin concave. HABITAT: In Arctic waters of the Atlantic and Pacific oceans and the Bering Sea. Northern Alaska to off northern Baja California. Rare south of Alaska. Also to Sea of Japan. Demersal over soft bottoms from about 45 to 6,560 ft. BIOLOGY: Largest specimens found in deep water. Sought commercially in northern waters.

ALSO IN THE AREA: Sakhalin Sole, *Limanda sakhalinensis*, page 351.

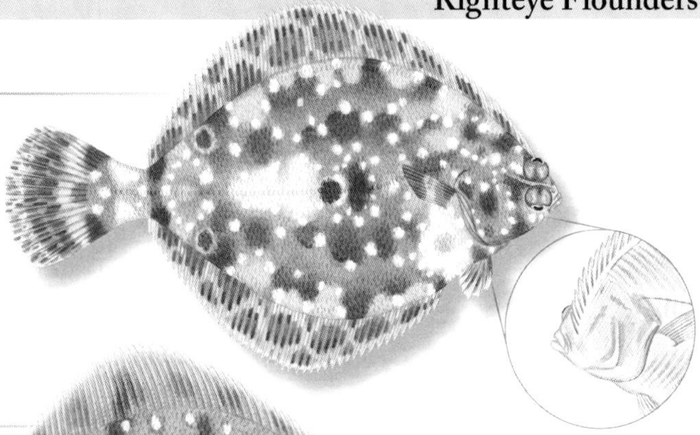

Spotted Turbot
to 11 in.

underside
of head

Hornyhead Turbot
to 14.5 in.

variants

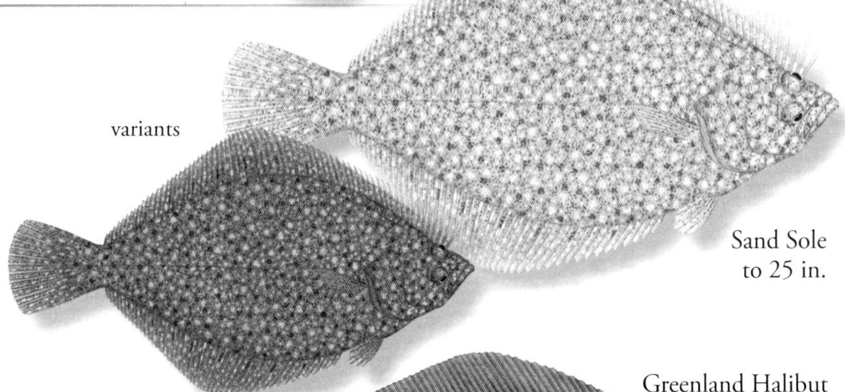

Sand Sole
to 25 in.

Greenland Halibut
to 4.3 ft.

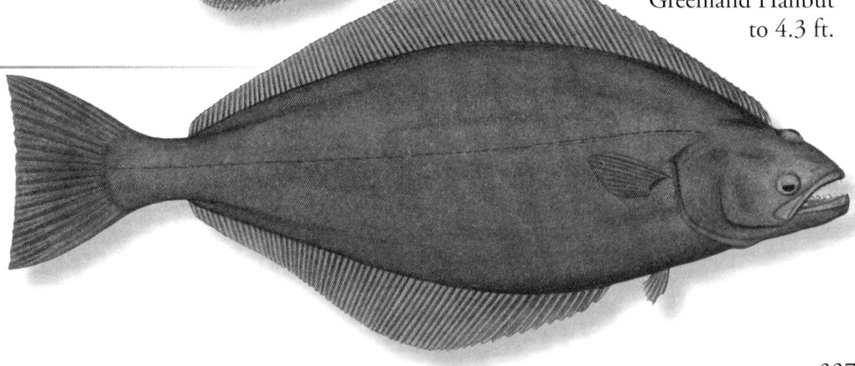

Cynoglossidae - Tonguefishes

California Tonguefish - *Symphurus atricaudus* (Jordan & Gilbert, 1880)

FEATURES: Eyed side shades of brown to gray and nearly uniformly colored or with numerous darker spots and/or irregular bars. Fins variably spotted. Body bars often become darker and tail area becomes black at night. Mouth small, nearly horizontal. Eyes very small and close set. Dorsal and anal fins confluent with tail. Pectoral fins absent. Body rounded anteriorly, tapered to a blunt tail, and forming a tongue shape. HABITAT: Barkley Sound, Vancouver Island, and Washington to Gulf of California. Demersal over soft bottoms of bays and estuaries, along open coast, and around eelgrass from shore to about 1,000 ft.

Balistidae - Triggerfishes

Finescale Triggerfish - *Balistes polylepis* Steindachner, 1876

FEATURES: Color variable. Shades of greenish brown to pale greenish or bluish dorsally and on sides. May have some blue around lips, head, and rims of fins. May also appear blotched. Paler on abdomen. Body otherwise unmarked. Mouth small, with deeply notched incisor-like teeth. First dorsal fin with three stout spines. Second dorsal fin with a tall anterior lobe. Caudal fin with elongate upper and lower rays. Scales small. HABITAT: Metlakatla, AK (rare), to central Chile, including Gulf of California. Most common south of CA. Found over rocky reefs and near adjacent sand or rubble from about 10 to 1,680 ft. Also along sandy beaches. Juveniles occur under drifting debris.

Black Durgon - *Melichthys niger* (Bloch, 1786)

FEATURES: Dark blackish brown to almost uniformly blackish. Cheeks may have yellowish to orange tinge, and blue lines may radiate from eyes. Scale margins often edged in blue, forming a distinct cross-hatched and zigzag pattern. All with a bright, pale blue band at dorsal- and anal-fin bases. Mouth small. Longitudinal ridges on cheeks absent. First dorsal fin with three stout spines. Second dorsal fin with a nearly straight margin. Caudal fin with elongate upper and lower rays. Scales behind gill opening enlarged, scales on posterior body keeled. HABITAT: Circumglobal in tropical seas. In eastern Pacific from San Diego, CA, and southern Gulf of California to Ecuador, including Galápagos Islands. Occur in clear water over reefs from surface to about 250 ft. BIOLOGY: Often seen in aggregations feeding at the surface.

Redtail Triggerfish - *Xanthichthys mento* (Jordan & Gilbert, 1882)

FEATURES: Color varies. Females and males shades of gray, tan, to yellow or green. Females with red on dorsal- and anal-fin margins and yellow on caudal-fin margins; sides of head may be paler or darker than body. Males with yellow on dorsal- and anal-fin margins and red on caudal-fin margin; sides of head may be dark green to blue or brown; dorsal area white during breeding. All with blue bands on cheeks and dark-margined scales. Mouth small. First dorsal fin with two visible spines. Dorsal and anal fins with moderately high anterior lobes. Caudal fin with elongate upper and lower rays. HABITAT: Ventura, CA, to Chile, including Galápagos Islands. Also scattered locations to the Philippines. Occur over rocky reefs from surface to about 430 ft.

Tonguefishes

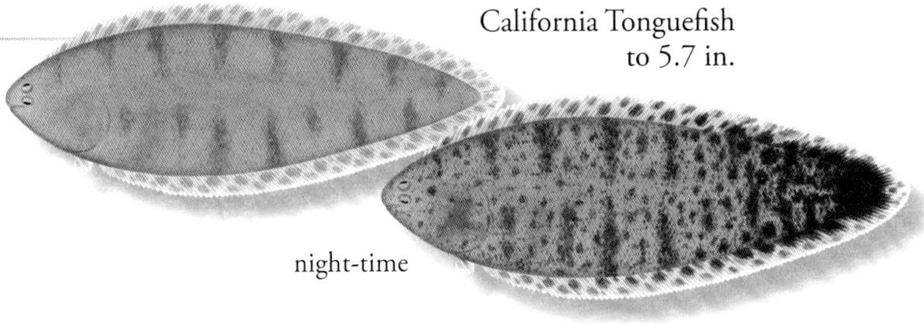

California Tonguefish
to 5.7 in.

night-time

Triggerfishes

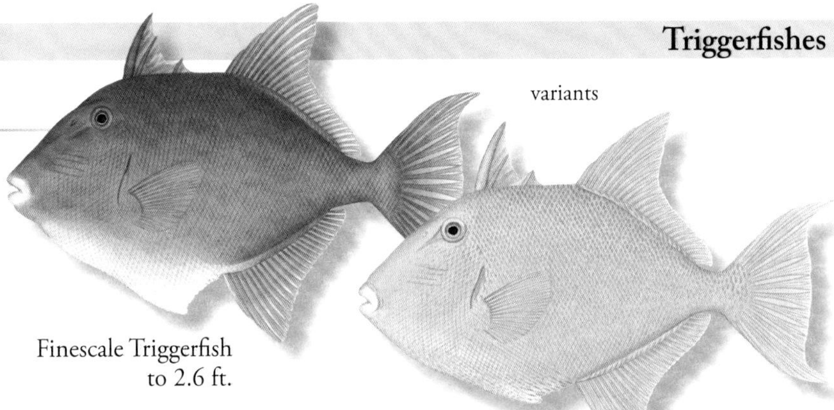

variants

Finescale Triggerfish
to 2.6 ft.

Black Durgon
to 20 in.

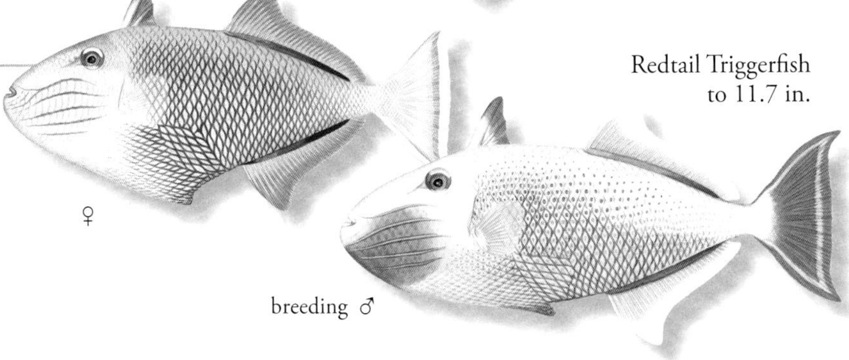

Redtail Triggerfish
to 11.7 in.

♀

breeding ♂

Ostraciidae - Boxfishes

Spiny Boxfish - *Lactoria diaphana* (Bloch & Schneider, 1801)

FEATURES: Color, pattern highly variable. Shades of tan to gray, with darker blotches, spots, and hexagonal pattern along plate edges. Scales over eyes and anterior to anal fin expanded into spine-like projections. Also with a dorsal spine. Small specimens with lateral spines. Pectoral rays number 10–11. HABITAT: Santa Barbara, CA, to Chile, including Galápagos. Scattered in Indo-Pacific. Juveniles pelagic; adults over reefs to about 160 ft.

Tetraodontidae - Puffers

Oceanic Puffer - *Lagocephalus lagocephalus* (Linnaeus, 1758)

FEATURES: Dark bluish black dorsally. Others described as greenish dorsally. Silvery on sides, white below. Dark and pale areas distinctly delineated. Upper portion of pectoral fins dark. Several rows of dark spots present under pectoral-fin base. Snout blunt, rounded. Lower caudal-fin lobe longer than upper lobe. HABITAT: Circumglobal in tropical to temperate seas. In eastern Pacific from Alder Creek Beach, CA, to Peru, including Gulf of California and Galápagos Islands. Oceanic and near surface to about 3,280 ft.

Bullseye Puffer - *Sphoeroides annulatus* (Jenyns, 1842)

FEATURES: Color variable. Dorsal area shades of brown to gray or blackish, with oblique pale yellow to white bands that form a bullseye shape on back. Back and sides with numerous blackish spots. Ventral area white. Area between eyes flat. Forehead, sides, and abdomen covered with small prickles. Cirri absent. HABITAT: Redondo Beach, CA, to Peru, including Gulf of California. Occur from surface to bottom over rocky reefs and adjacent sandy areas from shore to about 340 ft. BIOLOGY: Flesh may be toxic.

Longnose Puffer - *Sphoeroides lobatus* (Steindachner, 1870)

FEATURES: Color, pattern variable. Dorsal area shades of olive, with brownish mottling and numerous small irregular pale spots. Lower sides with a row of dark blotches. Ventral area white. Area between eyes concave. Back with a pair of flap-like cirri. Lower sides with scattered cirri. Most of body covered with small prickles. HABITAT: Redondo Beach, CA, to central Chile and southern and central Gulf of California. Occur over sandy, weedy, and rubble bottoms from shore to about 350 ft. BIOLOGY: Flesh may be toxic.

Diodontidae - Porcupinefishes

Spotted Burrfish - *Chilomycterus reticulatus* (Linnaeus, 1758)

FEATURES: Tannish dorsally and on sides. May have darker areas on upper body. Ventral area white. Relatively small, close-set, blackish spots on head, upper body, and fins. Anal fin may lack spots. Juveniles are blue dorsally, white below, with blackish spots dorsally and on sides. Eyes lack a tentacle. Head and spines covered with short, erect spines. Peduncle with one spine on dorsal surface. HABITAT: Circumtropical. In eastern Pacific from Long Beach Harbor, CA, to Chile and Galápagos Islands. Demersal in coastal waters from near shore to about 460 ft. BIOLOGY: Flesh may be toxic.

Boxfishes

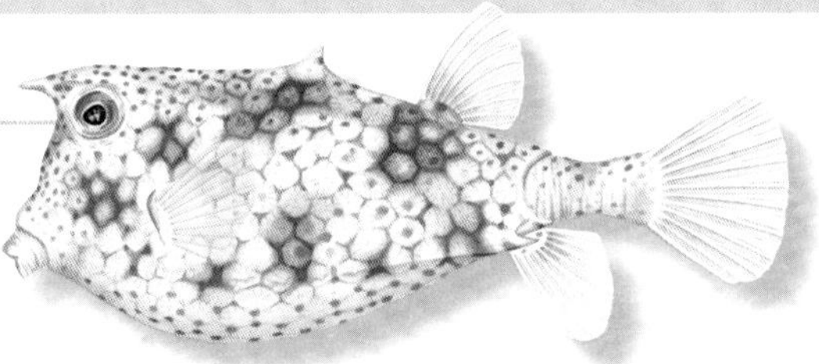

Spiny Boxfish
to 13.4 in.

Puffers

Oceanic Puffer
to 2 ft.

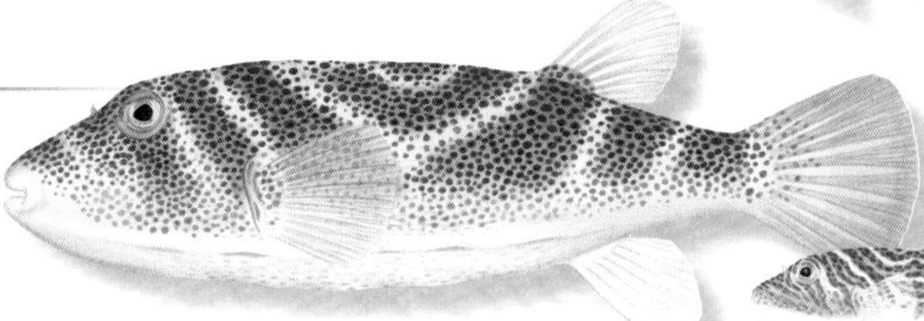

juvenile

Bullseye Puffer
to 18 in.

Longnose Puffer
to 12 in.

Porcupinefishes

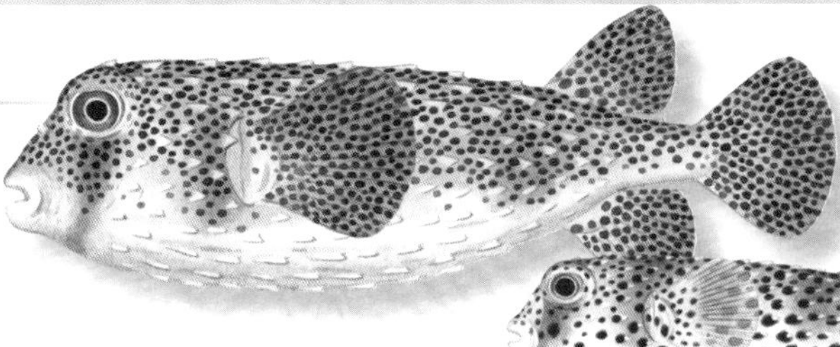

Spotted Burrfish
to 2.4 ft.

juvenile

Diodontidae - Porcupinefishes, *cont.*

Balloonfish - *Diodon holacanthus* Linnaeus, 1758

FEATURES: Color and pattern somewhat variable. Tannish dorsally, paler below. A dark bar extends through and below eyes. Body with several dark blotches. Head and body with scattered, moderately large blackish spots. Fins usually unspotted; some with spots at fin bases. Head and body covered with long spines; those on forehead longer than those on body. Spines absent on caudal peduncle. HABITAT: Circumglobal in tropical to warm temperate seas. In eastern Pacific from La Jolla, CA, to Easter Island off central Chile, including Gulf of California. Occur around mangrove areas and over shallow reefs and soft bottoms from near surface to about 330 ft. BIOLOGY: Inflate when disturbed. Flesh may be toxic.

Porcupinefish - *Diodon hystrix* Linnaeus, 1758

FEATURES: Color and pattern variable. Pale brownish yellow to olivaceous dorsally, whitish below. Numerous small, blackish to brownish spots on head, upper body, and all fins. May have a darkly pigmented ring around abdomen. Head and body covered with long spines; those on forehead shorter than those on body. Spines present on caudal peduncle. HABITAT: Circumglobal in tropical to warm temperate seas. In eastern Pacific from San Diego, CA, to Chile, including Gulf of California and Galápagos Islands. Reef associated from near shore to about 440 ft. BIOLOGY: Inflate when disturbed. Flesh and organs may be toxic.

Molidae - Molas

Ocean Sunfish - *Mola mola* (Linnaeus, 1758)

FEATURES: Color and pattern variable. Shades of pale to dark grayish to brownish dorsally, whitish to dusky below. Some fairly uniformly colored, others spotted. Snout blunt and rounded. Mouth and nostrils small. Gill openings round. Pectoral fins rounded and set at midbody. Dorsal and anal fins tall and nearly vertical. Posterior margin of modified tail rounded and vaguely to distinctly scalloped. Body round to oval in profile. HABITAT: Circumglobal in tropical to temperate seas. In eastern Pacific from Gulf of Alaska to central Chile, including Gulf of California and Galápagos Islands. Oceanic. Found near surface to about 2,110 ft. Deepest record off eastern Pacific coast is 1,286 ft. BIOLOGY: Ocean Sunfish are poor swimmers. Swim by waving dorsal and anal fins from side to side. Often observed lying horizontally at the surface.

Slender Mola - *Ranzania laevis* (Pennant, 1776)

FEATURES: Dark blackish blue dorsally, iridescent silvery on sides and below. Head and body with wavy silver bars and spots. Bars on head outlined with dark lines and spots. Snout bluntly pointed. Mouth small and vertically oriented. Gill openings round. Pectoral fins pointed, located on upper body. Dorsal and anal fins tall and nearly vertical. Posterior margin of modified tail nearly straight. Body at least twice as long as it is deep and laterally compressed. Small specimens very elongate. HABITAT: Circumglobal in tropical to warm temperate seas. In eastern Pacific from Oceano, CA, to Chile, including Gulf of California. Pelagic and oceanic. Reported from surface to about 660 ft.

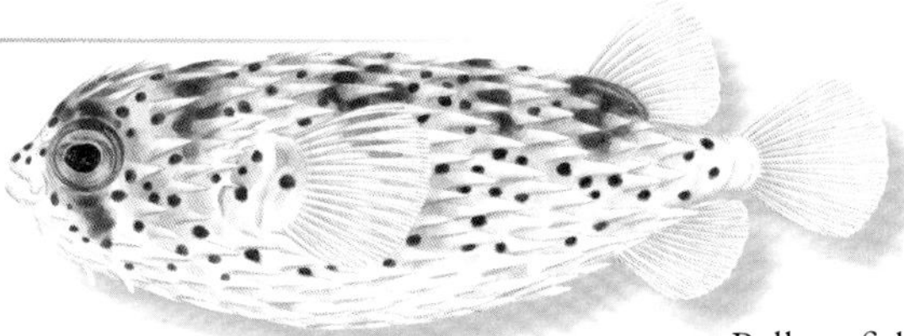

Balloonfish
to 23.6 in.

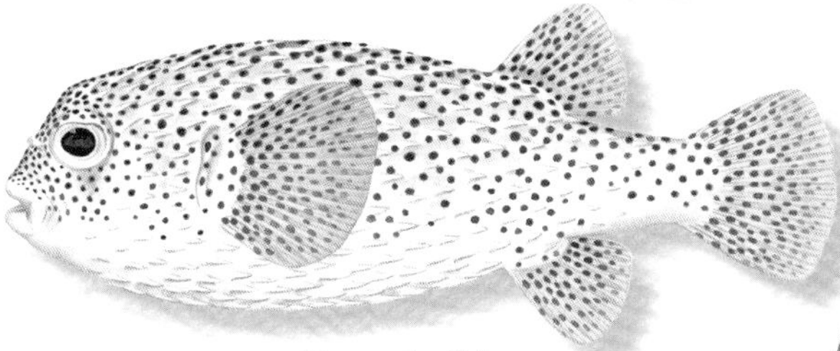

Porcupinefish
to 3 ft.

Molas

Slender Mola
to 2.7 ft.

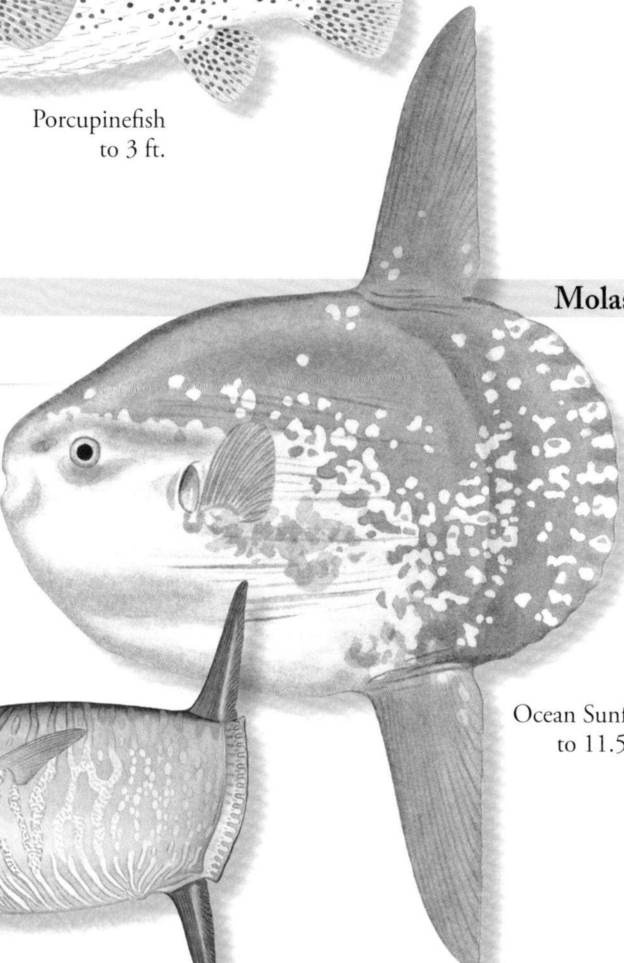

Ocean Sunfish
to 11.5 ft.

Glossary

Below are common technical terms used in this book to describe fishes and their environment. Many of the terms referring to color patterns, anatomical features, and topography are illustrated in the front of this book on pages 18–22.

Adipose eyelid: A fatty, transparent tissue that covers the eyes in some fishes.
Adipose fin: A fleshy, rayless fin on the dorsal midline between the dorsal and caudal fin in some fishes. In some species it is fused to the tail and separated from it by a slight notch.
Anadromous: A fish is anadromous when it lives primarily in the ocean and migrates into fresh water to spawn.
Anal fin: The fin on the ventral surface of fishes, located between the anus and the caudal fin.
Anterior: At or toward the front, or head, of a fish.
Band: A stripe or elongated area of a different color from the surrounding area.
Barbel: A fleshy, tentacle-like appendage usually on the chin, on the snout, or near the mouth of some fishes that is usually sensitive to chemicals or to touch.
Base: The point or area where a fin emerges from the body.
Brackish: Describes water that is a mixture of fresh and salt water and is less salty than ocean water and more salty than fresh water.
Buckler: A bony structure, often with a projecting spine, located on the sides, on the dorsal surface, or along the abdomen of some fishes.
Canine: A prominent, conical, and pointed tooth, usually larger than other teeth.
Catadromous: A fish is catadromous when it lives primarily in fresh water and migrates to the ocean to spawn.
Caudal fin: The fin at the rear end of the body, also called the tail fin.
Caudal peduncle: The portion of the fish body between the end of the anal fin and the base of the caudal fin.
Cephalic disk: A modified dorsal fin on the head of remoras that forms an oval-shaped adhesive disk.
Cephalic fin: A portion of modified pectoral fin on the head of manta rays that is elongated, flexible, and used to direct water into the mouth.
Chest: The ventral portion of the body in front of the pelvic fins; sometimes referred to as the breast.
Chevron: A diffuse, V-shaped mark that may be dark or pale.
Circumglobal: Occurring around the world, in all oceans.
Cirrus (pl. cirri): A fleshy filament or tab on the head, body, or fins; sometimes occurring as fringe-like, as branched, or as a series.
Clasper: An elongate modification of the pelvic fin of chimaeras, sharks, skates, and rays that is used for internal fertilization.
Commensal: Describes a relationship between two species in which one species benefits and the other species is neither helped nor harmed.
Compressed: Describes a fish body that is flattened from side to side.
Concave: Arched or rounded inward.
Confluent: Refers to fins that are joined or meet without a notch.
Continental shelf: The submerged portion of a continent that is relatively flat, gently sloping, and extends from shore to a depth of about 660 ft (200 m).

Continental slope: The submerged portion of a continent that is steeply sloping and extends from a depth of about 660 ft (200 m) to 13,200 ft (4,000 m).
Continuous: Unbroken; usually refers to fins that are joined without a gap or notch.
Convex: Arched or rounded outward.
Corselet: Region behind the head and surrounding the pectoral fins that is covered in specially modified scales.
Demersal: Living unattached at or near the bottom.
Denticle: A small tooth-like or thorn-like projection.
Depressed: Describes a fish body that is flattened from top to bottom.
Detritus: Disintegrating organic material—usually refers to small, particulate plant or animal debris.
Dorsal fin: A fin located on the back of a fish; often notched or separated into two or more fins.
Dorsally: Refers to the back or upper portion of a fish.
Estuary: An area partially enclosed by land where fresh water combines with salt water.
Filamentous: Refers to a long, slender, thread-like structure.
Filter-feed: Describes a method of feeding by removing plankton from the water using gill rakers.
Finlet: A short, isolated, separate fin segment between the dorsal and caudal fins and between the anal and caudal fins.
Flotsam: Floating wreckage or debris—usually man made.
Hermaphroditic: A fish that possesses both male and female reproductive organs either at the same time or sequentially.
Initial (phase): The color phase before sex change in a species that is hermaphroditic, with female and male reproductive organs at different periods in its life. *See also* **Terminal.**
Inshore: Refers to waters over the continental shelf.
Intertidal: The shore zone between high and low tide lines.
Keel: A raised ridge on a scale or on either side of the caudal peduncle.
Laminae: Thin, flat structures of the cephalic disk on remoras.
Lateral line: A sensory organ along the side of the body that perceives vibrations and is formed by a series of tubes, canals, or pores and motion-sensing cells. The lateral line is complete when it extends from the head to the caudal fin; it is incomplete when it does not reach the caudal fin.
Laterally: Refers to the side of a fish.
Lure: A fleshy appendage at the tip of a modified dorsal-fin spine that is used to attract prey.
Margin: The outermost edge of a fin.
Maxilla: The outermost or hindmost main bone of the two bones that form the upper jaw.
Nape: The part of the back immediately behind the head, or the area on the dorsal surface between the head and the dorsal-fin origin.
Ocellated: Having a pattern of a single ocellus or multiple ocelli.
Ocellus (pl. ocelli): A spot of color inside a ring of another color.
Offshore: Refers to waters beyond the edge of the continental shelf.

Omnivore: An animal that feeds on both animals and plants.
Opercle: The portion of the side of the head, typically composed of large, flattened bones, that covers the gills.
Origin: Refers to the front of a fin, where the first spine or ray emerges from the body.
Papillae: Small, fleshy projections.
Pectoral fins: The paired fins behind the gill opening on each side of the body.
Pelagic: Living in open waters or open ocean, away from the bottom.
Pelvic fins: The paired fins on the lower part of the body and in front of the anus. The position of the pelvic fins varies.
Peritoneum: The membrane lining the body cavity.
Planktivore: An animal that feeds exclusively on planktonic (small drifting) matter.
Posterior: At or toward the rear, or tail, of a fish.
Preopercle: The portion of the side of the head composed of a bone lying anterior to the opercle. The preopercle is often defined by a groove in the skin.
Profile: Side view; describes the outline of a fish body.
Protrusible: Refers to a mouth that is able to extend and lengthen forward to catch prey.
Ray: Supporting elements of the fins. Soft rays are segmented, flexible, and branched, while spiny rays are unsegmented, hard, and unbranched.
Reticulating: A pattern or network of irregular, repeating, and interconnecting dark or pale lines.
Rhomboid: A shape similar to a square, but with oblique angles.
Rostrum: The area of the snout that is sometimes flattened or elongate.
Saddle: A patch of color extending across the back and onto the upper sides that may form a Y, V, or U shape.
Saline: Containing salt.
Scute: A bony plate or modified scale that is usually keeled.
Serrated: Having saw-like notches along the edge.
Simple: Not divided or branched.
Snout: The region of the head in front of the eyes and above the mouth.
Spine: A sharp, unbranched, unsegmented, rigid fin ray; also a sharp bony projection on the head or body.
Spiracle: The circular opening between the eye and first gill opening in some sharks, rays, and sturgeons.
Subtropical: Describes the regions between the tropical and temperate latitudes.
Temperate: Describes the regions between the subtropical and polar latitudes.
Terminal (phase): The color phase after sex change in a species that is hermaphroditic, with female and male reproductive organs at different periods in its life. *See also* **Initial.**
Thorn: A sharp, bony projection.
Tropical: The world's warmest region, between the equator and the tropics.
Tubercle: A hardened, conical projection on the surface of the body.
Turbid: Refers to water that is murky due to suspended silt or sediment.
Ventrally: Refers to the abdominal or lower portion of a fish.
Vermiculation: Cluster of wavy lines or blotches.
Vomer: Median, unpaired bone toward front roof of the mouth; often bears teeth.

Rare Fishes Also in the Area

The fishes listed here have also been recorded within the area from Alaska to California. They were not illustrated because of their rarity, limited range, or deep-water habitat, or due to lack of specimen reference including photographs or video.

Shorthead Hagfish - *Eptatretus mcconnaugheyi* Wisner & McMillan, 1990
FEATURES: Life colors unavailable. Pairs of gill pouches usually number 13. A poorly documented species. To 19 in. HABITAT: Two reported populations: the first from Santa Monica Bay, CA, to Cedros and San Benito Islands; the other restricted to lower Gulf of California. Demersal from about 140 to 1,360 ft.

Okhotsk Skate - *Bathyraja violacea* (Suvorov, 1935)
FEATURES: Shades of lilac, with darker vermiculations dorsally; may be uniformly colored. Ventral surface mostly white, sometimes with dark blotches. Scapular or middorsal thorns absent. Thorns on tail unevenly spaced. Denticles along eye rims enlarged. Tail usually shorter than pre-tail length. To 2.4 ft. HABITAT: Northern and western Bering Sea, also to Japan. Other reports from AK not confirmed. Demersal from about 70 to 3,640 ft.

Pacific Worm Eel - *Myrophis vafer* Jordan & Gilbert, 1883
FEATURES: Shades of uniform brown; others described as pale gray to whitish, with darker speckling dorsally. Snout conical, depressed, and overhanging mouth. Gill opening small. Dorsal-fin origin far behind pectoral fins. Tip of tail covered with rays. To 18 in. HABITAT: San Pedro, CA, to Peru, including Gulf of California. Demersal from intertidal zone and tidepools and over shallow muddy and sandy bottoms from near shore to about 50 ft.

Spurred Hatchetfish - *Argyropelecus hemigymnus* Cocco 1829
FEATURES: Blackish on back, silvery on sides and below; body pigmentation distinct or diffuse. First dorsal fin with a tall posterior spine. Anal fin with 11 rays. Ventral photophores unevenly arranged. Body short and deeply keeled. To 1.8 in. HABITAT: Circumglobal in water column from about 330 to 2,620 ft and below about 45°N. In eastern Pacific, recorded from off WA and OR.

Tropical Hatchetfish - *Argyropelecus lychnus* Garman, 1899
FEATURES: Blackish on back, silvery on sides and below. Dark spots along posterior lateral line minute. Upper preopercular spine usually curved ventrally. Anal fin with 12 rays. Ventral photophores unevenly arranged. Body short and deeply keeled. To 3 in. HABITAT: Point Conception, CA, to Chile; possibly also in the Indian Ocean. Found in the water column from about 650 to 1,300 ft.

Bottlelight - *Danaphos oculatus* (Garman, 1899)
FEATURES: Life colors unavailable. Eyes large. Dorsal spines absent. Anal fin long-based, with 24–25 rays. Ventral photophores generally in a straight row. Body elongate and moderately deep anteriorly. To 2.3 in. HABITAT: Pacific and Indian oceans. In eastern Pacific from off Brooks Peninsula, BC, to Chile. Found in water column from about 170 to 3,000 ft.

Rare Fishes Also in the Area

Dusky Hatchetfish - ***Sternoptyx obscura*** **Garman, 1899**
FEATURES: Blackish on back, silvery on sides and below. First dorsal fin blade-like, with a stout posterior spine. Anal fin long-based, with 14–15 rays. Body hatchet shaped and deeply indented above anal fin. To 1.8 in. HABITAT: Pacific and Indian oceans. In eastern Pacific from central CA to Peru-Chile border. Found in water column from about 620 to 3,700 ft.

Longfin Dragonfish - ***Tactostoma macropus*** **Bolin, 1939**
FEATURES: Uniformly brownish black to black. Mouth large and curved upward with numerous sharp teeth. Chin with a short barbel. Pectoral fins absent. Pelvic fins located closer to tail than to the head. Dorsal and anal fins located just anterior to caudal fin. Lower body with two rows of small photophores. To 18 in. HABITAT: Bering Sea to southern CA. Also to Japan. Found in water column from about 80 to 6,560 ft.

Broadfin Lampfish - ***Nannobrachium ritteri*** **(Gilbert, 1915)**
FEATURES: Life colors unavailable. Head comparatively small. Opercular margin angular, with corner low on body. Pectoral fins small, not reaching past pelvic-fin base. Dorsonasal photophores absent. Ventral nasal photophores small. Body slender. To 5.5 in. HABITAT: NE Pacific from south of Aleutian Islands to southern Baja California. Found in water column from about 25 to 3,600 ft.

California Flashlightfish - ***Protomyctophum crockeri*** **(Bolin, 1939)**
FEATURES: Life colors unavailable. "Headlight" photophores absent or poorly developed. Head and eyes comparatively large. Anal fin long-based, with 20–23 rays. Photophores on lower caudal peduncle number 2. To 4.6 in. HABITAT: Southern British Columbia to Baja California. Also off Japan. Found from surface to about 3,120 ft.

Bigfin Lanternfish - ***Symbolophorus californiensis*** **(Eigenmann & Eigenmann, 1889)**
FEATURES: Life colors unavailable. "Headlight" photophores absent or poorly developed. Head and eyes comparatively large. Pectoral fins comparatively long. Photophores on lower caudal peduncle number 2. To 5 in. HABITAT: West of British Columbia to central Baja California. Also to Japan. Found from surface to about 2,500 ft.

Mexican Lampfish - ***Triphoturus mexicanus*** **(Gilbert, 1890)**
FEATURES: Silvery brown. "Headlight" photophores poorly developed. Opercular margin angular, with corner low on body. Pectoral fins minute. Body slender. To 3 in. HABITAT: British Columbia to off southern Mexico. In water column from about 65 to 3,700 ft.

Whiptail Ribbonfish - ***Desmodema lorum*** **Rosenblatt & Butler 1977**
FEATURES: Adults silvery, with red dorsal and caudal fins. Large juveniles with dark round spots. Body deep anteriorly, tapers to long, whip-like tail. Dorsal fin extends from over head to tip of tail. Anal fin absent. Caudal fin small. Pelvic fins fan-like in juveniles, reduce with age. To 3.7 ft. HABITAT: Temperate waters of the North Pacific. In eastern Pacific from central CA to southern Baja California. In water column from near surface to about 1,640 ft.

Rare Fishes Also in the Area

Scalloped Ribbonfish - *Zu cristatus* (Bonelli, 1820)
FEATURES: Silvery with vague dark bars; fins mostly red. Juveniles silvery with blackish spots or bars. First dorsal fin rays and pelvic fins elongate in juveniles, reduce with age. Abdominal profile scalloped in juveniles, becoming straight in adults. Body deep anteriorly, tapers to a slender tail. Dorsal fin extends from head to tip of tail. Caudal fin large. To 4 ft. HABITAT: Circumglobal in tropical to warm seas. In eastern Pacific from off Newport Beach, CA, to Chile. In water column from surface to about 2,300 ft.

Spottedtail Goosefish - *Lophiodes caulinaris* (Garman, 1899)
FEATURES: Spotted and blotched in shades of brown to gray. Caudal fin with a vertical series of pale spots on middle of rays. Mouth very large and gaping. Head very broad and flattened, tapering to narrow tail. First free dorsal-fin spine modified into a lure. To 15.7 in. HABITAT: Morro Bay, CA, to Peru, including Gulf of California. Demersal from about 50 to 1,250 ft.

Brickred Rockfish - *Sebastes baramenuke* (Wakiya, 1917)
FEATURES: Living color unavailable. After capture, shades of red, with a dark blotch on opercle and two dark bands below eyes. Head with three dark bars between snout and spiny dorsal fin. Head large, body deep and robust. To 20 in. HABITAT: Western Aleutian Islands off Kiska Pass. Also to Sea of Japan. Found at about 330 to 2,500 ft.

Gray Rockfish - *Sebastes glaucus* Hilgendorf, 1880
FEATURES: Living color unavailable. Described as gray, brown, or black and heavily washed and mottled in yellow to orange yellow. At least three dark bands radiate from eyes. Head spines reduced. First dorsal fin with 14 spines. Body oval in profile. To 23 in. HABITAT: Scattered in Bering Sea and off Atka Island. Also to Sea of Japan. Occur from about 10 to 1,800 ft.

Yellowmouth Rockfish - *Sebastes reedi* (Westrheim & Tsuyuki, 1967)
FEATURES: Living color unavailable. After capture, primarily red to orange red, with a dark blotch on opercle, dusky saddles on back. Mouth white with yellow, black, and/or red blotches. Head spines weak. Knob under tip of lower jaw weak to moderate. To 23 in. HABITAT: Off Sitka, AK, to near San Francisco, CA. Occur from about 230 to 1,200 ft.

Broadfin Thornyhead - *Sebastolobus macrochir* (Günther, 1877)
FEATURES: Reddish, pinkish, to brown. Spiny dorsal fin with one or two black blotches posteriorly. Head spines strong. Pectoral fins notched, lower rays elongate and free at tips. To 15 in. HABITAT: Scattered records from eastern Bering Sea and W Aleutian Islands. Also to Sea of Japan. Demersal over soft bottoms from about 330 to 4,900 ft.

Smootheye Poacher - *Xeneretmus leiops* Gilbert, 1915
FEATURES: Dusky olive, with dark blotches dorsally. Whitish below. First dorsal fin with black margin, second with dusky margin. Snout tip with a vertical spine and two small lateral spines. 1–4 small barbels under chin, and one barbel at end of maxilla. Eyes lack spines. Lower pectoral-fin rays elongate, free at tips. To 11 in. HABITAT: Forrester Island, AK, to off Santa Catalina Island, CA. Demersal from about 120 to 1,300 ft.

Rare Fishes Also in the Area

Greater Sand Perch - *Diplectrum maximum* **Hildebrand, 1946**
FEATURES: Dark grayish brown dorsally and on sides, fading to white below. Operculum with a large bluish gray blotch. Spiny dorsal fin banded in reddish brown and bluish gray. Lower preopercle rounded and projecting, with 4–8 bony spines. Anal fin short-based, with 7 rays. To 14.6 in. HABITAT: Newport Beach, CA, to northern Peru. Demersal over soft bottoms from near shore to about 425 ft.

Deepwater Serrano - *Serranus aequidens* **Gilbert, 1890**
FEATURES: Shades of tan dorsally and on sides with irregular darker bars. White below. Opeculum with a large bluish gray blotch. Spiny dorsal fin with blackish membranes and yellowish margin. Eyes large. Preopercle finely serrated. Anal fin short-based, with 7 rays. To 9.6 in. HABITAT: Whites Point, CA (rarely), to western Gulf of California and central Mexico to Colombia. Demersal over soft bottoms from about 250 to 850 ft.

Cocinero - *Carangoides vinctus* **(Jordan & Gilbert, 1882)**
FEATURES: Bluish gray, with golden or greenish reflections and about 8 pale bars dorsally. Silvery white below. Dorsal, caudal, and anal fins yellowish. Small blackish spot on upper opercular margin. Pectoral fins long, pointed. Lateral line arched anteriorly, followed by 46–53 scutes. To 15 in. HABITAT: San Diego Bay, CA (rarely), to Peru, including central and southern Gulf of California. Pelagic from surface to about 165 ft.

Whitemouth Jack - *Uraspis helvola* **(Forster, 1801)**
FEATURES: Dusky with silvery reflections, sometimes with 6 narrow, pale bars on upper sides. Mouth bluish black, with white tongue, roof, and floor. Pectoral fins moderately long and pointed. Lateral line arched anteriorly, followed by 23–40 scutes. To 20 in. HABITAT: Circumglobal in tropical seas. In eastern Pacific from Santa Catalina Island, CA (rarely), to Ecuador. Pelagic from about 30 to 980 ft.

Cottonmouth Jack - *Uraspis secunda* **(Poey, 1860)**
FEATURES: Dusky bluish gray with silvery reflections, sometimes with 6–8 narrow, pale bars on upper sides. Mouth and tongue bright white. Pectoral fins moderately long and pointed. Lateral line arched anteriorly, followed by 32–38 scutes. To 18.5 in. HABITAT: Circumglobal in tropical seas. In eastern Pacific form Santa Catalina Island, CA (rarely), to Costa Rica. Pelagic to about 120 ft. May be junior synonym of *U. helvola*.

Aleutian Pout - *Commandorella popovi* **Taranetz & Andriashev, 1935**
FEATURES: Uniformly olive brown to cherry red dorsally. Yellowish below. Smaller specimens with small dark reticulations or spots. Anal fin of mature males black. Dorsal fin often with a black ocellated spot. Dorsal-fin origin over pectoral fins. Pectoral fins large. To 6.4 in. HABITAT: Scattered from Kodiak and Aleutian islands. Demersal in seaweed above high tide mark and intertidal zone to about 7 ft. Also known as *Gymnelus popovi*.

Persimmon Eelpout - *Eucryphycus californicus* **(Starks & Mann, 1911)**
FEATURES: Rosy red. Pectoral fins bright orange and large and rounded. Pelvic fins absent. To 8.7 in. HABITAT: Monterey Bay to San Diego, CA. Demersal over kelp and eelgrass accumulations of submarine canyons from about 200 to 1,790 ft.

Rare Fishes Also in the Area

Polar Eelpout - *Lycodes turneri* **Bean, 1879**
FEATURES: Shades of purple, with 10–12 bluish white bands with purplish olive borders, or umber with cream bands and dark umber borders dorsally. Anterior bands broken. Underside of head and abdomen white. Head broad, depressed. Chin crests well developed. Pectoral fins fan-like. To 25 in. HABITAT: Beaufort and Chukchi seas to Bristol Bay, AK. Demersal over soft bottoms from about 33 to 410 ft.

Bearded Eelpout - *Lyconema barbatum* **Gilbert, 1896**
FEATURES: Living color unavailable. Dark spots on sides and possibly fins. Underside of lower jaw with minute barbels or cirri. Pelvic fins small. To 6.7 in. HABITAT: Southern OR to central Baja California. Demersal over muddy to sandy mud bottoms from about 150 to 1,310 ft.

Lined Clingfish - *Gobiesox eugrammus* **Briggs, 1955**
FEATURES: Shades of translucent cream, with orange or red marks. Dorsal and caudal fins black with white margins. Eyes comparatively large. Posterior nostrils at anterior edge of eyes. Dorsal fin with 14–15 rays (including 2–3 hidden). To 2.2 in. HABITAT: Bird Rock, San Diego County, CA, to central Baja California. Demersal from about 30 to 270 ft. Reported to live near or under sea urchins.

Channel Islands Clingfish - *Rimicola cabrilloi* **Briggs, 2002**
FEATURES: Living color unavailable. Preserved specimens pale and unmarked. Head relatively narrow, moderately compressed. Posterior nostrils above or just behind anterior edge of eyes. Males with 1–2 well-developed canine-like teeth in lower jaw. Dorsal fin with 7–8 rays. To 1.2 in. HABITAT: Islands off southern CA from intertidal and shallow subtidal waters.

Southern Clingfish - *Rimicola dimorpha* **Briggs, 1955**
FEATURES: Living color unavailable. Dorsal and anal fins short-based and low with 6–7 and 6–8 rays, respectively. Caudal peduncle shallow. Body slender. To 1.2 in. HABITAT: Northern Channel Islands, CA, to central Baja California. Demersal from intertidal to shallow subtidal zone.

Striped Escolar - *Diplospinus multistriatus* **Maul, 1948**
FEATURES: Silvery with narrow dotted lines along sides. Gill membranes black. Mouth large, toothy. Eyes large. Dorsal fin long-based and deeply notched at about ⅔ of length. Caudal fin forked. Anterior anal-fin rays very short, posterior rays low. Pelvic fins minute. Body elongate, with two lateral lines. To 12 in. HABITAT: Circumtropical in tropical to temperate seas. In eastern Pacific from central CA. Oceanic from about 165 to 3,280 ft.

Sakhalin Sole - *Limanda sakhalinensis* **Hubbs, 1915**
FEATURES: Eyed side uniformly brown or with vague, darker blotches. Blind side off-white. Fins brown. Maxilla extends to anterior edge of lower eye. Interorbital ridge moderately high. Lateral line arched over pectoral fin. Caudal-fin margin rounded. Scales on eyed side mostly rough. To 13 in. HABITAT: St. Lawrence Island, AK, to southeastern Bering Sea. Also to Sea of Okhotsk. Demersal from about 30 to 360 ft.

Select Bibliography and Additional Resources

References listed below will provide the reader with further information about the species included in this book and their natural history, ecology, biology, and environment. Out-of-print references may be found online or at public libraries.

Allen, G. R., and D. R. Robertson. 1994. *Fishes of the Tropical Eastern Pacific.* Honolulu: University of Hawaii Press.

Allen, L. G., D. J. Pondella, and H. H. Horn. 2006. *The Ecology of Marine Fishes—California and Adjacent Waters.* Berkeley: University of California Press.

Burgess, W. E., and H. R. Axelrod. 1984. *Fishes of California and Western Mexico.* Neptune City: T. F. H. Publications.

Butler, J. L., M. S. Love, and T. E. Ladig. 2012. *A Guide to the Rockfishes, Thornyheads, and Scorpionfishes of the Northeast Pacific.* Berkeley: University of California Press.

Campagno, L., and S. Fowler. 2005. *Sharks of the World.* Princeton: Princeton University Press.

Carpenter, K. E., ed. 2002. *FAO Species Identification Guide for Fishery Purposes; The Living Marine Resources of the Western Central Atlantic.* Vol. 1–3. Rome: Food and Agriculture Organization of the United Nations.

Castro, J. I. 1983. *The Sharks of North American Waters.* College Station: Texas A&M University Press.

Collette, B. B., and G. Klein-MacPhee. 2002. *Bigelow and Schroeder's Fishes of the Gulf of Maine.* 3rd ed. Washington: Smithsonian Institution Press.

Collette, B. B., and C. E. Nauen. 1983. *FAO Species Catalogue.* Vol. 2, *Scombrids of the World.* Rome: Food and Agriculture Organization of the United Nations.

Deans, N. L., and M. Rigsby, eds. 1997. *A Natural History of the Monterey Bay National Marine Sanctuary.* Monterey: Monterey Bay Aquarium Foundation.

Ebert, D. A. 2003. *Sharks, Rays, and Chimaeras of California.* Berkeley: University of California Press.

Gotshall, D. W. 1989. *Pacific Coast Inshore Fishes.* 3rd ed. Monterey: Sea Challengers.

Gotshall, D. W. 1998. *Sea of Cortez Marine Animals: A Guide to the Common Fishes and Invertebrates, Baja California to Panama.* Monterey: Sea Challengers.

Grove, J. S., and Lavenberg, R. L. 1997. *The Fishes of the Galápagos Islands.* Stanford: Stanford University Press.

Helfman, G. S. 2007. *Fish Conservation: A Guide to Understanding and Restoring Global Aquatic Biodiversity, Evolution, and Fishery Resources.* Washington, DC: Island Press.

Helfman, G. S., B. B. Collette, D. E. Facey, and B. W. Bowen. 2009. *The Diversity of Fishes: Biology, Evolution, and Ecology.* 2nd ed. West Sussex, UK: Wiley-Blackwell.

Helfman, G. S., and B. B. Collette. 2011. *Fishes: The Animal Answer Guide.* Baltimore: Johns Hopkins University Press.

Hart, J. L. 1973. *Pacific Fishes of Canada.* Fisheries Research Board of Canada, Bulletin 180.

Kramer, D. E., et al. 1995. *Guide to Northeast Pacific Flatfishes.* Fairbanks: University of Alaska Fairbanks.

Select Bibliography and Additional Resources

Kramer, D. E., and V. M. O'Connell. 2003. *Guide to Northeast Pacific Rockfishes, Genera Sebastes and Sebastolobus*. Fairbanks: University of Alaska Fairbanks.

Lamb, A., and P. Edgell. 2010. *Coastal Fishes of the Pacific Northwest.* 2nd ed. Madeira Park, BC: Harbour Publishing.

Long, J. A. 2011. *The Rise of Fishes*. Baltimore: Johns Hopkins University Press.

Love, M. S. 2011. *Certainly More Than You Want to Know About the Fishes of The Pacific Coast.* Santa Barbara: Really Big Press.

Love, M. S., et al. 2005. *Resource Inventory of Marine and Estuarine Fishes of the West Coast and Alaska: A Checklist of North Pacific and Arctic Species from Baja California to the Alaska-Yukon Border.* Seattle: US Department of the Interior, US Geologic Survey, Biological Resources Division.

Love, M. S., M. Yoklavich, and L. Thorsteinson. 2002. *The Rockfishes of the Northeast Pacific.* Berkeley: University of California Press.

McEachran, J. D., and J. D. Fechhelm. 2005. *Fishes of the Gulf of Mexico*. Vol. 1 and 2. Austin: University of Texas Press.

McGinnis, S. M. 1984. *Freshwater Fishes of California*. Berkeley: University of California Press.

Mecklenburg, C. W., T. A. Mecklenburg, and L. K. Thornsteinson. 2002. *Fishes of Alaska.* Bethesda: American Fisheries Society.

Miller, D. J., and R. N. Lea. 1972. *Guide to Coastal Marine Fishes of California.* California Dept. of Fish and Game, Fish Bulletin 157.

Nakamura, I. 1985. *FAO Species Catalogue.* Vol. 5, *Billfishes of the World.* Rome: Food and Agriculture Organization of the United Nations.

Nelson, J. S. 2006. *Fishes of the World.* 4th ed. Hoboken: John Wiley & Sons.

Page, L. M., et al. 2013. *Common and Scientific Names of Fishes from the United States, Canada, and Mexico.* 7th ed. Bethesda: American Fisheries Society, Special Publication 34.

Pollard, W. R., et al. *Field Identification of Coastal Juvenile Salmonids*. Madeira Park, BC: Harbour Publishing.

Stevenson, D. E., J. W. Orr, G. R. Hoff, and J. D. McEachran. 2007. *Field Guide to Sharks, Skates, and Ratfish of Alaska*. Fairbanks: University of Alaska Fairbanks.

The websites listed below have searchable databases that are useful sources of taxonomic and ichthyological information:

Biodiversity Heritage Library: http://www.biodiversitylibrary.org/

California Academy of Sciences, Eschmeyer's Catalog of Fishes: http://www.calacademy.org/scientists/projects/catalog-of-fishes

Encyclopedia of Life: http://www.eol.org/

Fishbase: http://www.fishbase.org/search.php

Integrated Taxonomic Information System: http://www.itis.gov/

IUCN Red List of Threatened Species: http://www.iucnredlist.org/

Smithsonian National Museum of Natural History, Research and Collections, Vertebrate Zoology: http://vertebrates.si.edu/fishes/

Smithsonian Tropical Research Institute: http://www.stri.org/sftep/

Index

Index

Index

Index

Index

Index

Index

Index

Index

S

Index

Index

Index

About the Authors

Val Kells is a widely recognized freelance Marine Science Illustrator. She received her B.S. in Environmental Studies and completed her formal training in Scientific Communication and Illustration at the University of California, Santa Cruz, in 1985 and 1986. She has worked with designers, educators, and curators to develop a wide variety of illustrations for interpretive and educational displays in public aquariums, museums, and nature centers. Her artwork has been displayed in over twenty aquariums, including the Long Beach Aquarium of the Pacific, Monterey Bay Aquarium, the North Carolina Aquariums, and the Texas State Aquarium, among others. Val also collaborates with editors, researchers, and scientists, and her work has been published in many books and periodicals. She recently illustrated and co-authored *A Field Guide to Coastal Fishes: From Maine to Texas* and illustrated *Field Guide to Fishes of the Chesapeake Bay*, both published by Johns Hopkins University Press. Val lives in Virginia with her family. She is an avid fisherman and naturalist and spends her off time exploring, fishing, and documenting the aquatic habitats along the coasts.

Luiz A. Rocha is the Follett Chair of Ichthyology at the California Academy of Sciences. He received his B.S. and M.S. at the Universidade Federal da Paraíba, Brazil, in 1996 and 1999, and his Ph.D. at the University of Florida in 2003. His major research interests include evolution, conservation, taxonomy, and community ecology of marine fishes. He frequently tries to combine these fields, linking ecology to evolution and using molecular tools to answer biogeographic and taxonomic questions. The overall objective of this interdisciplinary research is to explain what generates and maintains the extremely high biodiversity in coastal marine regions, both shallow and deep water. He has spent more than 5,000 hours underwater collecting and observing fishes, and published more than 70 peer-reviewed articles. In addition, his work has been featured in many popular media outlets, including the *New York Times*, *Scientific American*, and *National Geographic* magazine, and supported conservation efforts across the globe. Luiz lives with his family in Marin County, California, and spends his weekends hiking and mountain biking with his children and their very active dog.

About the Authors

Larry G. Allen is an ichthyologist who is Professor of Biology at California State University, Northridge. He received his B.A. and M.A. degrees at California State University, Fullerton, in 1974 and 1976, and his Ph.D. at the University of Southern California in 1980. He has mentored over 40 successful Master's degree students, published more than 65 scientific papers, and taught a wide variety of courses, including Pelagic Organisms, Ichthyology, and Ecology of Marine Fishes. His research efforts since the 1970s have included biogeography and ecology of bay/estuarine and coastal marine fishes off California. Dr. Allen was the senior editor of the book *Ecology of Marine Fishes: California and Adjacent Waters*, published in in 2006. In recent years, Dr. Allen has focused his research on the biology of important recreational and commercial species of the northeastern Pacific Ocean. He is currently serving as president of the American Society of Ichthyologists and Herpetologists. Larry lives in the Santa Monica Mountains near Los Angeles, California. He is a life-long angler and spends a lot of time catching and studying fishes.

About the Art. The illustrations were created with watercolor and gouache on 300 lb hotpress Arches® watercolor paper. Brushes included sizes 2 to 14, round, oval, and flat. Techniques included the use of liquid rubber mask, blotters, and gesso.

Notes

Notes

Notes

Notes

Notes

Also from Johns Hopkins University Press

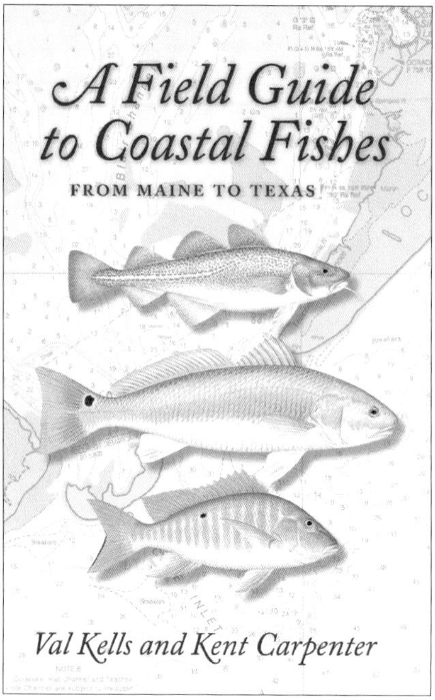

A Field Guide to Coastal Fishes is a comprehensive, current, and accurate identification guide to the more than 1,000 nearshore and offshore fishes that live in brackish and marine waters from Maine to Texas.

Reliable and up to date, this is the most complete book ever published on East and Gulf Coast fishes—perfect for boat, home, or classroom. Its beautiful design and accessible format make it an ideal guide for fishermen, divers, students, scientists, naturalists, and fish enthusiasts alike.

Paperback; 448 pages; 1079 color illustrations; 1 map

ISBN: 978-0-8018-9838-9

The only comprehensive field guide to the Chesapeake's fishes, this book is an indispensable resource for both anglers and students of the Bay.

Key features of this field guide include: full-color illustrations of more than 200 species; chacteristics, range, and occurrence in the Bay; reproduction, diet, and statistics from fisheries research; spot illustrations that highlight critical features of certain fish; illustrations of juveniles when they look different from adults; and appendices that include identification keys.

Paperback; 360 pages; 251 color illustrations; 2 maps

ISBN: 978-1-4214-0768-5